ACS SYMPOSIUM SERIES **567**

Formulation and Delivery of Proteins and Peptides

Jeffrey L. Cleland, EDITOR
Genentech, Inc.

Robert Langer, EDITOR
Massachusetts Institute of Technology

Developed from a symposium sponsored
by the Division of Biochemical Technology
at the 205th National Meeting
of the American Chemical Society,
Denver, Colorado,
March 28–April 2, 1993

American Chemical Society, Washington, DC 1994

Library of Congress Cataloging-in-Publication Data

Formulation and delivery of proteins and peptides / Jeffrey L. Cleland, editor; Robert Langer, editor.

    p.   cm.—(ACS symposium series, ISSN 0097–6156; 567)

"Developed from a symposium sponsored by the Division of Biochemical Technology at the 205th National Meeting of the American Chemical Society, Denver, Colorado, March 28–April 2, 1993."

Includes bibliographical references and indexes.

ISBN 0–8412–2959–7

1. Protein drugs—Dosage forms—Congresses. 2. Peptide drugs—Dosage forms—Congresses.

I. Cleland, Jeffrey L., 1964–   . II. Langer, Robert S. III. American Chemical Society. Division of Biochemical Technology. IV. Series.

RS431.P75F67   1994
615'.19—dc20   94–3789
                                                                 CIP

The paper used in this publication meets the minimum requirements of American National Standard for Information Sciences—Permanence of Paper for Printed Library Materials, ANSI Z39.48–1984. ∞

Foreword

THE ACS SYMPOSIUM SERIES was first published in 1974 to provide a mechanism for publishing symposia quickly in book form. The purpose of this series is to publish comprehensive books developed from symposia, which are usually "snapshots in time" of the current research being done on a topic, plus some review material on the topic. For this reason, it is necessary that the papers be published as quickly as possible.

Before a symposium-based book is put under contract, the proposed table of contents is reviewed for appropriateness to the topic and for comprehensiveness of the collection. Some papers are excluded at this point, and others are added to round out the scope of the volume. In addition, a draft of each paper is peer-reviewed prior to final acceptance or rejection. This anonymous review process is supervised by the organizer(s) of the symposium, who become the editor(s) of the book. The authors then revise their papers according to the recommendations of both the reviewers and the editors, prepare camera-ready copy, and submit the final papers to the editors, who check that all necessary revisions have been made.

As a rule, only original research papers and original review papers are included in the volumes. Verbatim reproductions of previously published papers are not accepted.

M. Joan Comstock
Series Editor

Contents

v

vii

Preface

THE THERAPEUTIC AND COMMERCIAL SUCCESS of drugs developed from biotechnology depends in part on the ability to formulate and deliver these drugs. During the past 20 years, many of the basic steps in the production of recombinant products have been well studied. Molecular biology, fermentation, cell culture, purification, and recovery of recombinant proteins are described in detail in several texts. However, the formulation and delivery of proteins and peptides is still somewhat of a "black box". The basic principles and underlying mechanisms for the successful formulation and delivery of proteins and peptides have provided some general rules for formulations. Continued studies on the mechanisms of degradation may lead to basic principles for formulation design based on the primary sequence of the protein. The delivery of proteins and peptides may be more difficult to reduce to general rules because the physicochemical properties of the drug and its behavior in vivo both play a critical role in its successful delivery. As our understanding of formulation and delivery improves, the probability of developing a viable biotechnology drug for human use increases, and the result will be a further evolution of the biotechnology industry.

Research on formulation and delivery issues for proteins and peptides has progressed rapidly in the past few years. Although several texts on different aspects of formulation and delivery are available, the technology continues to evolve such that updated texts are continually needed. This book is designed to provide an updated review of the recent research in formulation and delivery of proteins and peptides.

The book is divided into three sections: formulation, lyophilization, and delivery. Several major issues are involved in the formulation of proteins and peptides, addressed in the first section of the book. The degradation products formed during storage must be quantified and characterized. This process requires several analytical methods for each new drug (Chapter 2). These methods allow the researcher to assess the physical and chemical changes in the drug over time and may provide insight into the in vivo behavior of the drug. Chemical degradation, including deamidation and subsequent cleavage (Chapter 3), and oxidation by both chemicals (Chapter 4) and light (Chapter 5) are described along with the relationship between chemical and physical degradation such as oxidation and aggregation (Chapter 6). Finally, once these degradation mechanisms have been well studied for several model systems, a general scheme for predicting stability in a given formulation can be derived (Chapter 7).

Although storage of proteins and peptides in aqueous solutions is usually preferred for simplicity and reduced development cost, the instability of proteins and peptides often requires the use of lyophilization (Chapters 8–14). An overview of protein lyophilization addresses the major issues involved in development of these formulations (Chapter 8). Each component of a formulation can have a dramatic impact on the stability of the protein during freezing and subsequent drying. The mechanisms for excipient stabilization of proteins are analogous to cosolvent–protein interactions observed in aqueous solutions (Chapter 9). To understand further the role of excipients in stabilizing proteins during lyophilization, the use of Fourier transform infrared spectroscopy (FTIR) is described and may be applied to several proteins (Chapters 10 and 11). In FTIR studies of proteins after lyophilization (Chapter 10) and during the lyophilization process (Chapter 11), the secondary structure of the protein is assessed to measure the extent of denaturation. This technique provides correlations between the state of the protein and the role of the excipients in stabilizing the protein. Lyophilization may also be critical for some protein and peptide products that cannot be stored in an aqueous solution. Examples of proteins requiring lyophilization for a viable product are discussed in Chapters 12 and 13. In addition, the ability to generate a lyophilized formulation that is stable at room temperature would offer many advantages. The impact of sugars such as trehalose on the chemical stability of dried formulations is presented (Chapter 14).

The final section of the text focuses on the delivery of proteins and peptides. A number of potential routes and methods exist for the administration of drugs. The major routes include invasive methods such as subcutaneous or intravenous injections and noninvasive techniques such as pulmonary delivery. To reduce the frequency of administrations or to target the drug to the site of action, depot systems are described for several proteins and polymer systems (Chapters 15–18). Biodegradable polymer systems are often used to deliver proteins and peptides, and each polymer has unique characteristics. The behavior of each polymer system, including its mechanism of release and interaction with the entrapped protein, is presented in detail. Other potential depot systems such as liposomes and hydrogels are not discussed in this text. As an alternative to invasive delivery by depot formulations, proteins may be administered as an aerosol to the lung. The pulmonary delivery of recombinant human deoxyribonuclease I, Pulmozyme, is described; this protein is the first U.S. Food and Drug Administration approved protein administered as an aerosol formulation (Chapter 19). Other proteins may also benefit by pulmonary delivery, and new methods are required to characterize these formulations (Chapter 20). Finally, many other noninvasive delivery routes including transdermal and oral administration are not covered in the context of this book but are described in detail elsewhere.

Acknowledgments

We thank the Divisions of Biochemical Technology and Polymeric Materials: Science and Engineering, Inc., for their continuing support of this important area of research. Dhinakar Kompala should be commended for his efforts in making the organization of this meeting proceed smoothly. The assistance of Eliana DeBernardez-Clark during the meeting is also appreciated. We also appreciate the efforts of Anne Wilson of ACS Books in the publication process. The support and guidance as well as patience of several Genentech colleagues are greatly appreciated, and we especially thank Andrew J. S. Jones, Michael F. Powell, Rodney Pearlman, and Jessica Burdman. We also appreciate the assistance of the reviewers whose work ensured the submission of quality research. The authors did an excellent job of submitting a timely account of their recent research and updated reviews, and their efforts are evident in the quality of the work presented herein.

JEFFREY L. CLELAND
Genentech, Inc.
South San Francisco, CA 94080

ROBERT LANGER
Massachusetts Institute of Technology
Cambridge, MA 02139

June 7, 1994

Chapter 1

Formulation and Delivery of Proteins and Peptides

Design and Development Strategies

Jeffrey L. Cleland[1] and Robert Langer[2]

[1]Pharmaceutical Research and Development, Genentech, Inc., South San Francisco, CA 94080
[2]Department of Chemical Engineering, Massachusetts Institute of Technology, Cambridge, MA 02139

The success of most peptide and protein drugs is dependent upon the delivery of the biologically active form to the site of action. In the design and development of formulations to achieve this goal, the formulation scientist must consider the clinical indication, pharmacokinetics, toxicity, and physicochemical stability of the drug. The development of a stable formulation is a necessary step for each new protein or peptide therapeutic. The degradation pathways and their impact on stability should be systematically analyzed and competing degradation rates must be balanced to arrive at the most stable formulation possible. Several routes of administration should also be considered and future development of new formulations may expand the number of potential options. Formulations for each route of administration may be unique and, therefore, have special requirements. In the case of depot formulations, there are many potential matrices, each of which has distinct characteristics that affect its interactions with the drug and its behavior *in vivo*. The formulation characteristics may have a dramatic impact on the *in vivo* stability of the drug as well as the pharmacokinetics and pharmacodynamics. The optimization of formulations, the routes of delivery, the design of depot systems, and the correlation between physicochemical stability and *in vivo* behavior are discussed in detail with recent examples. For new biotechnology-derived drugs including nucleic acids (DNA vectors and antisense RNA) to reach commercialization, all of the issues involved in the design and development of a drug formulation must be considered at an early stage of the overall development process.

Many aspects of biopharmaceutical process development have been well studied over the past twenty years. Difficulties in fermentation, cell culture, and, to some extent, purification and recovery have largely been overcome and these process steps have been well characterized for the production of many protein pharmaceuticals. However, one important field lags behind these others in its development. The design and production of protein and peptide drug formulations is not well developed and many of the mechanisms for stabilization and delivery of these drugs have not been

0097–6156/94/0567–0001$08.00/0

determined. In many cases, companies may initially neglect formulation and stability issues, resolving to simply store proteins or peptides in phosphate buffered saline or other solutions that have not been optimized for stabilizing the drug. Several unknowns still exist when developing a stable dosage form for peptides and proteins. Each molecule has its own unique physical and chemical properties which determine its *in vitro* stability. The formulation scientist must also be concerned about the *in vivo* stability of the drug. Thus, the development of successful formulations is dependent upon the ability to study both the *in vitro* and *in vivo* characteristics of the drug as well as its intended application.

Effect of Formulation Design and Delivery on Drug Development

As shown in Figure 1, a formulation scientist is confronted with a complex decision in choosing a formulation for delivery of a therapeutic protein or peptide. In the literature, the most common discussions of protein and peptide formulations focus on the physicochemical stability of these molecules. Indeed, the properties of the drug molecule are critical in determining the appropriate formulation for successful delivery and stability. The vast majority of the literature on protein and peptide formulations describes the degradation pathways for the drug. Many degradation pathways have been well characterized and, in some cases, degradation may often be predicted from the primary sequence of the protein or peptide (see *1* for examples). Once the formulation scientist has found a set of conditions that provide extensive stability (>2 year shelf-life), the formulated drug is tested in animal models for toxicity and pharmacokinetics. In many cases, this testing phase does not occur until the drug has moved from research into development. At this stage, many problems can occur including poor bioavailability due to the instability of the drug *in vivo,* rapid clearance, or the distribution of the drug in the body. Furthermore, an attempt is often made to resolve these difficulties by administering excess drug to achieve the desired biological effect. However, excessive drug doses often lead to toxicity problems. By this stage, the development of the drug has reached a critical decision point. The tendency in most organizations is to reconsider the development of the drug, sometimes resulting in the 'death' of the development project. However, the formulation scientist has the unique opportunity to work with the scientists in pharmacokinetics and toxicology to 'save' the development of the drug. By altering the formulation or the route of delivery, a drug can often have another opportunity to reach the stage of an Investigational New Drug (IND) filing. Unfortunately, the formulation scientist may not become involved until the drug has already encountered difficulties in animal studies. Thus, it is essential for the formulation scientist to work closely with the discovery research team, the pharmacokinetics department, and the toxicology department prior to the decision to move the drug into full scale development.

After all the difficulties are resolved in the early development stages, many protein and peptide drugs can still encounter problems in the clinic. The major clinical hurdles may be similar to those observed in the pre-IND animal studies. However, the company may have filed an IND for a therapeutic indication that will encounter complex formulation and delivery problems. The route and frequency of administration and the bioactivity or potency of the drug in humans are critical issues that are often not addressed in the pre-IND animal studies. If difficulties in delivery or potency of the drug arise during clinical trials, the formulation scientist along with others on the development team must reconsider the design of both the drug formulation and the clinical plan. These pitfalls may often be avoided by testing the drug in a suitable animal model, if available, and an extensive analysis of the patient population including a marketing survey of the end users (physicians, nurses, and/or patients). By establishing early in the development stage (e.g. between research and Phase I clinical trials) the best route and formulation for the drug, the potential for a

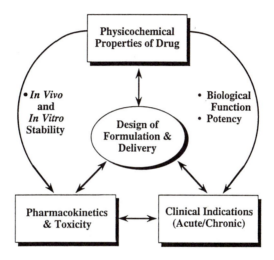

Figure 1: Key factors influencing the design of drug formulations and delivery. The physicochemical properties of the drug can affect the pharmacokinetics and toxicity as well as the clinical indication. The *in vitro* and *in vivo* stability of the drug determines its fate upon administration. The potential clinical utility of the drug is dependent upon the drug characteristics, biological function, and potency. To obtain the desired pharmocological response, a drug must be administered with a stable formulation. The design of a delivery system must also consider the clinical indication, pharmacokinetics, pharmacodynamics, toxicology, and drug properties.

clinically successful product and, ultimately, a marketed product increases dramatically.

The best route for delivery of a protein or peptide drug is often not investigated during the research stage or early in development. The protein or peptide is commonly administered systemically through an intravenous (i.v.) injection in initial animal testing. Thus, for indications that require a high local dose of the drug at the target site, high drug doses are required by i.v. injections. Due to toxicity problems, the efficacious dose may not be reached via i.v. administration. More recently, alternative routes of delivery have been studied. In particular, the therapeutic protein, recombinant human deoxyribonuclease I (rhDNAse), must be delivered directly to the lung of cystic fibrosis patients to degrade the DNA in the mucus. rhDNAse delivered systemically would clearly have little effect on the target site. While this example is an obvious candidate for an alternate delivery route (aerosol delivery of rhDNAse), many other proteins and peptides may also benefit from alternative routes of delivery for therapeutic or clinical reasons. It is therefore essential to investigate the site of action and assess any side effects before choosing a route of administration.

In addition, when companies are developing competitive products, the future sales of the product may rest upon the superior formulation and delivery of the drug, assuming that the efficacy of the competing products are similar. For example, many existing therapeutic proteins such as human growth hormone and insulin are administered chronically requiring daily injections. Competitors with superior drug formulations that release a sustained level of the protein and, thus, require less frequent injections would dominate the market. An example of competing products is the development of sustained release formulations for a luteinizing hormone-releasing hormone (LHRH) agonists. Takeda Pharmaceuticals developed an LHRH agonist (leuprolide acetate) - polylactide-coglycolide formulation that could be administered monthly and provided a continuous sustained therapeutic level of LHRH for one month (2-5). This product, Lupron Depot®, had a ¥57 billion (~$570 million) market in 1992 for prostate cancer, precocious puberty and endometriosis indications and competition from other types of LHRH agonist formulations, including daily injections and daily nasal delivery, have been insignificant (6). Similar competitive products also consist of controlled release systems using polylactide-coglycolide with different LHRH agonists (goserelin acetate, Zoladex®, 7, triptorelin, Decapeptyl®, 8). Overall, the clinical administration, patient compliance, pharmacokinetics, toxicity, and physicochemical properties of the drug must be considered to successfully develop a pharmaceutical protein or peptide drug.

Formulation Development Considerations

While development of novel delivery routes or systems is often necessary, the first step in development of any protein or peptide drug formulation involves the complete characterization of the drug properties and its stability in different formulations. Typically, a formulation scientist will begin by considering the physicochemical properties of the protein such as the isoelectric point, molecular weight, glycosylation or other post-translational modification, and overall amino acid composition. These properties along with any known behavior of the drug in different solutions (e.g. different buffers, cofactors, etc.) as well as its *in vivo* behavior should guide the choice of formulation components for testing in the initial screen of candidate formulations. The potential candidate formulations are composed of U. S. Food and Drug Administration (FDA) approved buffer components, excipients, and any required cofactors (e.g. metal ions). Often, the first choice of candidate formulations is based upon the previous experience of the formulation scientist with other proteins or peptides and, in many cases, a simple phosphate buffered saline solution may be one of the initial candidates.

A simplified approach to formulation development may proceed through the steps depicted in Figure 2. After obtaining all the available background information, one often evaluates several parameters in the initial screen of candidate formulations. One parameter that impacts all the major degradation pathways is the solution pH. Thus, the initial formulations also assess the pH dependence of the degradation reactions and the mechanism for degradation can often be determined from the pH dependence (9). The formulation scientist must quickly analyze the stability of the protein in each solution. Rapid screening methods usually involve the use of accelerated stability at elevated temperatures (e.g. 40° C; see references 10-13 for discussions of elevated temperature studies). Unfortunately, the FDA will only accept real time stability data for shelf life and accelerated stability studies may only serve as a tool for formulation screening and stability issues related to shipping or storage at room temperature. The degradation of the protein for both accelerated and real time studies is then followed by assays developed for analysis of degradation products (see reference 14 for detailed review). The most common degradation pathways for proteins and peptides are listed in Table I. Several recent reviews have analyzed these pathways as well as potential methods to prevent degradation (11, 15-18). In each case, the amount of degradation must be minimized to achieve greater than or equal to 90% of the original drug composition after 2 years (e.g. t $_{90} \geq 2$ years). The FDA usually requires that a pharmaceutical product is not more than 10% degraded and the company must demonstrate that the degradation products do not have any adverse effects on the safety or efficacy of the drug. Many proteins and peptides can degrade extensively without effecting either their safety or efficacy. For example, 70% deamidated recombinant human growth hormone (rhGH) is fully bioactive and non-immunogenic, but this extent of degradation is not acceptable by regulatory agency standards for a therapeutic protein (19). The effect of degradation on the safety and efficacy of a protein or peptide is difficult to ascertain without extensive testing. Thus, the more conservative standards of the FDA and other regulatory agencies may often provide a less expensive alternative if a stable formulation (> 2 year shelf-life) can be developed.

To fulfill the regulatory requirements for a stable formulation, the scientist must consider all of the major degradation routes and the potential conditions for optimization. In the case of aggregation, the addition of surfactants or sugars can prevent denaturation events that lead to irreversible aggregation. If the deamidation rate is the dominant degradation route, the use of amine buffers such as Tris, ammonium, or imidazole may slow the deamidation. Alternatively, a reduction in pH will also decrease the deamidation rate, but the reduced pH may also lead to cleavage or cyclization at Asp-X residues where X is usually a residue with a small side chain (e.g. Gly or Ser) and this degradation has been observed in several proteins (1). Proteins with Asp-X degradation must then be placed in a higher pH buffer to avoid cleavage or cyclization. High pH conditions (> pH 8) will however catalyze oxidation, thiol disulfide exchange, and β-elimination reactions. These degradation pathways may be inhibited by the addition of free radical and thiol scavengers such as methionine. In addition, the method used to prevent one type of degradation may influence another degradation pathway. For example, by adding surfactants or other polymers to prevent aggregation, the residual peroxide in the surfactant may cause a more rapid oxidation (20). In some cases, the formulation pH must be reduced to decrease the rate of deamidation. Reducing the pH may also alter the solubility of the protein since many proteins have isoelectric points at or near the optimal pH (pH 5-6) for minimizing the deamidation rate. For each protein formulation, all the degradation pathways must be evaluated and often a balance must be achieved between the different degradation pathways.

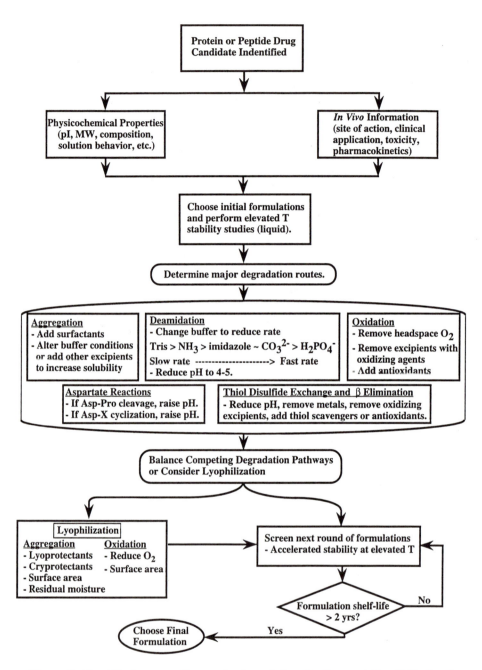

Figure 2: Simplified process diagram for formulation development (See Table I and text for detailed discussion).

Table I: Common Degradation Routes for Proteins and Peptides in Aqueous Solutions [a]

Degradation Route	Region Effected/Results	Major Factors
Aggregation	Whole protein; reversible or irreversible self-association	Shear, surface area, surfactants, pH, T, buffers, ionic strength
Deamidation	Asn or Gln; acidic product, isoform, or hydrolysis	pH, T, buffers ionic strength
Isomerization/ Cyclization	Asn-X, Asp-X (X= Gly or Ser); basic product	pH, T, buffers ionic strength
Cleavage	Asp-X; fragments (proteolysis also possible from trace proteases)	pH, T, buffers
Oxidation	Met, Cys, His, Trp, Tyr; oxidized forms	Oxygen (ions, radicals, peroxide), light, pH, T, buffers, metals, (surfactants), free radical scavengers
Thiol Disulfide Exchange	Cys; mixed disulfides: intermolecular or intramolecular	pH, T, buffers, metals, thiol scavengers
β Elimination	Cys; dehydroalanine, free thiol	pH, T, buffers, oxygen (ions, radicals, peroxide), metals

[a] This table lists degradation pathways commonly observed for proteins and peptides. However, this list is not comprehensive and many of these degradation routes may occur independently or in combination with one another.

The formulation scientist also has the option of developing a solid formulation such as a lyophilized powder. The removal of excess water from the formulation minimizes the degradation rates for deamidation and hydrolysis. The residual moisture in a solid protein or peptide formulation can contribute to the physical stability of the protein by preventing its denaturation and subsequent aggregation upon reconstitution. Recent studies on lyophilization of proteins have shown that in the absence of excipients, proteins require some residual water, usually referred to as bound water, for stability (21). In the presence of excipients such as sugars, the amount of remaining water may often be reduced to levels below the hydration layer (22). Two different theories are currently used to explain the excipient stabilization of proteins and peptides in the lyophilized state. The first theory is based upon the observed differences in crystallinity for each excipient in the dry state. The major differences in each excipient have been correlated to their glass transition temperature (23). This theory neglects the specific interactions between excipients during drying and in the final dried state. The second theory, often referred to as the water replacement hypothesis, contends that some excipients can substitute for water in the dried state and thereby provide stabilization (24, 25). However, this theory also has its faults since many excipients with similar hydrogen bonding characteristics provide different degrees of stabilization (e.g. mannitol versus trehalose). Recent work by Carpenter and coworkers has demonstrated that there may be two distinct mechanisms for excipient stabilization. These mechanisms include protection during freezing (cryoprotectants) and drying (lyoprotectants) (26). The specific interactions between proteins and excipients with these properties have not yet been determined, but it is probable that an understanding of these interactions will lead to more rational design of solid protein and peptide formulations. While these studies focus on the prevention of denaturation and aggregation of proteins in the lyophilized state, other degradation routes are also critical for solid protein and peptide formulations. For example, residual oxygen head space in a vial containing the solid formulation can affect the rate of oxidation (22). Therefore, the degradation issues for lyophilized formulations are comparable to liquid formulations, but deamidation and hydrolysis rates are usually slower in the solid state.

Once the decision has been made to narrow the formulation candidates or proceed with a lyophilized formulation, another round of screening at elevated temperatures is usually performed. If these studies account for the potential differences between real time data at the actual storage conditions (usually 2-8° C) as described elsewhere (10-13), then the formulation scientist can select two or three final candidate formulations that should have a greater than 2 year shelf-life for real time stability studies. Unfortunately, the formulation scientist is often asked to have a stable formulation within a very short time (< 6 months). Thus, the real time stability data is often limited, but these data are required for the IND filing since the FDA will not accept any accelerated stability data. Stability problems encountered later are treated as amendments during the clinical trials. Further changes after the IND filing may include scale-up of the manufacturing process including the formulation. The scale-up of formulations may include new processing and storage containers, bulk filling equipment, and new delivery systems or vial configurations. These modifications may require another evaluation of the formulation with additional optimization. Hopefully, any necessary changes in the formulation already in clinical trials will not alter the *in vivo* characteristics of the drug (e.g. clearance, immunogenicity or potency).

The Crucial Steps in Delivery

While considering the alternatives for formulating a protein or peptide drug, the formulation scientist must also consider the route of administration. As mentioned previously, the route of administration can often be a critical variable in determining the success of the final product. As shown in Table II, there are many potential routes for the delivery of protein or peptide drugs. For proteins, only the direct injection and pulmonary routes of delivery have been approved by the FDA. Due to their large size, susceptibility to proteolytic degradation and requirements for an intact tertiary structure, proteins are difficult to deliver by oral, topical, or transdermal routes. Oral delivery of peptides and proteins usually results in low bioavailability (< 10%) and attempts to make stable prodrugs of peptides, which are more readily absorbed have provided some increase in the bioavailability (*28*). Unlike oral delivery, topical and transdermal delivery routes may be efficient enough for a localized treatment of skin disorders or diseases and, perhaps, for other localized therapies. Another method for protein delivery is the use of depot systems. Several companies are investigating their potential application. The depot systems offer the opportunity to localize the drug to the target site, reduce the frequency of injections, extend the half-life of the drug, and enhance its *in vivo* stability. Unfortunately, these systems also have inherent disadvantages that have not yet been overcome for their successful development into commercial products. A major disadvantage is often the use of harsh conditions (e.g. organic solvents or high temperature) to produce the depot, resulting in denaturation of the protein (*29*). In addition, for a long term depot formulation, the protein must be stabilized in an aqueous environment under physiological conditions at 37° C. Significant degradation has been observed for proteins stored under conditions analogous to those encountered in the depot (*30, 31*). For most of the novel delivery routes or depot formulations, the environment at the site of administration must be considered and stability studies of the drug in the same environment (e.g. serum) should be performed to assure that complete degradation does not occur before the drug reaches the desired site of action (*32*). Thus, the successful development of these formulations requires the initial development of protein formulations that are stable to both the process conditions and the *in vivo* environment.

Design of Depot Systems for Delivery. After the development of a formulation that meets these criteria, the material for the depot system must be chosen. The material should be biodegradable, well characterized, and nontoxic. The depot should also not alter the pharmacological properties of the drug and should act only as an inert carrier. Several examples of materials tested for depot systems are listed in Table III. While the materials derived from natural sources may often be more biocompatible, these materials may vary in their physical characteristics and can be difficult to obtain in a highly purified form. Many of the natural materials are derived from recombinant or animal sources and may contain contaminants such as endotoxins. In addition, some natural materials such as collagen or other proteins may invoke an unwanted immune response. In contrast, the synthetic materials are usually well characterized, highly pure polymers. The synthetic polymers have different physical and chemical characteristics. The polyanhydrides, polyesters, polyiminocarbonates, and polycaprolactones require the use of organic solvents or high temperatures for drug formulation. These systems are however well characterized and can be made reproducibly. On the other hand, the hydrogels, polyamino acids, and polyphosphazenes can be used to encapsulated drugs in an aqueous environment without organic solvents or elevated temperatures. These polymers are usually less stable and additional development is required to completely characterize their properties for controlled release of proteins and peptides. Of the synthetic materials, only the polyesters, specifically polylactic-coglycolic acid (PLGA) and polylactic acid (PLA), are currently used in commercial depot systems, Lupron Depot®(*2-5*) and Zoladex® (*58, 59*). These products were recently approved as alternate formulations and the polymer matrix, PLGA, has been well characterized and has been used extensively in humans. These polymers have been used for over twenty

Table II: Routes of Delivery for Proteins and Peptides

Delivery Routes	Formulation and Device Requirements	Commercial Products[a]
Invasive		
Direct Injection	Liquid or reconstituted solid, syringe	
intravenous (i.v.)		Activase®
subcutaneous (s.q.)		Nutropin®
intramuscular (i.m.)		RecombiVax®
intracerebral vein (i.c.v.)		
Depot system	Biodegradable polymers, liposomes permeable polymers (not degradable), microspheres, implants	
LHRH analogs (s.q. or i.m.)		Lupron Depot® Zoladex® Decapeptyl®
Noninvasive (see reference 27 for review)		
Pulmonary	Liquid or powder formulations, nebulizers, metered dose inhalers, dry powder inhalers	Pulmozyme®
Oral	Solid, emulsion, microparticulates, absorption enhancers	
Nasal	Liquid, usually requires permeation enhancers	Synarel®
Topical	Emulsion, cream or paste (liposomes)	
Transdermal	Electrophoretic (iontophoresis), electroporation, chemical permeation enhancers, prodrugs, ultrasonic	
Buccal, Rectal, Vaginal	Gels, suppositories, bioadhesives, particles	

[a] Nutropin® (recombinant human growth hormone), Activase® (recombinant human tissue plasminogen activator), and Pulmozyme® (recombinant human deoxyribonuclease I) are all products of Genentech, Inc. RecombiVax® (recombinant Hepatitis B surface antigen) is produced by Merck & Co. Lupron Depot® (leuprolide acetate - PLGA) is a product of Takeda Pharmaceuticals. Zoladex® (goserelin acetate - PLGA) is produced by the Imperial Chemical Industries, Ltd. Decapeptyl® is a manufactured by Debiopharm. Synarel® (nafarelin acetate) is made by Syntex Corporation.

Table III. Biodegradable Materials for Controlled Delivery of Proteins or Peptides

Materials	Degradation Mechanism	Reference [a]
Natural		
Starch	Amylase	*33, 34*
Alginate [b]	pH, Enzymes	*35, 36*
Collagen (Gelatin)	Collagenase	*37-39*
Proteins (Cross-linked albumin)	Enzymes	*40, 41*
Tricalcium phosphate [c] or calcium carbonate (hydroxyapatite)	Dissolves over time	*42-44*
Synthetic		
Hydrogels	Chemical or enzymatic hydrolysis, solubilization in aqueous media	*45*
Polyanhydrides	Hydrolysis	*46-48*
Polyesters (polylactides)	Ester hydrolysis, esterases	*49, 50*
Poly (ortho esters)	Ester hydrolysis, esterases	*49*
Polyiminocarbonates	Hydrolysis	*51, 52*
Polycaprolactones	Hydrolysis	*53, 54*
Polyamino acids	Enzymes	*55*
Polyphosphazenes [b]	Hydrolysis, dissolution	*56, 57*

[a] Reviews describing the application of these materials for delivery of proteins or peptides. When recent reviews were not available, several examples were listed.

[b] Alginate polyphosphazene depot systems usually require a cross-linking agent such as calcium or polycations.

[c] These depot systems are often used for mechanical strength in bone formation and are usually used in combination with another carrier such as collagen or starch.

years in resorbable sutures and have well established Drug Master Files at the FDA. In addition, PLGA and PLA are readily degraded in water to lactic and glycolic acid by ester hydrolysis. They are also available in a range of compositions and molecular weights from commercial suppliers. Several studies have been performed to determine the *in vitro* and *in vivo* characteristics of these polymers and their release of proteins and peptides (*5, 60, 61*). The *in vivo* degradation and injection site histology of several PLGA and PLA preparations have also been studied in animals (*62-64*). These studies indicated that the polymer did not cause any abnormal inflammation or irritation and the surrounding tissues completely recovered without scaring after degradation of the polymer. Thus, while the other materials may offer some advantages in manufacturing or *in vivo* performance, these polymers will probably be used for the first commercially available, injectable depot system containing a protein.

The polylactides have been used to develop protein or peptide depot systems for different applications ranging from bone formation to vaccines. Several recent examples of PLGA or PLA formulations containing proteins or peptides are listed in Table IV. For bone indications, the formulations were used to maintain a local concentration of the protein at the site while another material such as calcium carbonate was used for mechanical strength (*65, 66*). Since peptide - PLGA formulations are proven products, many of the peptide indications may result in PLGA formulations for chronic therapies (*67-71*). Finally, the vaccine - PLGA formulations may ultimately provide a route for single administration of vaccines which are currently given in multiple doses (*72-76*). These new vaccine formulations would eliminate the common problem of the patient's failure to return for booster immunizations, especially in large urban areas with transient populations and underdeveloped countries. In addition to these applications, new methods of preparation and derivatives of the polylactides may provide new routes of administration. For example, clearance of the drug may be reduced by encapsulating the protein in PLGA nanospheres with PEG attached to their surface. PEG would act to shield the microspheres from clearance by the reticuloendothelial system. These 'stealth' nanospheres may allow for the continuous release of proteins and peptides due to the increased blood circulation half life (*79*). Overall, the polylactides may provide a more efficient method for delivery of drugs that require chronic or repeated administration.

Other potential depot delivery systems such as liposomes may also be able to reduce the administration frequency of drugs and provide improved *in vivo* drug characteristics. Liposomes may act as inert carriers of proteins, peptides, and DNA or RNA for gene therapy or antisense technology, respectively. While there are a few commercial liposome drug formulations on the market, these formulations do not consist of a protein (*80-82*). However, liposomes may find utility in solubilizing lipophilic peptides or proteins and as a result of their large size, they may act as a depot for the drug, preventing clearance or degradation. A protein or peptide is often presented on the surface of the liposome and can thus bind to its receptor or perform its biological function (*83*). Many of these formulations may only increase the duration of response for a few hours. The composition of liposomes directly impacts the amount of time that the liposome remains in the circulation. The common components of a liposome are phospholipids, cholesterol, and stabilizing components such as polyalcohols or sugars for protection during processing. By altering the liposome composition, the duration of bioactivity can be extended to several days. This effect was demonstrated by utilizing Stealth® liposomes and vasopressin (*84*). Stealth® liposomes consist of lipids with polyethylene glycol (PEG) covalently attached to their hydrophilic head groups. These PEG molecules act as a shield to prevent uptake by phagocytic cells in the reticuloendothelial system. Liposomes may

Table IV. Partial List of Recent Examples Using Polylactides for Controlled Release of Proteins or Peptides

Protein or Peptide	Polymer [a]	Application	Reference
Bone morphogenetic protein	PLA-PEG copolymer (650 Da PLA - 200 Da PEG)	Bone formation	*65*
Transforming growth factor-β_1	50:50 PLGA (40-100 kDa) (+ demineralized bone matrix)	Bone formation	*66*
Thyrotropin	75:25 PLGA (11 kDa)	Central nervous system dysfunction	*67*
Growth hormone releasing factor	75:25 PLGA (91 kDa)	Growth hormone deficiency	*68*
Somatostatin analogue	55:45 PLGA (23-76 kDa)	Acromegaly, tumors	*69*
Neurotensin Analogue	PLA (2-6 kDa)	Psychothropic	*70*
Cyclosporin A	50:50 PLGA (0.44 & 0.80 dL/g)	Immunosuppression	*71*
Colonizing factor antigen (*E. coli*)	PLGA (0.73 dL/g)	Oral vaccine	*72*
Cholera toxin B subunit	PLA (2 kDa)	Oral vaccine	*73*
Diphtheria toxoid formilin treated	PLA (49 kDa)	Vaccine	*74*
Ovalbumin	50:50 PLGA (22 kDa) 85:15 PLGA (53 kDa)	Vaccine	*75*
Tetanus toxoid	50:50 PLGA (100 kDa)	Vaccine	*76*
LHRH antagonists	50:50 PLGA 75:25 PLGA	Tumor suppression	*77*
Horse radish peroxidase Bovine serum albumin	75:25 PLGA (10 kDa)	Marker proteins Mechanistic studies	*78*

[a] X:Y PLGA indicates the mole fraction (%) of lactide (X) and glycolide (Y) in the copolymer. The polymer size is reported as either molecular weight in kilodaltons (kDa) or intrinsic viscosity in decaliters/gram. Some references did not provide complete descriptions of the polymer.

also have utility for presentation of antigens and adjuvants. Liposome presentation of proteins may mimic the normal antigen - membrane configuration and the liposome will be consumed by phagocytic cells. This approach has been applied to the development of vaccine formulations by using dehydration-rehydration vesicles (DRVs) which trap protein on their surface during the drying process (85). In addition, it may be possible to target liposomes to specific cell populations by placing a marker protein such as a monoclonal antibody or cytokine on the surface of the liposome (86, 87). Thus, liposomal formulations may provide unique methods for targeting and delivery of proteins and peptides.

The Role of Pharmacokinetics in Delivery. When choosing the type of delivery system and the route of administration, the pharmacokinetics of the formulation must be considered. It may often be easier to predict the pharmacokinetics of a depot formulation which provides a sustained therapeutic level of the drug than a solution of the drug administered at different sites or in different formulations. The impact of the formulation on the pharmacokinetics of a drug are well illustrated in the case of insulin as shown in Table V. Insulin has the ability to form multimers of the native protein and these multimers (zinc-hexamer) can reversibly form crystals. After subcutaneous injection, the multimers must dissociate into monomers to diffuse across the capillary membrane and bind to the receptor. By formulating insulin in different solutions, the pharmacokinetics are dramatically altered. The differences in time before the insulin reaches therapeutic levels and the duration of its therapeutic effect can result in severe consequences if the incorrect formulation is administered to a patient. In particular, a patient with severe side effects from hyperglycemia may require a rapid onset formulation and may be seriously injured if administered a slower acting form. In addition to the differences in pharmacokinetics due to the associated state of the protein, each formulation has a different stability (88, 90, 91). For example, the presence of rhombohedral crystals and free zinc can accelerate hydrolysis (88). The dominant degradation mechanism is hydrolysis in all the formulations except the isophane solutions where the formation of high molecular weight aggregates occurs at a much greater rate. These formulations also invoke a higher incidence of antibodies to insulin as well as the additive, protamine (90). The degradation of insulin by the solution conditions (additives, pH, etc.) is the result of the conformational flexibility of the insulin monomer (91). In the case of insulin, the formulation has an enormous impact on the stability and the pharmacokinetics of the drug, both of which determine its ultimate application. Many other protein and peptide drug formulations may directly alter the pharmacokinetics and pharmacodynamics of the molecule as well as its *in vitro* stability (92, 93).

Conclusions and Future Directions

The design and development of formulations and methods of delivery for proteins and peptides is dependent upon several variables. The relationship between the formulation, route of delivery, pharmacokinetics, toxicity, and clinical indication must be carefully balanced to successfully develop a protein or peptide drug. A formulation scientist should consider the possible options for each new therapeutic entity prior to the design of a formulation. Many new systems such as polymers and liposomes may offer attractive alternatives to traditional liquid or solid dosage forms. While these new delivery options often require invasive methods, development of new noninvasive delivery routes such as pulmonary or transdermal delivery may yield promising results for future applications. The formulation scientist will soon have to address new challenges in the delivery of DNA and RNA for gene therapy and antisense technology, respectively. The success of these future therapies may depend upon the ability to successfully deliver the nucleic acid polymer to the desired cell nucleus. Thus, the formulation scientist will play a key role in the advancement of these technologies. Further, the targeting of drugs to specific sites of action should

Table V. Effect of Insulin Formulation on the Pharmacokinetics after Subcutaneous Injection (*89, 90*)

Trade Name [a]	Formulation	Pharmacokinetics [b]
Humulin® R Novolin® R	Zinc-insulin crystalline suspension (Acid regular)	Rapid onset, short duration Start: 0.5 hr; Peak: 2.5-5 hr End: 8 hr
Humulin® N Novolin® N	Isophane suspension protamine, zinc crystalline insulin (buffer water for injection)	Intermediate-acting, slower onset, longer duration than regular insulin Start: 1.5 hr; Peak: 4-12 hr End: up to 24 hr
Humulin® 70/30 [c] Novolin® 70/30	70% isophane suspension 30% zinc crystalline	Intermediate-acting, faster onset, longer duration Start: 0.5 hr; Peak: 2-12 h End: up to 24 hr
Humulin® U	Extended zinc-insulin suspension - all crystalline	Slow-acting, slow onset, longer, less intense duration than R or N forms.
Humulin® L Novolin® L	70% zinc-insulin crystalline suspension, 30% amorphous insulin (cloudy suspension)	Intermediate-acting, slower onset, longer duration Start: 2.5 hr; Peak: 7-15 hr End: 22 hr
Humulin® BR	Zinc crystalline insulin dissolved in sodium diphosphate buffer	Rapid onset, short duration Use in pumps only.

[a] Humulin® products are recombinant human insulin derived from *Escherichia coli* and produced by Eli Lilly & Company. Novolin® products are recombinant human insulin derived from *Saccharamyces cervasiae* and produced by Novo Nordisk. Both companies also sell other forms of recombinant human insulin and may have additional forms (formulations or new drugs) in clinical trials.

[b] The pharmacokinetics of each formulation may vary greatly among different individuals. The values reported here are taken from reference *89*. The onset of therapeutic levels of insulin is referred to as the start of the effect. The maximum serum level of insulin is denoted as the peak and the time at which the insulin levels are below therapeutic levels is listed as the end of the therapeutic time course.

[c] Solution consists of 70% N form and 30% R form for both products.

increase the utility of many proteins and peptides that are potent and toxic. Low doses of these drugs can then be used and directed to the target site. Finally, future advances in the transport of proteins and peptides across biological barriers such as the blood brain barrier (*94-99*) and gastrointestinal epithelium (*100-104*) may result in new delivery options and successful therapies.

Acknowledgments

The authors would like to thank Dr. Andrew J. S. Jones and other Genentech reviewers for their comments and suggestions to improve the manuscript. We also appreciate the support and encouragement of Dr. Rodney Pearlman and the Pharmaceutical Research and Development Department at Genentech.

Literature Cited

1. Oliyai, C.; Schöneich, C.; Wilson, G. S.; Borchardt, R. T. In: *Topics in Pharmaceutical Sciences*, Crommelin, D. J. A.; Miha, K. K., Eds.; Med. Pharm. Scientific Publishers, Stuttgart, 1992, pp. 23-46.
2. Okada, H.; Inoue, Y.; Heya, T.; Ueno, H.; Ogawa, Y.; Toguchi, H. *Pharm. Res.* **1991**, 787-791.
3. Ogawa, Y.; Yamamoto, M.; Okada, H.; Yashiki, T.; Shimamoto, T. *Chem. Pharm. Bull. (Japan)* **1988**, *36*, 1095-1103.
4. Ogawa, Y.; Yamamoto, M.; Okada, H.; Yashiki, T.; Shimamoto, T. *Chem. Pharm. Bull. (Japan)* **1988**, *36*, 1502-1507.
5. Ogawa, Y.; Yamamoto, M.; Okada, H.; Yashiki, T.; Shimamoto, T. *Chem. Pharm. Bull. (Japan)* **1988**, *36*, 2576-2581.
6. *Pharma Japan* **1993**, *1355*, 2.
7. *FDC Reports: The Pink Sheet* **1991**, *53*, T&G 5.
8. *Scrip* **1992**, *1765*, 25.
9. Loudon, G. M. *J. Chem Ed.* **1991**, *68*, 973-984.
10. Yoshioka, S.; Aso, Y.; Izutsu, K.-I.; Terao, T. *J. Pharm. Sci.* **1994**, *83*, 454-456.
11. Cleland, J. L.; Powell, M.F.; Shire, S. J. *Critical Reviews in Therapeutic Drug Carrier Systems* **1993**, *10*, 307-377.
12. Pearlman, R.; Nguyen, T. *J. Pharm. Pharmacol.* **1992**, *44 (Suppl. 1)*, 178-185.
13. Gu, K. M.; Erdos, E. A.; Chiang, H.; Calderwood, T.; Tsai, K.; Visor, G. C.; Duffy, J.; Hsu, W.; Foster, L. C. *Pharm. Res.* **1991**, 8, 485-490.
14. Jones, A. J. S., In *Protein Formulations and Delivery*; Cleland, J. L.; Langer, R., Eds.; American Chemical Society Symposium Series, ACS Books, New York, NY 1994, (This Volume), Chapter 2.
15. Manning, M. C.; Patel, K.; Borchardt, R. T. *Pharm. Res.* **1989**, *6*, 903-919.
16. Wang, Y.-C. J.; Hanson, M. A. *J. Parenter. Sci. Technol.* **1988**, *42*, S2-S15.
17. Chen, T. *Drug Dev. Indust. Pharm.* **1992**, *18*, 311-325.
18. Volkin, D. B.; Klibanov, A. M. In: *Protein Function: A Practical Approach*, Creighton, T. E., Ed.; IRL Press, New York, NY, 1989, pp. 1-24.
19. Skottner, A.; Forsman, A.; Skoog, B; Kostyo, J. L.; Cameron, C. M.; Adamfio, N. A.; Thorngren, K. G.; Hagerman, M. *Acta Endocrinol.* **1988**, *118*, 14-19.
20. Hora, M. S.; Rana, R. K.; Wilcox, C. L.; Katre, N. V.; Hirtzer, P.; Wolfe, S. N.; Thomson, J. W. *Dev. Biol. Stand.* **1991**, *74*, 295-307.
21. Hsu, C. C.; Ward, C. A.; Pearlman, R.; Nguyen, H. M.; Yeung, D. A.; Curley, J. G. *Dev. Biol. Stand.* **1991**, *74*, 255-267.
22. Pikal, M. J.; Dellerman, K.; Roy, M. L. *Dev. Biol. Stand.* **1991**, *74*, 21-27.
23. Franks, F. *Cryo. Lett.* **1990**, *11*, 93-99.
24. Crowe, J. H.; Crowe, L. M.; Carpenter, J. F. *BioPharm* **1993**, *6 (4)*, 40-43.
25. Crowe, J. H.; Crowe, L. M.; Carpenter, J. F. *BioPharm* **1993**, *6 (3)*, 28-33.
26. Prestrelski, S. J.; Tedeschi, N.; Carpenter, J. F.; Arakawa, T. *Biophys. J.* **1993**, *65*, 661-669.

27. Wearley, L. L. *Critical Reviews in Therapeutic Drug Carrier Systems* **1991**, *8*, 331-394.
28. Oliyai, R.; Stella, V. J. *Annu. Rev. Pharmcol. Toxicol.* **1993**, *32*, 521-544.
29. Jones, A. J. S.; Nguyen, T. H.; Cleland, J. L.; Pearlman, R. In:*Trends and Future Perspectives in Peptide and Protein Drug Delivery;* Lee, V. H. L.; Hashida, M.; Mizushima, Y., Eds.; Drug Targeting & Delivery; Harwood Academic Publishers, Gmbh: Amsterdam, The Netherlands, 1994, *in press.*
30. Hageman, M. J.; Bauer, J. M.; Possert, P. L.; Darrington, R. T. *J. Agric. Food Chem.* **1992**, *40*, 348-355.
31. Costantino, W. R.; Langer, R.; Klibanov, A. *Pharm. Res.* **1994**, *11*, 21-29.
32. Powell, M. F.; Grey, H.; Gaeta, F.; Sette, A.; Colón, S. *J. Pharm. Sci.* **1992**, *81*, 731-735.
33. Stjarnkvist, P.; Degling, L.; Sjöholm, I. *J. Pharm. Sci.* **1991**, *80*, 436-440.
34. Arthursson, P.; Edman, P.; Laakso, T.; Sjöholm, I. *J. Pharm Sci.* **1984**, *73*, 1507-1513.
35. Wheatley, M. A.; Chang, M.; Park, E.; Langer, R. *J. Appl. Polymer Sci.* **1991**,*43*, 2123-2135.
36. Downs, E. C.; Robertson, N. E.; Riss, T. L.; Plunkett, M. L. *J. Cellular Physiology* **1992**, *152*, 422-429.
37. Lindholm, T.S.; Gao, T. J. *Ann. Chir. Gyn.* **1993**, *82 (S207)*, 3-12.
38. Horisaka, Y.; Okamoto, Y.; Matsumoto, N.; Yoshimura, Y.; Hirano, A.; Nishida, M.; Kawada, J.; Yamashita, K.; Takagi, T. *J. Biomed. Matrials. Res.* **1994**, *28*, 97-103.
39. Takaoka, K.; Koezuka, M.; Nakahara, H. *J. Orthopaedic Res.* **1991**, *9*, 902-907.
40. Santiago, N.; Milstein, S.; Rivera, T.; Garcia, E.; Zaidi, T.; Hong, H.; Bucher, D. *Pharm. Res.* **1993**, *10*, 1243-1247.
41. Levy, M. C.; Andry, M. C. *J. Microencapsulation* **1991**, *8*, 335-347.
42. Ripamonti, U., Ma, S.; Reddi, A. H. *Matrix* **1992**, *12*, 202-212.
43. Herr, G.; Wahl, D.; Kusswetter, W. *Ann. Chir. Gyn.* **1993**, *82(S207)*, 99-107.
44. Kenley, R. A.; Yim, K.; Abrams, J.; Ron, E.; Turek, T.; Marden, L. J.; Hollinger, J. O. *Pharm. Res.* **1993**, *10*, 1393-1401.
45. Kalpana, K. R.; Park, K. *Adv. Drug Del. Rev.* **1993**, *11*, 59-84.
46. Shieh, L.; Tamada, J.; Tabata, Y.; Domb, A.; Langer, R. *J. Controlled Release* **1994**, *29*, 73-82.
47. Ron, E.; Turek, T.; Mathiowitz, E.; Chasin, M.; Hageman, M.; Langer, R. *PNAS* **1993**, *90*, 4176-4180.
48. Langer, R. *Acc. Chem. Res.* **1993**, *26*, 537-542.
49. Heller, J. *Adv Drug Delivery Reviews* **1993**, *10*, 163-204.
50. Shah, N. H.; Railkar, A. S.; Chen, A. S.; Chen, F. C.; Tarantino, R.; Kumar, S.; Murjani, M.; Plamer, D.; Infeld, M. H.; Malick, A. W. *J. Controlled Release* **1993**, *27*, 139-147.
51. Arshady, R. *J. Controlled Release* **1991**, *17*, 1-21.
52. Pulapura, S.; Ki, C.; Kohn, J. *Bopmaterials* **1990**, *11*, 666-678.
53. Coffin, M. D.; McGinity, J. W. *Pharm. Res.* **1992**, *9*, 200-205.
54. Marchalheussler, L.; Sirbat, D.; Hoffman, M.; Maincent, P. *J. Francais Opthalmol.* **1991**, *14*, 371-375.
55. Li, C.; Yang, D. J.; Kuang, L. R.; Wallace, S. *Int. J. Pharm.* **1993**, *94*, 1-3.
56 Andrianov, A. K.; Cohen, S.; Visscher, K. B.; Payne, L. G.; Allcock, H. R.; Langer, R. *J. Controlled Release* **1993**, *27*, 69-77.
57. Crommen, J.; Vandorpe, J.; Schacht, E. *J. Controlled Release* **1993**, *24*, 1-3.
58. Furr, B. J. A.; Hutchinson, F. G. *J. Controlled Release* **1992**, *21*, 117-128.
59. Ahmed, S. R.; Grant, J.; Shalet, S. M.; Howell, A.; Chowdhury, S. D.; Weahterspoon, T.; Blacklock, N. J. *Br. Med. J. Clin. Res.* **1985**, *290*, 185-187.
60. Kenley, R. A., Lee, M. O., Mahoney, T. R.; Sanders, L. M. *Macromolecules* **1987**, *20*, 2398-2403.

61. Okada, H.; Inoue, Y.; Heya, T.; Ueno, H.; Ogawa, Y.; Toguchi, H. *Pharm. Res.* **1991**, *8*, 787-791.
62. Csernus, V. J.; Szende, B.; Schally, A. V. *Int. J. Peptide Protein Res.* **1990**, *35*, 557-565.
63. Visscher, G. E.; Robison, R. L.; Maulding, H. V.; Fong, J. W.; Pearson, J. E.; Argentieri, G. J. *J. Biomed. Mater. Res.* **1985**, *19*, 349-365.
64. Miller, R. A.; Brady, J. M.; Cutright, D. E *J. Biomed. Mater. Res.* **1977**, *11*, 711-719.
65. Miyamoto, S.; Takaoka, K.; Okada, T.; Yoshikawa, H.; Hashimoto, J. Suzuki, S.; Ono, K. *Clin. Orthopaedics Related Res.* **1993**, *294*, 333-343.
66. Gombotz, W. R.; Pankey, S. C.; Bouchard, L.S.; Ranchalis, J.; Puolakkainen, P. *J. Biomater. Sci. Polymer Edn.* **1993**, *5*, 49-63.
67. Heya, T.; Okada, H.; Ogawa, Y.; Toguchi, H. *Int. J. Pharm.* **1991**, *72*, 199-205.
68. Mariette, B.; Coudane, J.; Vert, M.; Gautier, J.-C.; Moneton, P. *J. Controlled Release* **1993**, 237-246.
69. Bodmer, D.; Kissel, T.; TRaechslin, E. *J. Controlled Release* **1992**, *21*, 129-138.
70. Yamakawa, I.; Tsushima, Y.; Machida, R.; Watanabe, S. *J. Pharm Sci.* **1992**, *81*, 899-903.
71. Sánchez, A.; Vila-Jato, J. L.; Alonso, M. J. *Int. J. Pharm.* **1993**, *99*, 263-273.
72. Reid, R. H.; Boedeker, E. C.; McQueen, C. E.; Davis, D.; Tseng, L. Y.; Kodak, J.; Sau, K. *Vaccine* **1993**, *11*, 159-167.
73. Almeidia, A. J.; Alpar, H. O.; Williamson, D.; Brown, M. R. W. *Biochem. Soc. Trans.* **1992**, *20*, 316S.
74. Singh, M.; Singh, O.; Singh, A.; Talwar, G. P. *Int. J. Pharm.* **1992**, *85*, R5-R8.
75. Jeffery, H.; Davis, S. S.; O'Hagen, D. T. *Pharm. Res.* **1993**, *10*, 362-368.
76. Alonso, M. J.; Cohen, S.; Park, T. G.; Gupta, R. K.; Siber, G. R.; Langer, R. *Pharm. Res.* **1993**, *10*, 945-953.
77. Stoeckemann, K.; Sandow, J. *J. Cancer Res. Clin. Oncology* **1993**, *119*, 457-462.
78. Cohen, S.; Yoshioka, T.; Lucarelli, M.; Hwang, L. H.; Langer, R. *Pharm. Res.* **1991**, *8*, 713-720.
79. Gref, R.; Minamitake, Y.; Peracchia, M. T.; Trubetskoy, V.; Torchilin, V.; Langer, R. *Science* **1994**, *263*, 1600-1603.
80. Talsma, H.; Crommelin, D. J. A. *Pharm. Tech.* **1992**, *16*, 96-106.
81. Talsma, H.; Crommelin, D. J. A. *Pharm. Tech.* **1992**, *16*, 52-58.
82. Talsma, H.; Crommelin, D. J. A. *Pharm. Tech.* **1993**, *17*, 48-59.
83. Maruyama, K.; Mori, A.; Bhadra, S.; Subbiah, M. T. R.; Huang, L. *Biochem. Biophys Acta* **1991**, *1070*, 246-252.
84. Woodle, M. C.; Strom, G.; Newman, M. S.; Jekot, J. J.; Collins, L. R.; Martin, F. J.; Szoka, F. C. *Pharm. Res.* **1992**, *9*, 260-265.
85. Gregoriadis, G.; David, D.; Davies, A. *Vaccine* **1987**, *5*, 145-151.
86 Bakouche, O.; Brown, D. C.; Lachman, L. B. *J. Immunol.* **1987**, *138*, 4256-4262.
87. Crommelin, D.J.A.; Eling, W.M.C.; Steerenberg, P. A.; Nässander, U. K.; Storm, G.; DeJong, W. H.; Van Hoesel, Q. G. C. M.; Zuidema, J. *J. Controlled Release* **1991**, *16*, 147-154.
88. Brange, J.; Langkjaer, L.; Havelund, S.; Vølund, A. *Pharm. Res.* **1992**, *9*, 715-725.
89. *Physicians' Desk Reference* **1992**, 46th Edition, pp. 1267-1274; 1625-1629.
90. Brange, J.; Havelund, S., Hougaard, P. *Pharm. Res.* **1992**, *9*, 727-734.
91. Brange, J.; Langkjaer, L. *ActaPharm. Nord.* **1992**, *4*, 149-158.
92. Talmadge, J. E. *Adv. Drug Delivery Rev.* **1993**, *10*, 247-299.
93. Breimer, D. D. *J. Controlled Release* **1992**, *21*, 5-10.
94. Pardridge, W. M. *Amer. J. Pharm. Ed.* **1993**, *57*, 439-440.

95. Bickel, U.; Yoshikawa, T.; Pardridge, W. M. *Adv. Drug Delivery Rev.* **1993**, *10*, 205-245.
96. Menei, P.; Daniel, V.; Monteromenei, C.; Bouillard, M.; Pouplardbartgelaix, A.; Benoit, J. P. *Biomaterials* **1993**, *14*, 470-478.
97. Vanbree, J. B. M. M.; Deboer, A. G.; Danhof, M.; Breimer, D. D. *Pharm. World Sci.* **1993**, *15*, 2-9.
98. Pardridge, W. M. *Pharm.col. Tox.* **1992**, *71*, 3-10.
99. Brem, H.; Domb, A.; Lenartz, D.; Dureza, C.; Olivi, A.; Epstein, J. I. *J. Controlled Release* **1992**, *9*, 325-329.
100. Jimenezcastellanos, M. R.; Zia, H.; Rhodes, C. T. *Drug Dev. Indust. Pharm.* **1993**, *19*, 143-194.
101. Bai, J. P. F.; Amidon, G. L. *Pharm. Res.* **1992**, *9*, 969-978.
102. Touitou, E. *J. Controlled Release* **1992**, *21*, 139-144.
103. Muranishi, S.; Murakami, M.; Hashidzume, M. Tamada, K.; Tajima, S.; Kiso, Y. *J. Controlled Release* **1992**, *19*, 179-188.
104. Smith, P. L.; Wall, D. A.; Gochoco, C. H.; Wilson, G. *Adv. Drug Del. Rev.* **1992**, *8*, 253-290.

RECEIVED June 9, 1994

FORMULATION ISSUES

Chapter 2

Analytical Methods for the Assessment of Protein Formulations and Delivery Systems

A. J. S. Jones

Pharmaceutical Research and Development, Genentech, Inc., South San Francisco, CA 94080

The characterization of a protein drug is a complex undertaking, requiring the use of a wide range of methods to establish the properties of the drug substance. The development of a formulation, and especially a novel delivery system, begins with the information and analytical methods from the characterization and initially extends that information by evaluating the ranges of storage conditions or processing operations under which the protein remains in its initial state in terms of the rates of degradation over these ranges. The complexity of protein molecules means that there are many potential degradation pathways, each with their individual dependences on such parameters as pH, ionic strength, temperature and so on. Further, each protein represents a unique combination of such pathways and dependences. It is therefore critical that a broad array of methods be used to evaluate the effects of processing or storage to assure optimal maintenance of the safety and efficacy of the drug. This chapter surveys the methods that are commonly used, in terms of their principles, utility and limitations in the study of protein degradation.

The task of the protein formulation scientist begins when significant quantities of a highly purified protein preparation are available. The assessment of the formulation and delivery systems is therefore not concerned with proteins other than the active molecule. Consequently, the main tools required are those that are capable of resolving different forms of the active component from each other and from their degradation products. Frequently, the protein preparation can be purified to the point of showing only one component by many analytical methods. However, the application of mammalian cell culture methods has also allowed the production of large and complex proteins as well as glycoproteins. The larger a protein molecule, the more difficult it is to separate one form from minor degradation products during large-scale production and purification. In such cases, the starting material for formulation development will already be heterogeneous. In addition, glycoproteins frequently exhibit "microheterogeneity", a multiplicity of very similar molecules which differ only in the details of their glycan structures which may be present even though the amino acid sequence is the same for all. In such cases of product

0097–6156/94/0567–0022$08.72/0

heterogeneity, those purifying and characterizing the product must face regulatory issues regarding the properties of the individual forms and the reproducibility of the heterogeneity from batch to batch. This heterogeneity also dramatically complicates the work of the formulation scientist in evaluating the suitability or stability of the formulation. To be useful for studying degradation, the analytical methods must therefore have high resolution to afford separation between altered macromolecules which may differ only slightly in their physical properties from the parent molecule. For example, the loss of an amide group or the addition of an oxygen atom is an extremely slight change for a molecule with a mass of 20kD-150kD!

A major portion of the effort in developing these analytical methods is devoted to optimizing the assay conditions for maximal resolving capability. A thorough understanding of the principles of each method and the influence of the operating parameters is crucial to such an effort (1). It is beyond the scope of this chapter to cover all of these in depth, so the present goal will be to discuss the major methods of analysis of proteins, primarily from a physical and chemical viewpoint. The Table lists many of the common alterations found in modified or degraded forms of proteins observed during formulation or stability studies, along with methods which are expected to be useful in detection and characterization of such changes. The comments listed in the Table are caveats or restrictions that should be borne in mind when these methods are contemplated and they are illustrated in the text sections where the specific methods are discussed. Methods for evaluating non-product-related protein impurities will not be addressed, nor will potency assays, which, although critical for stability confirmation, are usually product-specific, frequently of low precision and cannot resolve different forms of the product.

Electrophoresis.

The acidic and basic groups of proteins make them polyelectrolytes and they can therefore be separated in an applied electric field. This phenomenon, known as electrophoresis, is one of the most common methods employed to separate mixtures of proteins on an analytical scale. High resolution "disc" (discontinuous) electrophoresis was developed by Ornstein (2) and Davis (3). In this format, the buffer systems used for the gel, sample and electrolyte chambers were discontinuous, resulting in a phenomenon known as "stacking", in which the sample proteins are initially concentrated into a narrow starting zone and consequently are sharp bands after separation. A practical text covering the many forms of electrophoresis has been published (4) and a recent review details the theoretical principles behind electromigration techniques (5).

Native Electrophoresis. In native electrophoresis, performed in the absence of denaturing agents, the major factors controlling the electrophoretic mobility of a macromolecule are its net charge (at the pH of the separation) and its Stokes' radius, a hydrodynamic parameter determined primarily by the size and, to a lesser extent, shape of the molecule. The basis for separation is therefore mainly by differences in mass-to-charge ratio. The use of an anti-convective matrix, such as a gel, can enhance the resolving power of the method. If the pores of the gel matrix are comparable to the dimensions of the proteins, they will present resistance to the movement of the molecules in a size-dependent fashion. Thus proteins with the same mass-to-charge ratio will be separated if they are of different sizes. Electrophoretic separations in native gel electrophoresis are therefore seen to produce information on both size and charge and independent data are required to resolve these two factors. If data are available to show that all the components present in the sample have the same mass, then this method does provide information on the heterogeneity of

Table of Common Protein Alterations Observed During Formulation Development

ALTERATION	METHOD*	COMMENTS**
Chemical		
Any mass change	Mass spectrometry	Used alone or in combination with listed methods
Charge Alteration (e.g. deamidation)	Isoelectric focusing Ion-exchange chromatography Native electrophoresis Reversed-phase-HPLC	Group must be charged at pI for detection Group must be charged at assay pH Group must be charged at assay pH Using peptide map
Neutral Alteration (e.g. met oxidation)	Reversed-phase-HPLC Hydrophobic interaction chromatography Amino acid analysis	Especially in peptide map May also detect ionic changes Adduct must be stable to hydrolysis conditions
Polypeptide Cleavage (Proteolysis or terminal processing))	N- and C-terminal sequencing Size exclusion chromatography Reversed-phase-HPLC SDS-polyacrylamide gel electrophoresis Isoelectric focusing	May require disulfide reduction and/or denaturation "Tryptic" map will not detect trypsin-like proteolysis May require disulfide reduction New termini may alter pI
Disulfide alterations	Reversed-phase-HPLC Reversed-phase-HPLC SDS-polyacrylamide gel electrophoresis	Run ± reduction of disulfides Peptide map run ± reduction of disulfides Run ± reduction of disulfides
Physical*		
Altered secondary structure	Far UV circular dichroism Infrared spectroscopy	Possible aromatic amino-acid interference Can be done in dry state etc.
Altered tertiary structure	Near UV circular dichroism UV absorption spectroscopy Fluorescence spectroscopy	Only aromatic residues sensitive to change " and difference and 2nd derivative modes helpful Only fluorescent residues sensitive to change
Aggregation	Size exclusion chromatography Light scattering Analytical Ultracentrifugation	Confirmation of area recovery essential Corrections needed for UV protein assay Static and dynamic light scattering methods Measure self- or ligand associations

* Methods frequently useful for analysis of alteration - only general description noted: e.g. many exist in capillary format also.
** Comments relate to caveats or restrictions. See text for details of principles of method, examples and references.
*** Effects of these may be detected in methods listed above for chemical changes.
Note that combinations of these methods are usually needed for unambiguous identification.

charges of those components (6). This heterogeneity of charge can arise from differences in deamidation or processing (of either the peptide or glycans) or occasionally by mutation.

SDS Polyacrylamide Gel Electrophoresis. The most common form of electrophoresis of proteins employs the denaturing agent sodium dodecyl sulfate (SDS). Reduced proteins tend to bind a relatively constant amount of SDS on a weight basis : approximately 1.4 gm of SDS per gram of protein (7). The SDS molecule carries a negative charge and complexes of proteins with SDS have very similar mass to charge ratios and therefore free electrophoretic mobility (8). When a mixture of SDS-saturated proteins is electrophoresed in a gel matrix with the correct pore size, the major factor determining their migration rate is their effective size. SDS polyacrylamide gel electrophoresis is commonly used for assessing purity and as a tool for determining apparent molecular weight (9,10). A widely used system is the discontinuous buffer system with SDS published by Laemmli (11). This system may be used for proteins from around 10,000 to 300,000 molecular weight by varying the acrylamide gel concentration and the cross-linker, bisacrylamide, concentration, to vary the pore size (8 ,12). With a fixed pore size only a limited range of polypeptide molecular weights can be well resolved; however, a gradient of acrylamide concentration can be used, resulting in a the resolution of a wider range of molecular weights in one analysis (13). The most reliable estimate of molecular weight is obtained from analyses in which the disulfide bonds have been reduced, and the non-helical portions of the polypeptide chains become truly random (8).

A common use for SDS-PAGE is the analysis of proteins over time during stability evaluation. This allows the detection of proteolytic degradation events and the development of covalent aggregates, either reducible or non-reducible forms. Proteins containing disulfide bonds can rearrange these bonds over time to yield dimers and higher aggregates linked by new intermolecular disulfides, such as those observed in human growth hormone (14), and tissue plasminogen activator (15).

Interpretation of SDS-PAGE separations is not always straightforward. The effect of reagents used to block the free sulfhydryl groups (if used, for example, to prevent re-oxidation during the analysis) can be complex. Thus, t-PA, which has 35 cysteine residues out of 527 total amino acids, runs to approximately the correct weight when simply reduced. However, if it is carboxymethylated with iodoacetic acid it has an apparent molecular weight closer to 95,000, with a similar increase observed for bovine serum albumin (16). Another example of anomalous migration comes from the derivatization of IL-2 with PEG in which molecules with an average of either 2 or 3 PEG (MW 6 kD) groups per molecule (MW 17 kD) ran to the same place on SDS-PAGE, even though their masses were quite different (17). In this case, differences in effective size or SDS binding may be the result of attachment of the PEG groups at different locations on the molecule. It is also commonly found that proteins with significant carbohydrate content will run as diffuse bands, probably attributable to a combination of heterogeneity of molecular weight, charge (as a result of variable extents of sialylation) and SDS binding (18). Occasionally proteins are very resistant to denaturation by SDS and unfold only slowly, so sample treatment must be carefully evaluated (19).

Isoelectric Focusing. As noted above, the mobility of a native protein is dependent on, among other things, its charge. This charge is determined by the pH of the solution. At the isoelectric point (pI) the protein's net charge will be zero and the mobility in an electric field will be zero. Thus, if the electric field is also a pH gradient, the protein will migrate to the point where the pH is the same as the pI and migration will stop. This phenomenon is known as isoelectric focusing (IEF) (20).

The key to the method was the development of a system for producing the pH gradient that is stable in the applied electric field (21) by preparing "carrier ampholytes" of appropriate pI and buffering capacity. Such gradients, when run with large-pore anti-convective agents such as agarose or acrylamide (22) can separate protein species which differ in pI by as little as 0.02 pH units (21). This method has been used to assess deamidation during stability studies of human growth hormone by quantitative densitometry (23) and the heterogeneity of recombinant tissue plasminogen activator caused by differences in the degree of sialylation of the complex carbohydrates (15,24). One significant disadvantage of isoelectric focusing results from the fact that most proteins exhibit minimum solubility around their pI and neutral surfactants or urea may be required to allow them to remain in solution while they are focusing.

The pH gradient in conventional isoelectric focusing has limited resolving power and eventually degenerates over time. A more recent and higher resolution method has been developed (25) in which the pH gradient is covalently immobilized on the acrylamide. This method allows resolution of species whose pI's differ by as little as 0.001 pH units in a suitable pH gradient (26). This method has been used to separate two forms of hemoglobin whose pI's differ by 0.003 pH units as a result of an alanine to glycine substitution (27).

Two-dimensional Gel Electrophoresis. For complex mixtures of proteins, the resolution afforded by any one-dimensional analysis yields limited information and the combination of isoelectric focusing and SDS-PAGE represents a powerful high resolution method for analysis of such mixtures (28). In this method, the sample is first separated by IEF, usually in a thin tube with urea and neutral surfactant present to ensure optimal focusing and resolution; the gel is then removed from the tube and equilibrated briefly in SDS sample buffer before being laid horizontally across a slab-gel which is then run as a normal one dimensional SDS-PAGE. After the separation in the second dimension, the protein zones (spots) are spread throughout the gel and X-Y coordinates can be correlated with pI and molecular weight through the use of standards. The utility of 2D gel electrophoresis in the study of highly purified proteins is not significantly greater than the individual single dimensions, although it may be useful confirmation (29).

Protein Detection After Electrophoresis. After the proteins have been separated by the desired method, they must be visualized. After fixation, the proteins can be detected by staining techniques, enzymatic assays, chemical reactivity, ligand binding or immunoreagents. For detection of enzymes, the gel system can be chosen so that the enzyme remains active (or is regenerated after a denaturing method of analysis), and the band(s) of activity can be detected by the use of suitable substrates (30).

For simple detection of the locations of proteins after electrophoresis Coomassie blue is one of the most sensitive stains (31). Quantitation of the amount of protein present, for example by densitometry or elution and spectrophotometry (for a recent review, see 32) using these dye-binding methods must take into account the wide variation in dye binding capacity from protein to protein (33) and purity estimates of a final product pharmaceutical must remain somewhat imprecise as a result.

The introduction of silver staining (34) dramatically improved the sensitivity routinely achievable. Many improvements and variations on the method attest to its utility and widespread use. Several commonly used methods have appeared with claimed improvements in cost (35), sensitivity (36) or convenience and uniformity (37).

Silver staining methods are mainly used in situations where sensitivity is a major issue, such as in the detection of impurities in a pharmaceutical protein preparation. However, the "staining" phenomenon is not stoichiometric and attempts to develop wide range standard curves have shown that the band intensities are not simply related to the amount of protein loaded (*38*). There is also a wide range of staining intensity per unit mass of protein loaded, as with the dye-binding methods. Thus, quantitation after silver staining is even more problematic than with the dye-based methods. Silver staining is consequently not a useful tool in stability studies in which the effect of observed changes (often estimated to be on the order of 0.01%) cannot be meaningfully evaluated.

Capillary Electrophoresis. In electrophoretic separations the higher the applied voltage, the faster the separation, while resolution also increases with the square root of the voltage (*39*). The narrow diameter (typically about 50µm) of the capillaries used in capillary electrophoresis not only allows dissipation of the Joule heat generated by increased voltages, but also eliminates the need for an anticonvective medium due to the rapid diffusion of analytes within the small sample zone (*40*). The separation principles remain the same as in conventional electrophoresis but now the detection methods employed in chromatography systems can be used. As a result, the method is still undergoing rapid development in the field of protein analysis (*41,42,43,44*).

The capillaries used are usually made of fused silica and at pH values above 3-4, the silanol groups are ionized to produce a negatively charged surface. Thus bulk flow is observed as a result of electroendosmosis. Consequently, analytes will migrate with a velocity determined by the sum of the electrophoretic mobility and the electroendosmotic flow (*40*). By coating the column, for example with trimethylchlorosilane (*40*), many of the charges of the silanols can be shielded and electroendosmotic flow is significantly reduced. Early work used capillaries over 1 meter long, but more recently, shorter capillaries of 25-50 cm have been used since "the ideal situation is to apply as high a voltage as is available to capillaries as short as possible, yielding the highest separation efficiency in the shortest possible time"(*40*).

Selection of the optimal pH during the electrophoresis of proteins and peptides is as important in capillary electrophoresis as it is in conventional methods, primarily because it is the mass-to-charge ratio of the analytes that determine differences in their behavior. Thus, pH was found to be critical in an evaluation of the heterogeneity of a variety of recombinant proteins (*45*) and in the separation of deamidation products of the growth-hormone releasing factor (*46*). Frenz et al. (*47*) compared the separation of native and deamidated forms of hGH by ion-exchange HPLC and capillary electrophoresis. Both methods gave good separation and while variable recovery of hydrophobic peptides is not uncommon in HPLC, the electrophoretic method suffers uncertainty in the amounts of different peptides loaded by electrophoretic means.

Nonetheless, separation of peptides from a proteolytic digest based on electrophoretic principles complements the chromatographic (reversed-phase HPLC) methods normally used. This increases the confidence that all components have been separated and detected, because co-elution of the same peptides in both methods is extremely unlikely. This complementation improves the quality of the information concerning identity and purity of the protein under investigation as shown by two groups using the tryptic digest of hGH (*47,48*). The ability to obtain such high-quality information may be critical in investigations of the mechanisms of degradation pathways: if the mechanism is understood, the formulation scientist is in a position to rationally design a formulation in which the degradation is minimized.

Capillary Isoelectric Focusing. Coating the capillary walls was essential for the development of isoelectric focusing in capillaries *(49)* since the proteins end up stationary in this method. Hjerten and Zhu employed two methods to elute the focused bands: pumping and electrophoretic mobilization *(49)*. Both methods were successful, but the technical simplicity of mobilization (the addition of salt to either electrolyte *(50)*) has resulted in its wide adoption. Capillary IEF of monoclonal antibodies and comparison with slab-gel patterns showed that essentially the same patterns were obtained in both methods, although the benefits of on-line UV detection resulted in a more accurate quantitation than with the densitometry of the stained slab for reasons discussed previously *(51)*. The use of urea, detergents or solubilizers to maintain solubility and mobility was fcritical in a capillary IEF investigation of the heterogeneity of the glycoforms of recombinant tissue plasminogen activator. This separation demonstrated the high resolving power and utility of the method in a major area : biotechnology production of glycoproteins *(52)*.

Size Analysis by Capillary Electrophoresis. SDS-PAGE has become popular in large part due to the simple interpretation of the results in terms of apparent molecular weight. It has also been used in a capillary format and has the potential for rapid, automated analysis of samples (see *(41)* for discussion). However, technical difficulties, such as preparation of a bubble-free gel and the need for an open column zone beyond the end of the gel for sensitive detection, in the far-UV region, are being solved more slowly for this mode of capillary electrophoresis *(53)*. The addition of non-crosslinked polymers to the solution in capillary electrophoresis is technically easier and has been shown to cause separation of proteins on the basis of size by a "tangled polymer" mechanism, in which pores are generated by the overlap of tangled (but unbranched) polymer chains, rather than by cross-linking *(54)*.

Detection Methods for Capillary Electrophoresis. One of the major advantages of capillary electrophoretic methods is that on-line detection methods can be applied. For a summary of the instrumentation for these methods, the reader is directed to reference *55*. The absorption of the peptide bond in the far UV (around 200 nm) affords the most sensitive direct absorption method, while specific information on aromatic amino acid containing peptides or proteins uses the more traditional wavelengths around 280 nm.

The most powerful detection method so far developed for capillary separation methods (both electrophoretic and chromatographic) is on-line mass spectrometry. This area is clearly one in which there is much activity and rapid progress in the development of extremely powerful analytical methods for analysis of proteins. This is addressed below in the section on mass spectrometry.

Chromatography

Until the recent application of on-line detection methods to capillary electrophoresis, much of the quantitative analysis of peptides and proteins was performed with HPLC methods. The use of HPLC methods for macromolecules has recently been comprehensively reviewed *(56)* as have their applications in the pharmaceutical industry *(57)*.

Gel Filtration Chromatography. The apparent molecular weight of a protein can be determined by gel filtration on a column which has been calibrated with molecular weight standards *(58)*. This method is of relatively low resolution *(59)* primarily because it is based on the hydrodynamic properties of the protein and consequently gives accurate estimates only for spherical proteins (60) (which are usually used for calibration). Careful calibration of the column can be used to determine the effective radius of the protein of interest (61) and to evaluate this parameter as a function of a variable such as pH (or net charge) to measure the expansion and contraction of a protein due to charge repulsion effects (62). Since separation is based on apparent molecular weight, this method is frequently used to assess aggregation status and quantitate dimers, trimers, tetramers and other aggregates (63,64).

Denaturation of the sample and running of the column in denaturant can render this technique a measure of polypeptide length, in much the same way as SDS polyacrylamide gel electrophoresis, described above (65). While yielding lower resolution than SDS-PAGE, this method does allow the quantitation of polypeptide distribution independent of the (variable) dye-binding properties of the individual proteins. For example, the extent of cleavage of rt-PA to its two-chain form was determined in this manner(16). Detection of components eluting from a column can be accomplished in many ways, such as UV absorption or fluorescence. More recently light scattering methods have been successfully applied to increase the information content of the experiment. Photon correlation spectroscopy (PCS) is normally performed on static samples but has been found applicable to liquid chromatographic separations (66). Low angle laser light scattering has also been coupled to chromatographic detection (see , for example, 67) to yield molecular weight directly, without the need for any assumptions about shape that are required for PCS-based detection.

Ion Exchange Chromatography. This is a powerful technique on the preparative scale for the purification of proteins and its analytical counterpart is equally powerful (68). A protein with a net positive charge tends to bind to a matrix with a net negative charge by ionic interactions and vice versa, provided the ionic strength is sufficiently low.. The passage of a salt gradient over a column to which a mixture of proteins has been bound will cause the elution of each protein at its own critical salt concentration where the protein binds less tightly than the salts. Thus the proteins are separated from each other and the components of a mixture can be separately quantitated. Depending on the distribution of the charges on the surface , it is possible for the same protein to bind to both anion and cation exchangers at the same pH. For example, a selenium-containing thiolase was purified, after being adsorbed in the same loading buffer, by salt gradient elution from either a DEAE-cellulose column (69) or a CM-Sephadex column (70). Any factor which affects either the net charge or its distribution will affect elution behaviour. Many of these factors have been discussed in the of electrophoresis section above but in this case we are concerned with something more subtle than "net" charge.

One special factor needs to be considered here and that is the pH of the separation. If two forms of the same protein which differ by one charged amino acid are analysed at a pH where that group is uncharged, then this difference may not be detected by the ion exchange resin. Thus, a single homogeneous peak on ion exchange is not a guarantee that the sample is homogeneous. Analysis at two widely differing pH values would be needed to increase the certainty of that conclusion.

Chromatofocusing. As the name implies this method is somewhere between isoelectric focusing and ion exchange chromatography. This method was introduced as a refinement of ampholyte displacement chromatography (71) and was extensively developed by Sluyterman (72,73). It is performed under low salt conditions and can use the same ampholytes as used in isoelectric focusing. The sample is loaded onto an ion exchange column at a pH where it binds. The column is then eluted with ampholyte or buffer mixture selected to generate a pH gradient that gradually flows down the column. At some point the pI of a bound protein is reached and that protein therefore is released from the resin, at approximately its pI. The high resolving power of this method results from the focusing nature of the elution mechanism : if peak broadening begins (by diffusion as, for example, in gel filtration chromatography) the molecules that run ahead of the center of the peak will find themselves at a pH where they will spend more time bound and will slow down, and if they get left behind the altered pH will cause them to spend less time bound to the resin and consequently speed up their elution rate. The net result is tightly focused peaks eluting from the column. This method was used to separate hGH from a two-chain version, due to the introduction of ionizable groups at the extra N- and C-terminals and the alteration in pI caused by the chain cleavage (*74*).

Reversed Phase Chromatography. If a hydrophobic compound is coupled to a resin in a column, then proteins can bind through their hydrophobic regions to these columns (*75*). These hydrophobic interactions are weakened by decreasing the water concentration in the eluting mobile phase. This is readily accomplished by adding an organic modifier such as acetonitrile or propanol. Thus stronger interactions will require more organic solvent to displace the protein and a gradient elution with increasing organic modifier will release proteins in the order of their hydrophobic interaction strengths.

This method was initially developed for the separation of small organic molecules but was quickly adopted in amino acid and peptide separations (*76*). A major application of this methodology is in the separation of peptides from proteolytic digests of proteins in a technique known as peptide mapping (*77*). This is based on the observation that the composition generally determines the retention time of small peptides on a suitable column(*78, 79*). As a result, peptide mixtures can be resolved into a characteristic profile or "map". This method afforded a simple way of distinguishing met-hGH from hGH of pituitary origin due to the altered elution behavior of the N-terminal peptide (*80*). In general, chemical changes in peptides will alter their behavior and post-translational or degradation changes can often be detected by alterations in the map, although the nature of the change must be determined independently. This method has been successfully applied on a routine basis to molecules as large as t-PA (*81*), although not all peaks were completely resolved in a single system and ambiguities or co-elutions were determined by rechromatography in a different solvent system. The applications of reversedphase chromatography for the analysis of rDNA derived proteins has been recently reviewed (*82*).

Peptide Mapping. This is a very powerful and widely applicable tool if performed carefully. The high specificity of certain proteolytic enzymes in cleaving polypeptide chains only at certain residues results in a characteristic set of peptides (*83*). For example, if trypsin is used to generate a mixture of peptides by cleaving at lysine and arginine residues, all the peptides should end in lysine or arginine (except the C-terminal peptide) and the same pattern will be observed for each digest of a given protein. The HPLC pattern of this mixture is therefore a "fingerprint". Peptides can then identified by sequencing, amino acid analysis or mass spectrometry or some combination of these methods. By monitoring all the peptides (including the N- and C-terminal peptides specifically), the integrity of the whole chain can be assessed for each batch.   In some cases heterogeneity of peptides (such as sites of glycosylation/deamidation) can also be confirmed to be consistent from batch to batch and over time (*81*). This method has the capability of detecting changes at the 5-10% level for large proteins such as t-PA (*81*) and below 5% for smaller molecules such as bGH (*84*). Accordingly, with a well developed fingerprint method, it would be possible to validate the genetic stability of the cells being used for production by analysing the consistency of the structure of the product produced over the life of a master or working cell bank under production condition ranges. More recently, the information that can be obtained from this method has increased with the development of mass spectrometry interfaces and the elution position of specific peptides can be detected by selected ion monitoring. By rapidly scanning an appropriate mass range repetitively, it is possible to generate a three dimensional map (the three dimensions being elution time, mass and intensity)and potentially identify everything in the elution profile (at least by mass) even when substantial co-elution occurs(*85*).

Chemical Degradation.

During stability evaluation of proteins, two commonly observed types of chemical changes are the deamidation of asparagine sidechains and the oxidation of methionine. A review focusing on deamidation and oxidation of proteins from a formulation viewpoint has recently covered this area in detail (*86*). The peptide map is often used to identify which amino acids, if any, have been affected by these chemical degradation processes under defined storage conditions.

Deamidation. Deamidation is the loss of the amide group usually from asparagine and less frequently from glutamine side-chains. It occurs through a cyclic intermediate with the neighboring peptide group and appears to be controlled by both local sequence and local structural effects (*87,88*). The opening of the cyclic imide intermediate can yield either the normal peptide bond or the iso-peptide bond resulting in a new path for the peptide backbone. The amount of this new form can be quantitated by incorporation of labeled methyl groups that are transferred specifically to such a structure by a methyl transferase (*89*). The presence of deamidated forms of proteins is frequently manifested in heterogeneity on isoelectric focusing or ion exchange chromatography but peptide mapping by reversedphase HPLC is usually the tool employed to identify the site of the deamidation. For, example, ion exchange chromatography was used to isolate a deamidated form of soluble CD4 (after removal of sialic acid charge heterogeneity by neuraminidase treatment) and examined by peptide mapping either using trypsin or Asp-N, which selectively cleaves at the N terminal side of aspartic acid residues (*90*). This study showed that most of the deamidated product was deamidated at residue 52 and was in the isopeptide form and had reduced activity, consistent with its location in the binding site of this molecule.
　　Deamidation of a protein frequently has no apparent effect on its activity. For example, hGH has long been known to have sites which deamidate more readily than others. Deamidated hGH is consequently a frequently encountered form, even in material rapidly isolated from pituitaries (*91*), but it is fully biologically active (*92*). The sites of deamidation in recombinant hGH have been studied using a variety of methods (*93,94*).
Oxidation. The other frequently encountered degradation product is the result of the reactivity of the thioether linkage in methionine residues with oxidizers to yield methionine sulfoxides (*95*), often with loss of activity, either through loss of active site reactivity or through conformational alterations. Methionine oxidation of hGH has been studied, but in this case the major degradation products (with sulfoxides at positions 14 and/or 125) appear to have full biological activity (*96*).

Protein Chromatography in Reversed Phase HPLC. The magnitude of the effect of chemical changes on the elution behavior of intact proteins is smaller than for peptides and the utility of RP-HPLC for resolving mixtures of proteins diminishes as the size of the proteins increases. However, conditions can sometimes be found in which subtle changes can be detected after much optimization. For example met-hGH can be resolved from hGH (a difference of one amino acid in a protein of 191 amino acids) under certain conditions (82). The detection systems described above for capillary electrophoresis systems are also applicable for identifying some of the properties of peptides as they are eluted.

As in ion exchange chromatography, the behavior of a protein may be dominated by a surface "patch" (in this case a hydrophobic patch) which interacts much more strongly than the rest of the protein molecule. However, the existence of the patch may be affected by the method itself : the conformation of the protein in such organic phases will not necessarily be the same as in aqueous solution. Nonetheless, the aim of these separations is often analysis for heterogeneity of any kind and any loss of activity during such analyses is not critical.

The pH of the separation can be controlled and the ionization status of the protein will affect its hydrophobic/hydrophilic balance and may thereby reveal heterogeneity only at some pH values (46). The same caveats apply here as they do in gel filtration chromatography and recovery of very hydrophobic proteins or peptides can be very variable. The conclusions of such an analysis are based on the properties of those molecules which eluted and unless recovery studies are performed, they are not necessarily the conclusions about the sample that was loaded.

Hydrophobic Interaction Chromatography. As mentioned above, hydrophobic interactions are generally strengthened by increasing the salt concentration (witness the salting out of proteins by such agents as ammonium sulfate). In this form of chromatography (97) a (generally small) hydrophobic (e.g. propyl or phenyl) group is coupled to the column and the protein is loaded in very high salt (several molar). Under these conditions, the surface hydrophobic patches will interact with the column and the protein will bind. By running a reverse salt gradient (i.e. decreasing salt concentration) the stabilizing effect will be decreased until the interaction is not strong enough to retain the protein and it will be eluted. The stronger the hydrophobic interaction, the lower the salt concentration at which the protein will elute. This method was particularly useful in separating the 4 major forms of rt-PA : the single and double chain versions of the Type I and Type II glycoforms (98).

Primary Structure

Sequencing. Peptide sequencing is often a convenient method to identify the site of a chemical or enzymatic cleavage in degradation products generated during storage, allowing the formulation scientist to design ways of minimizing the degaradation. Full sequence determination is required in order to be able to describe the final drug product appropriately. While primary data are generally obtained from the cDNA sequence, determination of the complete protein sequence is required to confirm the cDNA sequence and may detect other polypeptide chains or new amino termini arising from proteolysis, a common degradation pathway if proteases have been active during the purification process (74) or are present as trace impurities in the final product.

Direct N-terminal sequencing has long used a chemical procedure known as Edman degradation, which derivatizes the amino-terminal amino acid in such a way as to release the amino acid and expose the amino terminal of the next amino acid (99). Mass spectrometry is playing an increasingly important role in peptide sequencing and has no problem with N-terminally blocked sequences (see, for example, *100*).

For final product evaluation, this method has been applied to the intact protein for a limited number of residues, primarily to confirm the integrity of the N- terminus of the chain but this is not very useful for stability evaluation. The confirmation of batch-to-batch consistency of the primary structure of the product and its stability over time generally comes from the peptide map. Confirmation of the identity/integrity and stability of the carboxy terminus for final products also usually comes from tracking of peptides from the peptide map (*16*).

Carbohydrate Analysis. Heterogeneity of glycoproteins frequently complicates the interpretation of the results from the analytical methods described in this chapter. The assembly and addition of the carbohydrate moieties to a protein is under the control of enzymes and this generally results in heterogeneity of the structures attached. For example, many carbohydrate structures contain sialic acids which are ionic and heterogeneity in the number of sialic acids per molecule can therefore result in the appearance of several bands on isoelectric focusing or ion exchange (*52,90*).

The glycan structures are generally quite stable and do not generally change during storage under the conditions commonly used for formulating protein pharmaceuticals. However, it may occasionally be beneficial/necessary to remove the (analytical heterogeneity caused by) carbohydrates in order to analyse for underlying degradation processes.

While the amino acids in a polypeptide are linked in a linear fashion, the saccharide units of a glycan may be linked in complex branching patterns and determination of their complete structure, including the definition of the configurations of the anomeric linkages, is a technically demanding undertaking. A common approach is to remove the glycans with enzymes and characterize their saccharide composition: for example N-glycanase has the potential to remove all the N-linked glycans (*101*). The sites of attachment of the glycans may be deduced by comparing peptide maps before and after glycanase treatment (*81*) or by lectin affinity chromatography and reanalysis (*82*).

Mass Spectrometry Techniques

Since the early 1980's mass spectrometry has made spectacular progress in applications for protein, peptide and glycopeptide analysis. The sensitivity and information-generating capability of mass spectrometry (MS) have been discussed above for electrophoretic and chromatographic separations.

A basic requirement for the analysis of a compound in a mass spectrometer is a the ability to generate an ion of the compound in the vacuum of the instrument (139). The recent advances for protein analysis have come largely from the development of methods that allow large molecules to be introduced into the spectrometer without destruction, along with the extension of measurable mass ranges resulting from technical improvements in analyzer design. (For an excellent overview of the role of MS in analytical biotechnology, the reader is directed to reference *102* and the references therein). The use of a beam of fast argon atoms to irradiate a liquid matrix with low vapor pressure (such as glycerol) caused the ejection of protonated molecules which could be successfully analysed in a mass spectrometer (*103*). This method, known as FAB (fast atom bombardment) MS was widely adopted. General strategies for analysing the primary structure of recombinant DNA produced proteins were quickly developed (see for example refs.*104* and *105*)), and the sequence of recombinant human interleukin-2 was verified by FAB-MS analysis of CNBr and proteolytic digests (106).

The lack of fragmentation of peptides in FAB-MS results in the major ion ejected being $(M + H)^+$, allowing its use in precise molecular weight determination. However, other methods must be employed if more detailed structural information is required. In a single MS experiment, an unfractionated mixture of peptides (e.g. from an enzyme digest) is injected and the masses of the major ions are compared with those predicted from the cDNA sequence. It should be remembered that MS is not capable, in this mode, of distinguishing between residues that have the same mass but different structures (such as Lys and Gln or Leu and Ile). An additional limitation of the method is due to the nature of the ionization mechanism : in a mixture of peptides in the matrix, not all peptides are ejected with equal efficiency, with some being suppressed. It is therefore difficult to use the method as a way of quantitatively analysing mixtures.

Mass Spectrometer Interfaces. The utility of MS method was further increased by the development of the LC/MS interface and the MS/MS interface. In the former, methods were developed for allowing LC methodology to separate peptides before entry into the MS, such as the continuous flow FAB interfaced with microbore HPLC (107). The latter "interface" allowed the sequencing of individual peptides from complex mixtures.. This sequencing was achieved by fragmentation of the peptides selected by the first MS followed by passage through a cloud of ions in a collision cell : this is known as CID (collision induced dissociation). In this technique, recently reviewed in ref. 108, the collision results in a characteristic set of fragments, from which the sequence may be deduced,without the need for prior information (such as the cDNA sequence). Thus, by 1987 one of the first sequences of a protein (thioredoxin, Mr 11750.2) had been almost completely determined by this method (109). The utility of the MS-MS (or tandem MS) method for determination of structural modifications was summarized in a review by Biemann and Scoble (110). Clearly, the combination of enzymatic digestion, chemical derivatization, LC/MS and tandem MS is an extremely powerful one, circumventing some of the limitations of previous methods of structure determination (110). For example, this combination allowed the detailed structural characterization of recombinant soluble CD4 receptor, confirming over 95% of the primary sequence of this 369 amino acid glycoprotein, demonstrating the intact nature of both N- and C-termini, the positions of attachment of the glycans, the structures of the glycans and the correct assignment of the disulfide bridges (111). The use of MS methods to address characterization issues associated with peptides (112) and recombinant proteins has been extensively reviewed (102,113,114).

Electrospray Ionization Mass Spectrometry (ESI-MS). Two additional ionization methods have extended the applications of MS to protein characterization : ESI-MS and MALDI-MS. Electrospray ionization mass spectrometry (ESI-MS) has extremely high precision because it generates multiply-charged species of even very large proteins (reviewed in 115). In this method, an aerosol of protein solution is introduced into the MS through a needle to which a high voltage is applied, resulting in the generation of a family of multiply charged peaks (of the same molecule) differing by single charges. The major advantages of this method are that the mass-to-charge ratio of the ions is within the range of existing spectrometers and the precision determination of proteins with very high molecular weights. This precision arises from the fact that each peak in the family of peaks generated by the differently charged species generates an estimate of the MW and these estimates can be combined to increase the overall precision of the MW estimate. This approach was used to characterize recombinant γ-interferon and its C-terminal degradation products simultaneously, yielding an estimate of the MW of 16908.4 ± 1.2, while the theoretical mass is 16907.3 (*116*). A major benefit of ESI-MS results from the fact that a flowing stream of liquid is required, making it relatively simple to develop interfaces to LC instruments.

MALDI-MS (Matrix Assisted Laser Desorption Ionization MS). This was pioneered by Hillenkamp and Karas (*117,118*) who showed that if a high concentration of a chromophore is added to the sample, a high intensity laser pulse will be absorbed by the matrix and the energy absorbed volatilizes a portion of the matrix and carries the protein sample with it into the vapor phase essentially intact. The resulting ions are then analysed in a time of flight MS. The "gentle" nature of the ionization may be responsible for the ability of the method to provide information on quaternary structure (*118*). A major extension of this method was developed by Beavis and Chait (*119*) who showed that the method is relatively insensitive to large amounts of buffer salts and inorganic contaminants. This type of methodology may have wide utility for several reasons : it requires only pmol amounts of sample, the sample can be a crude mixture of proteins, it is very fast (less than 15 min from start to finish), does not fragment the molecules and the result is in principle as easy to interpret as (and indeed resembles) a densitometric scan of an SDS-PAGE gel, with a mass range up to well over 100 kDa.

MS is still rapidly developing tools that are of enormous benefit in protein analysis in a wide range of applications. With the exquisite ability to resolve and identify peptides with changes in mass of 1 amu (atomic mass unit) and the development of more affordable commercial instruments, MS will soon be an indispensable complement to HPLC and electrophoretic methods.

Secondary And Tertiary Structure

The formulation scientist must consider all aspects of the structure of the protein pharmaceutical when designing formulations for optimal stability. So far, methods for evaluating chemical changes only have been discussed, but changes in the conformation of the polypeptide chain can be equally important in the maintenance of biological activity and native structure. Altered conformation or misfolding can lead to altered activity and, in some cases, the development of immunogenicity of the product. The assessment of such structural alterations is complex and the data generally do not define the structure but rather some aspects of it. It is nonethtless a prudent course of action to confirm that these aspects are not altered during processing or storage of the product.

Methods for evaluating secondary structure are primarily physical, spectroscopic methods. An overview of optical spectroscopic methods has recently summarized the application of these techniques using bovine growth hormone as an example (*120*). It should be noted that these methods do not identify which amino acids are in which conformation or environment and the methods generally provide indirect information on the structure. In some cases, however, a unique amino acid, such as tryptophan, is available and some specific information about it can be obtained.

These spectroscopic methods are generally only capable of characterizing some physical parameter of a sample in terms of the overall average property, such as helical content, as opposed to being able to detect changes in a small proportion of the sample. It is not possible to determine by these methods whether all the molecules changed by the same amount or whether a small fraction changed by a large amount after exposure to a particular set of conditions. However, these methods are frequently used as "handles" to follow important changes in the structure, such as unfolding or aggregation.

It is quite common for proteins expressed directly in bacterial cells to precipitate during synthesis (or during cell inactivation before harvest) and to form inclusion bodies or refractile bodies (see *121* for a review) and solubilization with denaturants followed by *in vitro* refolding is required to yield the correct structure of the native molecule (*122, 123*). The exhibition of the appropriate biological properties of such molecules are generally taken to indicate that the major structural or functional features have been formed during this refolding. However, biological activity is often difficult to measure precisely and additional confirmation that the structure is native is therefore sought (*122*).

Some of these spectroscopic methods are not readily amenable to analysis of large numbers of samples from stability or formulation screens but may be more useful in adding confirmation that the physical properties of the material have not been altered after, for example, a claimed shelf-life. An understanding of the effects of ranges of pH, salts or potential stabilizers on the structure of the protein can be very useful in selection of suitable conditions for maintaining structural integrity. They can also provide additional information on any physical changes which may accompany a specific chemical degradation event, such as oxidation, if the degradation product can be isolated for study.

Absorption Spectroscopy. The aromatic amino acids (Phe, Tyr and Trp) have ultraviolet absorption spectra in the 240 to 320 nm range and this provides a convenient way to measure their concentration, typically by the absorbance at around 280 nm (*124*). These absorption spectra are sensitive to their environment (*125*) and both difference spectroscopy (*126*) and derivative spectroscopy (*127*) can reveal these effects. In some proteins or peptides, the absorption of disulfide bonds is also detectable and can show a peak wavelength up to 340-350 nm (*128*).

The hydrophobic nature of the aromatic chromophores frequently results in their being folded into the interiors of proteins, (which have a different polarizability than aqueous solution) and consequently their spectra change compared to those observed in aqueous solution of the chromophores (*124*). This change is generally seen as a red-shift and an increase in absorptivity of 10-15% (*126*).

One of the most fundamental measurements of proteins is the determination of concentration. In many cases, and especially for unknown proteins, convenient methods such as the Lowry (*129*) and Bradford methods (*130*) provide sufficient information, if used with a defined standard such as BSA, to allow data to be reported in a way that can be compared between labs. The most convenient method for purified proteins is to use UV absorption spectroscopy to determine the concentration. This requires an accurate estimate of the absorption spectrum of a solution of known protein concentration. From this is derived the "extinction coefficient" usually meaning the absorbance of a 1 mg/mL solution at the wavelength maximum near 280 nm. An accurate measurement of the protein concentration (for the sample whose spectrum is recorded) is a critical part of the determination of the extinction coefficient. It is often measured by quantitative amino acid analysis, nitrogen assay or dry weight measurements.

However, since the sequence of the protein is known at this stage of product development, it is possible to calculate the UV absorption spectrum of the component aromatic amino acids. This will not be the spectrum of the protein, however, because it does not take into account the effects of the protein's structure on the spectra. Bewley (*131*) has shown that the magnitude of these effects can be directly measured in the following manner. The spectrum is recorded and the protein is digested with enzyme(s) until the spectrum stabilizes to that of the free aromatic amino acids in the resulting peptides, and their concentrations can be determined from their known extinction coefficients. This spectroscopic method avoids the errors associated with dry weight analysis and amino acid analyses and is very simple with modern computer-assisted spectrophotometers. The effects of protein structure on absorption spectra have been used by Brems (*132*) to follow the equilibrium unfolding of bovine growth hormone by guanidinium hydrochloride, in which the resulting blue shift and decrease in hyperchromicity combined to yield a change in absorbance of approximately 25% at 290 nm.

Fluorescence Spectroscopy. Many proteins exhibit fluorescence in the 300 to 400 nm range when excited at 250 to 300 nm. due to the presence of the aromatic amino acids. These spectra can tell us something about the microenvironments (of these reporter groups) that are generated by the folding of the protein (see, for example, *133*). Thus, a buried tryptophan is usually in a hydrophobic environment and will fluoresce with a wavelength maximum in the 325 to 330 nm range, while an exposed residue (or free amino acid) fluoresces at around 350 to 355 nm (*134*). A recent example of the use of fluorescence spectroscopy is a study of the conformational change in G-CSF induced by acid (*135*). At neutral pH, the fluorescence emission spectrum is dominated by tryptophan and significant energy transfer from tyrosine is present, while at acidic pH, the tryptophan fluorescence is decreased, apparently due to the conformational change which disrupts the energy transfer and allows significant fluorescence from the tyrosines, which consequently dominate the emission. These fluorescence studies also confirmed the presence of intermediates in the guanidine-induced unfolding transition of the molecule.

In addition to the solvent perturbation absorption spectroscopic method of determining solvent accessibility of absorbing chromophores, certain ions can quench the fluorescence, but only if they can approach sufficiently closely to the fluorophore. Thus Cs^+ and I^- ions can be used to probe the accessibility of fluorescent tryptophans and the quenching constants can be related to the environment of the residue in the protein (*136,137*).

Circular Dichroism Spectroscopy. Secondary structure can be evaluated by a technique known as circular dichroism (*138*). The arrangement of peptide bonds into regular, constrained structures such as α-helices and β sheets, gives rise to alterations in their side chain absorption spectra. This can be measured as a circular dichroism signal and the shape and intensity of this signal as a function of wavelength (in the 190 to 250 nm range) can be related to the secondary structure (*139,140*). Early methods of estimating secondary structure content from CD spectra employed spectra from model polypeptides (*139*) but an approach based on the known secondary structures of proteins (measured from X-ray structures) results in a definition of the spectral contributions of secondary structural elements in typical globular proteins (*141*). While these methods do not identify which parts of the chain are in which conformation, they allow an estimate of the overall time-averaged content of α-helix. This information can be tracked over time and in different buffers, pH's etc. to evaluate the stability of the overall secondary structure in an empirical manner (see, for example, *142*). It is frequently used in evaluating the development of secondary structural elements in folding studies (*143,144*).

It is generally accepted that most proteins are denatured by concentrations of guanidine hydrochloride in the range of 6-7M (*23*). Thiocyanate is also occasionally used but, as with SDS denaturation noted above, there are resistant proteins. Cockle et al. showed that incubation in 6M guanidine hydrochloride and 0.1M mercaptoethanol at 70 °C for 6 hrs was insufficient to denature lipophilin, a proteolipid which retained 85 % of its initial helical content after this treatment (*145*). In this study, even using 6M guanidine thiocyanate did not produce much more denaturation as assessed by fluorescence spectroscopy.

Aromatic amino acids (Trp, Tyr and Phe) can also give rise to CD spectra (in the 240 to 320 nm range). This is often referred to as "near-UV CD". These "reporter" groups, can therefore be used to monitor changes in their environment due to changes in tertiary structure. Strong signals are generally interpreted as indicating a high degree of constraint on the movement of the sidechains, since free amino acids (or freely rotating sidechains on proteins) give weak signals (*146*). Disulfide bonds are also chromophores which can also give rise to CD bands in this region, while free -SH groups do not (*138,146*). The near UV CD spectrum is commonly used as a "fingerprint" of the tertiary structure to provide confirmation that the correct structure is present (*29*) or that refolding efforts were successful when the product was isolated from insoluble starting material(*122*).

Infrared Spectroscopy. This measures the most rapid motions of atoms in proteins, due to the bending and stretching of the bonds themselves (37). It therefore represents another opportunity to obtain information about the environments of atoms. Fourier techniques have successfully been applied to IR spectroscopy yielding enhanced resolution and the ability to obtain secondary structure information by the deconvolution of the absorption envelope of the amide vibrations(*147*). In this method, the spectrum is related to the subtle effects of regular secondary structure on the energetics (i.e. vibrational frequency) of the amide group in the peptide linkage. IR spectroscopy has an advantage in that it can be applied to samples in various states (such as powders, films, solutions, etc.) (*120*). However, some of the semi-empirical interpretations can sometimes be misleading; the large loops in recombinant interleukin-2 were found to have an IR absorption peak characteristic of α-helix (*148*).

Recent application of FTIR to protein pharmaceuticals has resulted in the ability to evaluate aspects of protein structure in the solid state (*149*). The effects of additives and excipients on structural changes during freezing or lyophilization have been studied using this method (*150*) and calorimetry (*151*). These tools look very promising for use by the formulation scientist in the development of frozen or lyophilized formulations. With these techniques it is possible to study the structure of the protein at each step through the freezing, drying, storage and reconstitution cycles and separately optimize the excipients to maintain the correct structure throughout.

Analytical Ultracentrifugation. In this non-invasive method, protein solutions are subjected in a centrifuge to forces that are comparable in magnitude to those that cause the diffusion of proteins. The technique has been available for many years and the classical texts appeared quite some time ago (*152,153*). However, the recent development of a modern, computerized, simple and convenient instrument is anticipated to lead to a renewed interest in the utility of the technique : "Analytical ultracentrifugation is still the best method for quantitative studies of the interactions between macromolecules" (*154*).

Sedimentation velocity experiments use a high g force which results in the net migration of the solute towards the outside of the rotor and the rate at which the boundary migrates is related to the molecular weight of the molecule (*152*). If two non-interacting solutes of sufficiently different molecular weights are present, it is possible to obtain individual sedimentation rates for them, since their boundaries will be distinguishable. However, if they are close in molecular weight or there are interactions, these experiments become harder to interpret. Thus, in a preparation of highly purified recombinant human α–interferon, Shire observed highly asymmetric boundaries in velocity experiments, demonstrating an associating system, and was able to fit the data reasonably well with modeled association constants (*155*).

If diffusion forces and centrifugal forces are balanced, as in sedimentation equilibrium studies, the solute is not uniformly distributed throughout the cell and the concentration profile is determined by the molecular weight and the g force (*156*). If two or more species are present, in a non-interacting system, the data can be fit in a relatively straightforward fashion. However, if there are interactions, the equilibrium is not only between diffusion and centrifugation, but also between the interacting species (*157*). Association constants could be obtained for subunits of HIV reverse transcriptase by this method (*158*) using mathematical treatments developed by Lewis (*159*). One of the first publications using the new XLA model from Beckman reports the interaction of recombinant tumor necrosis factor with the soluble portion of its receptor, demonstrating that the TNF trimer binds to 2 or 3 molecules of the receptor to form a 3:2 or 3:3 stoichiometric complex (*160*). Application of ultracentrifugation methods to the study of protein protein interactions will become increasingly important in strategies for rational drug design. The new equipment, non-destructive nature and low sample requirements and sensitivity (as little as 0.02 mL at 0.1 mg/mL or less) add to the attractiveness of the method.

While the method is not sufficiently precise to measure very small changes in molecular weight, it can be useful in assessing degradation products in terms of critical properties such as interaction stoichiometey and association constants, such as for antigen-antibody or receptor-ligand complexes. Such absolute (in contrast to relative) information may be important in confirming that a small chemical change, such as a deamidation at an irrelevant site on a macromolecule, has no effects on the mechanism of action of the molecule.

Light Scattering. All matter scatters light to some extent, with larger particles scattering more than small ones. As a result, aggregated proteins, commonly encountered in stability studies, often make their presence known in the form of haziness or opalescence of a formulation which is intended to be clear. Aggregation in protein formulations has recently been critically reviewed (*86*). If precipitation is observed, it must be accounted for if a protein concentration measurement is being made by absorbance at 280 nm (*161*). Final product pharmaceutical solutions are inspected for clarity and qualitative descriptions are required; the European Pharmacopeia has defined reference preparations which are defined as "opalescent", "slightly cloudy" etc., but the test remains subjective. A relatively simple spectrophotometric evaluation has been developed which reduces the subjectivity of this test (*162*). This method was useful in the evaluation of the effects of freezing on the aggregation of hGH (*163*).

An unaggregated monomeric protein solution still scatters light and light scattering can therefore be used to measure the molecular weight (see the extensive treatment of light scattering in ref. *164*). For most proteins, which are much smaller than the wavelength of the light used in scattering studies (around 500 nm), the dependence of scattered intensity upon scattering angle is not strong. Consequently, simple right angle scattering has been shown to be a valid way of determining molecular weights up to 10^6 in eluents from HPLC columns (*165*). This method requires much simpler (and cheaper) instrumentation than that used in low angle laser light scattering studies, which can yield accurate molecular weights with no size constraints (*166,167*). This latter method has been applied to reversed-phase HPLC (*168*), HIC (*169*) and SEC (*170*). The two parameters which directly determine the accuracy of the molecular weight estimate are the extinction coefficient (which is used to determine the concentrations at each point in the elution profile) and the square of the refractive index increment (dn/dc) of the protein. This latter term is of particular importance in the evaluation of glycoproteins (67). This knowledge of molecular weight of a protein in a specific eluent can be particularly useful to detect aggregation in separation methods which are not directly based on size, or to identify situations where the protein interacts with sizing matrices and therefore yields erroneous values in such chromatographic separations.

Laser light scattering (or photon correlation spectroscopy, PCS) can be used to rapidly determine the diffusion coefficient of a protein by autocorrelation methods (171,172). For example, this method was used to show the pH dependent expansion of several proteins and to show an increase in net charge of lysozyme induced by denaturation (173). With powerful lasers and fast computers these measurements can now be performed rapidly enough to be used as an on-line detection system (66). Current methods have difficulty in quantitating heterogeneous mixtures , primarily the result of the presence of many exponentials in the autocorrelation function. However, populations of monomers and dimers can be quantitated and PCS was key to elucidating the mechanism of cosolvent-assisted refolding of proteins (174) to improve yield from recombinant DNA processes (175).

Conclusions

This chapter has covered the major methods of analysing proteins which can be used to guide the selection of conditions and components for optimal formulation and to study their degradation. Clearly, the selection of the methods used for an individual protein will be determined, at least in part, by the properties of the protein and the design criteria for the formulation. The challenge for the formulation scientist will lie in understanding the principles of these analytical methods, determination of the factors controlling the major degradation pathways and selecting appropriate strategies for minimizing them.

Literature Cited

1   Jones, A.J.S. **1993** *Advanced Drug Delivery Reviews* 10, 29-90.
2   Ornstein, L. **1964** *Ann. N.Y. Acad. Sci.* 121, 321-349.
3   Davis, B.J. **1964** *Ann. N.Y. Acad. Sci.* 121, 404-427.
4   *Gel electrophoresis of proteins : A practical approach*, Hames,B.D.; Rickwood,D., Eds.; **1981** IRL, Oxford.
5   Kleparnik, K.; Bocek, P. **1991** *J. Chromatogr.* 569, 3-42.
6   Fryklund,L.; Brandt,J.; Eketorp,G.; Fholenhag,K.; Skoog,B.; Skottner-Lundin,A.; Wichman,A. In *Hormone Drugs*; Gueriguian, J.L.; Bransome, E.D.;Jr.; Outschoorn, A.S., Eds.; USP, Rockville, MD. **1982** ; pp. 319-326
7   Reynolds, J.A.; Tanford, C. **1970** *J. Biol. Chem.* 245, 5161-5165
8   See,Y.P.; Jackowski, G. **1989** In : Creighton, T.E.(Ed.) *Protein structure : a practical approach,* IRL Press, Oxford, pp. 1-21.
9   Shapiro, A.L.; Vinuela, E.; Maizel, J.V. **1967** *Biochem. Biophys. Res. Commun.* 28, 815-820.
10   Weber,K and Osborne, M. **1969** *J. Biol. Chem.* 244, 4406-4412.
11   Laemmli, U.K. **1970** *Nature* 27, 680-685.
12   Neville, D.M.;Jr. **1971** *J. Biol. Chem.* 246, 6328-6334.
13   Hames,B.D. In: *Gel electrophoresis of proteins : A practical approach*; Hames,B.D.; Rickwood,D., Eds.; IRL, Oxford.**1981** ; pp. 1-91.
14   Hageman, M.J.; Bauer, J.M.; Possert, P.L.; Darrington, R.T.; **1992** *J. Agric. Food Chem.* 40, 348-355.

15   Nguyen,T.H.; Ward, C. In *Stability and characterization of protein and peptide drugs : case histories*; Wang, J.Y.; Pearlman, R., Eds.; Plenum Press, NY. **1993**; pp. 91-134.

16   Jones, A.J.S. and Garnick, R.L. In *Large-scale mammalian cell culture technology*, Lubiniecki, A.S. Ed.; Dekker, New York.**1990** pp. 543-566.

17   Kunitani, M.; Dollinger, G.; Johnson, D.; Fresin, L. **1991** *J. Chromatogr.* 588, 125-137.

18   Leach, B.J.; Collawn J.F.; Jr.; and Fish W.W. **1980** *Biochemistry* 19, 5734-5741.

19   Arakawa, T.; Hung, L.; Narhi, L.O. **1992** *J. Prot. Chem.* 11, 111-118.

20   Righetti, P.G. **1983** *Isoelectric Focusing : Theory, Methodology and Applications*. Elsevier, Amsterdam.

21   Vesterberg, O.; Svensson, H. **1966** *Acta. Chem. Scand.* 20, 820-834.

22   Righetti, P.G. In *Protein Structure : a practical approach,* Creighton, T.E., Ed.;IRL Press, Oxford.**1989** pp. 23-63.

23   Pearlman, R.; Nguyen,T.H. In *Therapeutic peptides proteins : Formulation, delivery and targeting*, Marshak,D.; Liu,D., Eds.; Cold Spring Harbor Laboratory.**1989** ; pp. 23-30

24   Canova-Davis, E.; Teshima, G.M.; Kessler, T.J.; Lee, P.-J.; Guzzetta, A.W.; Hancock, W.S.; **1990** *Am. Chem. Soc. Symp.* 434, 90-112.

25   Righetti, P.G. **1990** *Immobilized pH gradients : Theory and methodology*. Elsevier, Amsterdam.

26   Bjellqvist,B.; Ek,K.; Righetti, P.G. Gianazza, E.; Gorg, A.; Postel, W.; Westermeier,R. **1982** *J. Biochem. Biophys. Methods* 6, 317-339.

27   Cossu, G.; Righetti, P.G. **1987** *J. Chromatogr.* 398, 211-216.

28   O'Farrell, P. **1975** *J. Biol. Chem.* 250, 4007-4021.

29   Jones, A.J.S.; O'Connor, J.V. In *Hormone Drugs*; Gueriguian, J.L.; Bransome, E.D.;Jr.; Outschoorn, A.S., Eds.; USP, Rockville, MD. **1982** ; pp. 335-351.

30   Gabriel, O.; Gersten, D.M. **1992** *Anal. Biochem.* 203, 1-21.

31   Wilson, C.M. **1979** *Anal. Biochem.* 96, 263-278.

32   Syrovy, I.; Hodny, Z. **1991** *J. Chromatogr.* 569, 175-196.

33   Pierce, J.; Suelter, C.H. **1977** *Anal. Biochem.* 81, 478-480.

34   Switzer, R.C.; III, Merril, C.R.; Shifrin, S. **1979** *Anal. Biochem.* 98, 231-237.

35   Oakley, B.R.; Kirsch, D.R.; Morris, N.R. **1980** *Anal. Biochem.* 105, 361-363.

36   Merril, C.R.; Goldman,D.; Sedman, S.A.; Ebert, M.H. **1981** *Science* 211, 1437-1438.

37   Morrissey, J.H. **1981** *Anal. Biochem.* 117, 307-310.

38   Poehling, H.-M.; Neuhoff, V. **1981** *Electrophoresis* 2, 141-147.

39   Jorgenson, J.W.; Lukacs, K.D. **1981** *Clin. Chem.* 27, 1551-1553.

40   Jorgenson, J.W.; Lukacs, K.D. **1983** *Science* 222, 266-272.

41   Karger, B. L.; Cohen, A.S.; Guttman, A. **1989** *J. Chromatogr.* 492, 585-614

42   Deyl, Z.; Struzinsky, R. **1991** *J. Chromatogr.* 569, 63-122.

43   *New Directions in Electrophoretic Methods*. Jorgenson, J.W.; Phillips, M.,Eds.; Am. Chem. Soc. Symp. Series, ACS, Washington, D.C.**1987** Vol 335.

44   *Analytical biotechnology : Capillary Electrophoresis and Chromatography*. Horvath, Cs.; Nikelly, J.G., Eds.; Am. Chem. Soc. Symp. Series, ACS, Washington, D.C.**1990** Vol 434.

45   Wu, S.-L.; Teshima,G.; Cacia, J.; Hancock, W.S. **1990** *J. Chromatogr.* 516,115-122.

46   Bongers, J.; Lambros, T.; Felix, A.M.; Heimer, E. P. **1992** *J. Liq. Chromatogr.* 15, 1115-1128.

47   Frenz,J.; Wu, S.-L.; Hancock, W.S. **1989** *J. Chromatogr.* 480, 379-381.

48   Nielsen, R.G.; Riggin, R.M.; Rickard, E.C. **1989** *J. Chromatogr.* 480, 393-401.
49   Hjerten, S.; Zhu, M. **1985** *J. Chromatogr.* 346, 265-270.
50   Hjerten, S.; Liao, J.L.; Yao, K. **1987** *J. Chromatogr.* 387, 127-138.
51   Costello, M.A.; Woititz,C.; De Feo, J.; Stremlo, D.; Wen, L.-F.; Palling, D.J.; Iqbal,K.; Guzman, N. **1992** *J. Liq. Chromatogr.* 15, 1081-1097.
52   Yim, K.W. **1991** *J. Chromatogr.* 559, 401-410.
53   Cohen, A.S.; Karger, B.L. **1987** *J. Chromatogr.* 397, 409-417.
54   Zhu, M.; Hansen, D.L.; Burd, S.; Gannon, F. **1989** *J. Chromatogr.* 480, 311-319.
55   Guzman, N.A.; Hernandez, L.; Terabe, S. **1990** *Am. Chem. Soc. Symp.* 434, 1-35.
56   *HPLC of biological macromolecules.* Gooding, K.M.; Regnier, F.E., Eds.;Dekker, N.Y. **1990**
57   *HPLC in the pharmaceutical industry*, Benedek, K.; Swadesh, J.K. In : Fong, G.W.; Lam, S.K., Eds.; Dekker, N.Y. **1991** pp. 241-302.
58   Whitaker, R. **1963** *Anal. Chem.* 35, 1950-1953.
59   Gooding, K.M.; Regnier, F.E. In *HPLC of biological macromolecules.* Gooding, K.M.; Regnier, F.E., Eds.; Dekker, N.Y.**1990** pp.47-75
60   Andrews, P. **1970** *Methods Biochem. Anal.* 18, 1-53.
61   Potschka, M. **1987** *Anal. Biochem.* 162, 47-64.
62   Martenson, R.E. **1978** *J. Biol. Chem.* 253, 8887-8893.
63   Van Liedekerke, B.M.; Nelis, H.J.; Kint,J.A. Vanneste, F.W.; De Leenheer, A.P. **1991** *J. Pharmaceut. Sci.* 80,11-16.
64   Townsend, M.W.; DeLuca, P.P. **1991** *J. Pharmaceut. Sci.* 80, 63-66.
65   Mann, K.G.; Fish, W.W. **1972** *Methods Enzymol.* 26, 28-42.
66   Carr, R.J.G.; Rarity, J.G.; Stansfield, A.G.; Brown, R.G.W.; Clarke, D.J.; Atkinson, T. **1988** *Anal. Biochem.* 175, 492-499.
67   Arakawa, T.; Langley, K.E.; Kameyama, K.; Takagi, T. **1992** . *Anal. Biochem.* 203, 53-57.
68   Henry, M.P. In *HPLC of macromolecules A practical approach*; Oliver, R.W.A., Ed. IRL, Oxford. **1989** pp. 9-125
69   Hartmanis, M.G.N.; Stadtman, T.C. **1982** *Proc. Nat. Acad. Sci. USA* 79, 4912-4916
70   Sliwkowski, M.X.; Hartmanis, M.G.N. **1984** *Anal. Biochem.* 141, 344-347.
71   Leaback, D.H.; Robinson, H.K. **1975** *Biochem. Biophys. Res. Commun.* 67, 248-254.
72   Sluyterman, L.A.Ae.; Elgersma, O. **1978** *J. Chromatogr.* 150, 17-30.
73   Sluyterman, L.A.Ae.**1982** *Trends Biochem. Sci.* 7, 168-170
74   Canova-Davis, E.; Baldonado, I.P.; Moore, J.A.; Rudman, C.G.; Bennett, W.F.; Hancock, W.S. **1990** *Int. J. Peptide Protein Res.* 35, 17-24
75   Cowan, P.H. **1989** In *HPLC of macromolecules A practical approach.* Oliver, R.W.A. , Ed.; IRL, Oxford pp. 127-156
76   *Handbook of HPLC for the separation of amino acids , prptides and proteins,* Hancock W.S. Ed.; **1984** CRC Press Boca Raton, FL. ;Vol. I and II.
77   Fullmer, C.S.; Wasserman, R.H. **1979** *J. Biol. Chem.* 254, 7208-7212
78   Meek, J.L. **1980** *Proc. Natl. Acad. Sci. USA* 77, 1632-1636
79   Mant, C.L.; Hodges, R.S. **1989** *J. Liq. Chromatogr.* 12, 139- 172
80   Kohr, W.J.; Keck, R.; Harkins, R.N. **1982** *Anal. Biochem.* 122, 348-359.
81   Chloupek, R.C.; Harris, R.J.; Leonard, C.K.; Keck, R.G.; Keyt, B.A.; Spellman, M.W.; Jones, A.J.S.; Hancock, W.S. **1989** , *J. Chromatogr.* 463, 375-396.
82   Frenz, J.; Hancock, W.S.; Henzel, W.J.; Horvath, Cs. In *HPLC of biological macromolecules.* Gooding, K.M.; Regnier, F.E., Eds.; Dekker, N.Y.**1990** pp 145-177

83   Hancock, W.S.; Bishop, C.A.; Hearn, M.T.W. **1979** *Anal. Biochem.* 92, 170-173
84   Dougherty, J.J.; Snyder, L.M.; Sinclair, R.L.; Robbins, R.H. **1990** *Anal. Biochem.* 190, 7-20
85   Ling, V.; Guzzetta, A.W.; Canova-Davis, E.; Stults, J.T.; Hancock, W.S. , Covey, T.R.; Shushan, B.I.**1991** *Anal Chem.* 63, 2909-2915.
86   Cleland, J.L.; Powell, M.F.; Shire, S.J. **1993** *Crit. Revs. Therapeut. Drug Carrier Systems* 10, 307-377.
87   Wright, H.T. **1991** *Prot. Engineering* 4, 283-294
88   Kossiakoff, A.A. **1988** *Science* 240, 191-194.
89   Aswad, D. **1984** *J. Biol. Chem.* 259, 10714-10721
90   Teshima, G.; Porter, J.; Yim, K.; Ling, V.; Guzzetta, A. **1991** *Biochemistry* 30, 3916-3922
91   Lewis, U.J.; Singh, R.N.P.; Tutweiler, G.F.; Sigel, M.B. VanderLaan, E.F.; VanderLaan, W.P.**1980** *Rec. Prog. Hormone Res.* 36, 477-508
92   Skottner, A.; Forsman, A.; Skoog, B.; Kostyo, J.L. Cameron, C.M.; Adamafio, N.A.; Thorngren, K.G.; Hagerman,M. **1988** *Acta Endocrinol.* 118, 14-21
93   Hancock, W.S.; Canova-Davis, E.; Chloupek, R.C.; Wu, S.-L.; Baldonado, I.P.; Battersby, J.E. , Spellman, M.W.; Basa, L.J.; Chakel In *Therapeutic Peptides and proteins : Assessing the new technologies.* Marshak, D.; Liu, D., Eds.; Banbury Report 29 Cold Spring Harbor **1988** pp.95-118.
94   Becker, G.W.; Tackitt, P.M.; Bromer, W.W.; Lefeber, D.S.; Riggin, **1988** *Biotechnol. Appl. Biochem.* 10, 326-337
95   Brot, N and Weissbach, H. **1983** *Arch. Biochem. Biophys.* 223, 271-281
96   Teh, L.-C.; Murphy, L.J.; Huq, N.L.; Surus, A.S.; Friesen, H.G.; Lazarus, L.; Chapman, G.E. **1987** *J. Biol. Chem.* 262, 6472-6477
97   Shansky, R.E.; Wu, S.-L.; Figueroa, A.; Karger, B.L. In *HPLC of biological macromolecules.* Gooding, K.M.; Regnier, F.E., Eds.; Dekker, N.Y.**1990** pp 95-144.
98   Wu, S.-l, **1992** *LC.GC* 10, 430-434.
99   Edman, P.; Begg, C. **1967** *Eur. J. Biochem.* 1, 80-91.
100 Gibson, B.W.; Yu, Z.; Gillece-Castro, B.; Aberth, W.; Walls, F.C, ; Burlingame, A.L. In *Techniques in protein chemistry ,* Hugli, T.E., Ed.; **1989** Vol 3, 135-151.
101 Hirani, S.; Bernasconi, R.J.; Rasmussen, J.R. **1987** *Anal. Biochem.* 162, 485-492.
102 Carr, S.A.; Hemling, M.E.; Bean, M.F.; Roberts, G.D. **1991** *Anal. Chem.* 63, 2802-2824.
103 Morris, H.R.; Panico, M.; Barber, M. Bordoli, R.S.; Sedgwick, R.D.; Tyler, A.N. **1981** *Biochem. Biophys. Res. Commun.* 101, 623-631.
104 Morris, H. R.; Panico, M.; Taylor, G. W. **1983** *Biochem. Biophys. Res. Commun.* 117, 299-305.
105 Gibson, B. W.; Biemann, K.. **1984** *Proc. Natl. Acad. Sci. USA* 81, 1956-1960.
106 Fukuhara, K.; Tsuji, T.; Toi, K.; Takao, T.; Shimonishi, Y. **1985** *J. Biol. Chem.* 260, 10487-10494.
107 Caprioli, R.M.; DaGue, B.; Fan, T.; Moore, W. **1987** *Biochem. Biophys.Res. Commun.* 146, 291-299.
108 Biemann, K. **1990** *Methods Enzymol.* 193, 455-479.
109 Johnson, R. S ; Biemann, K. **1987** *Biochemistry* 26, 1209-1214.
110 Biemann, K.; Scoble, H. A. **1987** *Science* 237, 992-998.
111 Carr, S. A.; Hemling, M.E.; Folena-Wasserman, G.; Sweet, R.W.; Anumula,K.; Barr, J.R.; Huddleston, M.J.; Taylor, P. **1989** *J. Biol. Chem.* 264, 21286-21295.

112 *Mass spectrometry of peptides,* Desiderio, D.M., Ed.; CRC Press, Boca Raton, FL.**1991**
113 Scoble, H. A.; Martin, S. A. **1990** *Methods Enzymol.* 193, .519-536
114 *Methods in Enzymology* McCloskey, J.A. Ed.; Academic Press, NY.**1990** Volume 193.
115 Smith, R. D.; Loo, J. A.; Loo, R.R.O.; Busman, M.; Udseth, H.R. **1991** *Mass Spectrom. Revs.* 10, 359-451.
116 Maquin, F.;Schoot, B.M.; Devaux, P. G .; Green, B. N. **1991** *Rapid Commun. Mass Spectrom.* 299-302.
117 Hillenkamp, F.; Karas, M. **1990** *Methods Enzymol.* 193, 280-294
118 Karas, M.; Bahr, U.; Giessmann, U. **1991** *Mass Spectrom. Revs.* 10, 335-357.
119 Beavis, R. C.; Chait, B. T. **1990** *Proc. Natl. Acad. Sci. USA* 87, 6873-6877.
120 Havel, H.A.; Chao, R.S.; Haskel, R.J.; Thamann, T.J. **1989** *Anal. Chem.* 61, 642-650
121 Wetzel, R. In *Stability of protein pharmaceuticals* .Ahern, T.J.; Manning M.C., Eds.; Vol. 3 , **1992** ; pp.43-88.
122 Davio, S.R.; Hageman, M.J.; In *Stability and characterization of protein and peptide drugs : case histories*; Wang, J.Y.; Pearlman, R., Eds.; Plenum Press, NY. **1993**; pp. 59-90.
123 Geigert, J.; Solli, N.; Woehlke, P.; Vemuri, S.; In *Stability and characterization of protein and peptide drugs : case histories*; Wang, J.Y.; Pearlman, R., Eds.; Plenum Press, NY. **1993**; pp. 249-262.
124 Wetlaufer, D.B. **1962** *Adv. Prot. Chem.* 17, 303-390
125 Demchenko, A.P. **1986** *Ultraviolet spectroscopy of proteins.* Springer-Verlag Berlin.
126 Donovan, J.W. **1973** *Methods Enzymol.* 27, 497-548
127 Balestrieri, C.; Colonna, G.; Giovanne, A.; Irace, G.; Servillo, L. **1978** *Eur. J.Biochem.* 90, 433-440.
128 Woody, R.W. **1973** *Tetrahedron* 29, 1273-1283.
129 Lowry, O.H.; Rosebrough, N.J.; Farr, A.L.; Randall, R.J. **1951** *J. Biol. Chem.* 193, 265-275
130 Bradford, M.M. **1976** *Anal. Biochem.* 72, 248-254.
131 Bewley, T.A. **1982** *Anal. Biochem.* 123, 55-65.
132 Brems, D.N. **1988** *Biochemistry* 27, 4541-4546.
133 Kronman, M.J.; Holmes, L.G. **1971** *Photochem. Photobiol.* 14, 113-134.
134 Burstein, E.A.; Vedenkina, N.S.; Ivkova, M.N. **1973** *Photochem. Photobiol.* 18, 263-279.
135 Narhi, L.O.; Kenney, W.C.; Arakawa, T. **1991** *J. Prot. Chem.* 10, 359-367.
136 Eftink, M.R.; Ghiron, C.A. **1976** *Biochemistry* 15, 672-680.
137 Davis, J. M.; Arakawa, T.; Strickland, T.W.; Yphantis, D.A. **1987** *Biochemistry* 26, 2633-2638
138 Strickland , E.H. **1974** *C.R.C. Crit. Rev. Biochem.* 2, 113-175
139 Chen,Y.-H.; Yang,J.T.; Chau, K.H. **1974** *Biochemistry* 13, 3350-3359.
140 Johnson, W.C. **1988** *Ann. Rev. Biophys. Biophys. Chem.* 17, 145-166
141 Provencher, S.W.; Gloeckner, J. **1981** *Biochemistry* 20, 33-37
142 Otto, A.; Seckler, R. **1991** *Eur. J. Biochem.* 202, 67-73
143 Shirley, B.A. In *Stability of protein pharmaceuticals* .Ahern, T.J.; Manning M.C., Eds.; Vol. 3 , **1992** ; pp 167-194.
144 Chen, B.-L.; Baase, W.A.; Nicholson, H.; Schellman, J.A. **1992** *Biochemistry* 31, 1464-1476
145 Cockle, S.A.; Epand, R.M.; Moscarello, M.A. **1978** *J. Biol. Chem* 253, 8019-8026.

146 Bewley, T.A. **1979** *Rec. Prog. Hormone Res.* 35, 155-213
147 Susi, H.; Byler, D. M. **1986** *Methods Enzymol.* 130, 290-311.
148 Wilder, C.L.; Friedrich, A.D.; Potts, R.O.; Daumy, G.O.; Francoeur, M.L. **1992** *Biochemistry* 31, 27-31.
149 Prestrelski, S.J.; Arakawa, T.; Carpenter, J.F.; **1993** *Arch. Biochem. Biophys.* 303,465-473.
150 Arakawa ,T; Prestrelski, S.J.; Kenney, W.C.; Carpenter, J.F.; **1993** *Advanced Drug Delievry Revs.* 10, 1-28.
151 Carpenter, J.F.; Prestrelski, S.J.; Arakawa, T.; **1993** *Arch. Biochem. Biophys.* 303, 456-464.
152 Schachman, H.K. **1959** *Ultracentifugation in Biochemistry.* Academic Press, NY.
153 Williams,J.W. **1972** *Ultracentifugation of macromolecules.* Academic Press, NY.
154 Schachman, H.K. **1989** *Nature*, 341, 259-260.
155 Shire, S.J. **1983** *Biochemistry* 22, 2664-2671
156 Teller, D.C. **1973** *Methods Enzymol.* 27, 346-441
157 Minton, A.P. **1990** *Anal. Biochem.* 190, 1-6.
158 Becerra, S. P.; Kumar, A.; Lewis, M.S.; Widen, G.S.; Abbotts, J.; Karawy, E.M.; Highes, S.H.; Shiloach, J.; Wilson, S.H. **1991** *Biochemistry* 30, 11707-11719.
159 Lewis, M.S. **1991** *Biochemistry* 30, 11707-11719.
160 Pennica , D.; Kohr, W.J.; Fendly, B.M.; Shire, S.J.; Raab, H.E.; Borchardt, P.E.; Lewis, M.; Goeddel, D.V. **1992** *Biocemistry* 31, 1134-1141.
161 Winder, A.F.; Gent, W.L.G. **1971** *Biopolymers* 10, 1243-1252.
162 Eckhardt, B.M.; Oeswein, J.Q.; Bewley, T.A **1994** *J. Parenteral Sci. Technol.* (In press)
163 Eckhardt, B.M.; Oeswein, J.Q.; Bewley, T.A **1991** *Pharm. Res.* 8, 1360-1364
164 Tanford, C. **1961** *Physical chemistry of macromolecules,* Wiley, New York
165 Dollinger, G.; Cunico, B.; Kunitani, M.; Johnson, D.; Jones, R. **1992** *J. Chromatogr.* 592, 215-228.
166 Takagi, T. **1990** *J. Chromatogr.* 506, 409-416
167 Stuting, H.H.; Krull.I.S.; Mhatre,R.; Krzysko, S.C.; Barth, H.G. **1989** *LC.GC* 7, 402-404
168 Mhatre, R.; Krull, I.S.; Stuting, H.H. **1990** *J. Chromatogr.* 502, 21-46
169 Krull, I.; Stuting, H.H.; Krzysko, S.C **1988** *J. Chromatogr.* 442, 29-52
170 Stuting, H.H.; Krull, I.S.**1990** *Anal. Chem.* 62, 2107-2114
171 Bloomfield, V.A. **1981** *Ann. Rev. Biophys. Bioeng.* 10, 421-450.
172 Phillies, G.D.J. **1990** *Anal. Chem.* 62, 1049A-1057A
173 Nicoli, D.F.; Benedek, G.B. **1976** *Biopolymers* 15, 2421-2437.
174 Cleland, J.L.; Wang, D.I.C. **1990** *Biochemistry* 29, 11072-11078.
175 Cleland, J.L.; Builder, S.E.; Swartz, J.R.; Chang, J.Y.; Wang, D.I.C. **1992** *Biotechnol.* 10, 1013-1019.

RECEIVED June 10, 1994

Chapter 3

Solution and Solid-State Chemical Instabilities of Asparaginyl and Aspartyl Residues in Model Peptides

Cecilia Oliyai[1,2] and Ronald T. Borchardt[1]

[1]Department of Pharmaceutical Chemistry, University of Kansas, Lawrence, KS 66045

The kinetics and mechanisms of the degradation of asparagine (Asn) and aspartic acid (Asp) residues in model hexapeptides were examined in aqueous and solid states. Specific chemical reactions that affect the stability of Asn and Asp residues in polypeptides include deamidation of the Asn side chain, Asp-X and/or X-Asp amide bond hydrolysis, and Asp-to-isoAsp interconversion. The exogenous parameters which influence the product distribution and the rates of degradation in solution and in solid state were studied. Under lyophilized conditions, the nature of the excipient most significantly affected the rate of decomposition of the Asp-hexapeptide, whereas the pH of the pre-lyophilized solution determined the product distribution. In contrast, the pH of the starting solution dictated both the rate and extent of degradation of the freeze-dried Asn-hexapeptide. It was observed that pH, temperature, and buffer concentration played important roles in determining the chemical lability and degradation routes in aqueous medium.

Although the pharmacological properties of polypeptides have long been recognized, the realization of their use as therapeutic agents came only recently with the advent of biotechnology, which permits the production of proteins and peptides on a commercial scale. As a result, the number of recombinant products is widespread and ever increasing (1). Unlike small molecules, polypeptides possess not only their primary sequence but also higher order structure (i.e. secondary, tertiary, and quaternary). Thus, the development of stable formulations for proteins as pharmaceuticals is difficult, and success often depends on an understanding of the physical and chemical instability of the molecule. This chapter is not intended to address the broad topic of physical and chemical instability of polypeptides since there are many appropriate reviews and books which provide comprehensive treatments of these subjects (2-4). The primary objective of this review is to focus on the effects of exogenous factors on the chemical instability of proteins, specifically, the instability of asparagine (Asn) and aspartic acid (Asp) residues in solution and in the solid state.

[2]Current address: Formulation Research and Development, Chiron Corporation, 4560 Horton Street, Emeryville, CA 94608

0097–6156/94/0567–0046$08.00/0

Instability of Aspartic Acid Residues

In solution Several studies have indicated that the chemical stability of Asp residues in proteins and peptides is dependent on the local conformation and primary sequence around the potentially reactive Asp residue *(5, 6)*. Two chemical reactions that are known to affect specifically Asp-containing proteins and peptides include Asp-to-isoAsp interconversion via cyclic imide formation and Asp-X and/or X-Asp amide bond hydrolysis. Several investigators have illustrated the magnitude of these non-enzymatic degradation pathways, which can substantially reduce the biological activity of Asp-containing proteins *(5-12* and Oliyai, C.; Borchardt, R. T., University of Kansas, unpublished data).

Asp residues in polypeptides tend to undergo cyclization to produce the cyclic imide. Such rearrangement has been observed in classical solution and solid phase syntheses, recrystallization *(13)*, and hydrogenolysis *(14)* of peptides which contain β-alkylaspartyl and free β-carboxyl groups. In addition, peptide bonds of Asp residues are cleaved in dilute acid at a rate at least 100 times greater than other peptide bonds *(15)*. The enhanced hydrolysis, occurring at Asp-X and/or X-Asp locations, is attributed to intramolecular catalysis by the carboxyl group of the Asp side chain.

Our laboratory has attempted to delineate the relevant factors which modulate the chemical activity of Asp residues in polypeptides by studying the kinetics and mechanism of degradation of a model hexapeptide (Val-Tyr-Pro-Asp-Gly-Ala; Asp-hexapeptide) whose primary sequence contains a single Asp residue. In solution, the apparent rates of degradation of this Asp-hexapeptide were determined as a function of pH, buffer concentration, and temperature *(16)*. The major degradation pathways for the Asp-hexapeptide consisted of hydrolysis at the Asp-Gly amide bond and the isomerization of Asp to isoAsp via the cyclic imide intermediate, the Asu peptide. The extent and routes of degradation were pH-dependent.

Under highly acidic conditions (pH 0.3–2.0), the Asp-hexapeptide decomposed predominantly via intramolecular hydrolysis of the Asp-Gly amide bond, forming a tetrapeptide, Val-Tyr-Pro-Asp, and a dipeptide, Gly-Ala (Figure 1a). Two plausible mechanisms of the Asp-Gly amide hydrolysis were postulated based on the kinetic and solvent isotope experiments *(16)*. Both mechanisms, which possibly involve intramolecular nucleophilic and general base catalyses, are catalyzed by hydrogen ions *(16)*. In addition, the starting peptide could also cyclize to generate an Asu-hexapeptide (cyclic imide) which remained stable under acidic conditions and constituted approximately 7% of the total degradation products. The ring closure is known to be catalyzed by both acids and bases *(17-20)*. The Asp-hexapeptide favored cyclization over the peptide cleavage reaction as the pH was increased. The Asu-hexapeptide formed in cyclization further degraded to give rise to isoAsp and Asp peptides at higher pH values (pH $\geq$ 4) as a result of base-catalyzed hydrolysis. For example, at pH 4.0 the majority of the peptide (60%) rearranged to produce the Asu-hexapeptide, which was further hydrolyzed to form the isoAsp-hexapeptide while approximately 10% underwent Asp-X hydrolysis (Figure 1b). The overall kinetic scheme at pH 4.0 to 5.0 was complicated by this transition from predominantly amide hydrolysis to cyclic imide formation, which eventually led to isomerization. The Asp-hexapeptide exhibited apparent minimum chemical stability at pH 0.3 and pH 4.0 (Figure 2). In this pH-rate profile, the observed rate constants (k_{obs}) for the disappearance of the Asp-hexapeptide at pH 0.3 to 3.0 were determined from the slopes of linear plots of the logarithm of peptide peak area versus time. At pH 4.0 and 5.0, k_{obs} were estimated by adding the pseudo-first-order rate constants for the parallel formation of the Asu-hexapeptide and -tetrapeptide. These individual rate constants were generated by fitting the data to Laplace MicroMath (Salt Lake City, UT) and nonlinear least-squares regression (MINSQ). For pH 6 to 10.0, the loss of the Asp-hexapeptide was characterized by pseudo-first-order reversible kinetics whereby $k_{obs} = k_f$ (apparent rate constant for the formation of the isoAsp peptide) + k_r (apparent rate constant for the regeneration of the Asp peptide), where k_f and k_r were generated from the best fit obtained.

The contribution from the Asp-Gly amide hydrolysis to the overall degradation was negligible above pH 6.0 (Figure 1c). Furthermore, since the Asu-hexapeptide was extremely unstable at neutral and alkaline pH, hydrolyzing rapidly to form the isoAsp-hexapeptide and to regenerate the Asp-hexapeptide, only the isoAsp hexapeptide product was observed (Figure 1c). The apparent rate of degradation followed pseudo-first-order reversible interconversion kinetics *(16)*. Buffer catalysis did not occur to any significant extent for the forward and reverse reactions, which were also essentially pH-independent at pH 8.0 and above (Figure 2). It may bereasonable to suggest that the formation of Asu-hexapeptide from the Asp-hexapeptide involves the nucleophilic attack by the deprotonated amide nitrogen on the free carboxylic acid species. Thus, as the pH of the solution is increased, the equilibrium concentration of the nucleophilic amide ion increases but the fraction of protonated species decreases. These two effects exactly offset each other, and the observed rate becomes independent of pH.

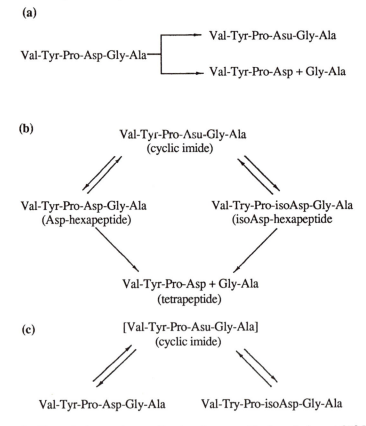

Figure 1. Degradation pathways for Asp-hexapeptide in solution at 37°C (0.5 M). (a) at pH 0.3 to 3 (pH 0.3, 1.1, 1.5, and 2.0 HCl; pH 3.0 formate); (b) at pH 4 to 5 (acetate); (c) at pH 6 to 10.0 (pH 6 and 7.4 phosphate; pH 8.0 tris; pH 9.0 and 10.0 borate). (Adapted from ref. 1).

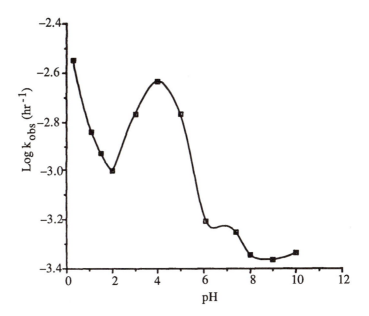

Figure 2. pH-rate profile for the degradation of the Asp-hexapeptide at 37°C (k_{obs} is the observed rate constant for the loss of the Asp-hexapeptide). (Adapted from ref. 1).

In the Solid State. Proteins often have poor and erratic oral bioavailability *(21)* and may require lyophilization to achieve adequate shelf-life stability *(22)*. However, development of freeze-dried formulations presents a new set of stability variables such as the effect of excipients, residual moisture content, and lyophilization cycle.

The nature of the excipients used in pharmaceutical formulations has been reported to influence the general stability of freeze-dried drugs. Excipient crystallinity, which can be modulated by the choice of lyophilization cycle, can affect the stabilization of proteins during lyophilization. Inclusion of certain amorphous additives has been shown to minimize freeze-drying-induced aggregate formation *(23-25)*. In another study, a partially amorphous excipient system provided the greatest protection against aggregation and chemical decomposition via methionine oxidation and Asn deamidation *(26)*. Crystallization of excipients, which leads to a decrease in Tg (glass transition temperature), has been observed to occur when the formulations are stored at 25°C *(27)*. As a result, the degradation rate increased sharply with increasing temperature and with increasing moisture level. Additives were also observed to preserve the native structure of proteins by acting as protectants during freezing and dehydration stresses *(28-31)*. Optimal recovery of activity of proteins upon freeze-drying and rehydration was correlated with the use of such additives in the formulations.

Water facilitates spontaneous chemical degradation by increasing molecular mobility and chain flexibility to encourage intermolecular and intramolecular rearrangements and by directly participating in chemical reactions. Moreover, water can interact with the excipient to render physical transformation of solids (i.e. excipients and drug), compromising the overall stability of the active drug *(32)*. Thus, the residual water level in freeze-dried products represents another critical parameter in the chemical stability of drugs. While the concept of pH may not have meaning in the solid state, the pH of the bulk solution determines the extent of ionization of the peptide and buffer species both in solution and in the solid state *(26)*. In some cases, the rate constant of degradation in solid state can be comparable to or higher than that in solution at a given pH *(33, 34)*. Also, depending on the choice of buffers, the pH can change during the freezing process as a result of crystallization of buffers *(35)*. Thus, it is important to confirm whether the same pH dependence for the rate of degradation generated from solution stability studies applies in the solid state. Assessment of these variables is imperative for designing the most stable dosage forms with minimal lot-to-lot shelf-life variability.

We have recently conducted a study which evaluated the individual and inter-active effects of formulation variables such as the pH of pre-lyophilized solution, moisture level, temperature, and type of bulking agent on the chemical stability of the model Asp-hexapeptide in the lyophilized state (Oliyai, C.; Patel, J. P.; Carr, L.; Borchardt, R. T., University of Kansas, unpublished data). All bulk solutions were prepared by dissolving appropriate amounts of peptide and excipients in 0.01 M of pH-adjusted dibasic sodium phosphate/citrate buffer solutions. These bulk solutions were then lyophilized and rehydrated with 0.8 µl (medium moisture level) and 1.6 µl (high moisture level) of H_2O. The samples that were not rehydrated following lyophilization were designated as having the lowest moisture level. The moisture-loaded samples were then stored at temperatures ranging from 40° to 60°C. The vials were removed at designated time intervals to be analyzed by reversed-phase HPLC. The degradation products were identified by coinjecting authentic samples.

The degradation pathways of the Asp-hexapeptide in the lyophilized state were dependent on the pH of the bulk solutions and the moisture content of the freeze-dried formulations. In general, the kinetics of the disappearance of Asp-hexapeptide followed pseudo-first-order reversible kinetic behavior under all experimental conditions (Figures 3a-b). This type of kinetic profile was justified by the product distribution observed. Under acidic conditions (pH 3.5 and 5.0), the lyophilized Asp hexapeptide predominantly decomposed to generate the Asu-hexapeptide, irrespective of the type of excipient present in the formulation (Table I). The hydrolysis of the Asp-Gly amide bond constituted a much less significant pathway under these conditions (Figure 4, Table I). At pH 6.5 and 8.0, the parent hexapeptide exclusively isomerized via formation of the Asu-hexapeptide to produce the isoAsp-hexapeptide (Figure 4, Table I). The extent of hydrolysis of the Asu-hexapeptide intermediate at pH 8.0 exceeded that at pH 6.5, rendering the isoAsp-hexapeptide the major product of degradation in the basic environment. Although the type of excipient (amorphous vs. crystalline) did not influence the degradation routes, the choice of excipient substantially affected the mean rate constant of peptide decomposition. Consistently, the amorphous lactose/peptide formulations were significantly more chemically stable at all temperatures, pH values, and moisture levels than formulations containing crystalline mannitol. These results are consistent with the view that an amorphous or partially amorphous matrix usually imparts more stabilization to the freeze-dried protein drugs than does the crystalline excipient system *(23-26)*.

Upon examination, the product distribution in the solid state was significantly different from that in solution. Evidently the hydrolysis of the Asu-hexapeptide intermediate and the Asp-Gly peptide bond (formation of the tetrapeptide) was suppressed in a water-deficient environment. Thus, at lower pH values (3.5, 5.0 and 6.5), the major product observed was the Asu-hexapeptide, and only trace amounts of the tetrapeptide and isoAsp-hexapeptide were detected at pH 3.5 and 6.5, respectively (Figure 4, Table I). At pH 8.0, the base catalysis component of the hydrolysis of the Asu-hexapeptide intermediate compensated for the nearly inoperative water catalysis term. Consequently, the rate of decomposition of the Asu-hexapeptide was sufficient to afford the isoAsp-hexapeptide as a major degradation product at pH 8.0 (Figure 4, Table I).

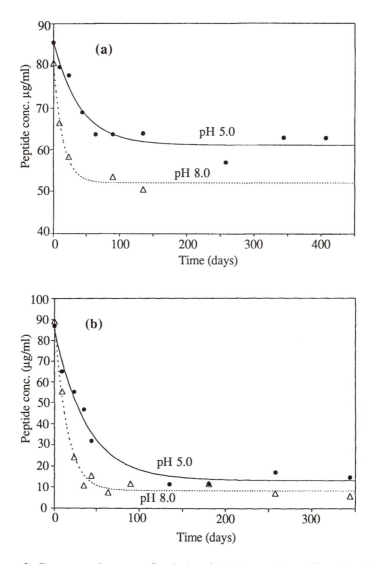

Figure 3. Representative curve-fitted plots for the loss of lyophilized Asp-hexa-peptide (50°C) in formulations containing (a) lactose (average moisture content = 2.7 % ± 0.2% H_2O) and (b) mannitol (average moisture content = 0.4% ± 0.05% H_2O).

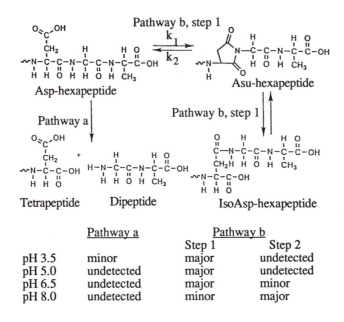

	Pathway a	Pathway b	
		Step 1	Step 2
pH 3.5	minor	major	undetected
pH 5.0	undetected	major	undetected
pH 6.5	undetected	major	minor
pH 8.0	undetected	minor	major

Figure 4. Degradation pathways for the Asp-hexapeptide in the solid state.

**Table I. Representative Peptide Distribution at 60°C
for the Lyophilized Asp-Hexapeptide in the Presence of
Mannitol and Lactose at Different pH Values (six months)**

| | % of Total Initial Concentration | | | |
Mannitol Formulations Average H_2O content (0.3% ± 0.08%)	Asp-hexa-peptide	IsoAsp-hexapeptide	Asu-hexa-peptide	Tetra-peptide
pH 3.5	15	ND[a]	45	2.5
pH 5.0	8	ND	57	ND
pH 8.0	9	23	18	ND
Lactose Formulations Average H_2O content (1.9% ± 0.7%)				
pH 3.5	55	ND	12	4
pH 5.0	73	ND	9	ND
pH 8.0	54	17	7	ND

[a]Not detectable

Instability of Asparagine Residues

In Solution. Deamidation of Asn residues accounts for a high incidence of chemical instability in proteins. The prevalence of this degradation pathway is illustrated by the ever-increasing number of polypeptides which experienced deamidation and, in some cases, loss of their biological activity as a result of the chemical transformation *(6, 12, 36-43)*. During the deamidation reaction of Asn residues, the side chain amide of the Asn residue either is hydrolyzed directly to an Asp residue or proceeds through a five-membered cyclic imide intermediate to Asp and isoAsp residues (Figure 5).

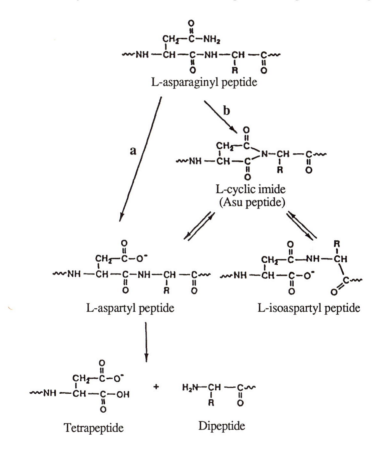

Figure 5. Deamidation pathways for the Asn-hexapeptide via (a) direct hydrolysis of the Asn amide side chain or (b) formation of the cyclic imide intermediate. (Adapted from ref. 20).

Previous studies have demonstrated that non-enzymatic deamidation is by no means a randomly programmed, post-translational modification. Several determinants such as primary sequence *(44, 45)*, local conformation *(40, 46, 47)*, and exogenous factors *(38, 48)* dictate the deamidation of native proteins. Therefore, it is important to understand the factors that affect deamidation in order to design rational strategies for protein stabilization.

In our laboratory, we have studied the effects of exogenous factors such as pH, buffer concentration, and temperature on the kinetics of deamidation and on the degradation product ratio, isoAsp/Asp peptides. A model Asn-containing hexapeptide, whose primary sequence coincides with residues 22-27 of adrenocorticotropic hormone (Val-Tyr-Pro-Asn-Gly-Ala), was selected for this study since the degradation products are easily separated from the parent peptide.

The solution degradation of this model hexapeptide, Asn-hexapeptide, was pH-dependent and followed pseudo-first-order kinetics (20). Under acidic conditions (pH 1-2) at 37°C, the Asn-hexapeptide degraded to produce the normal Asp-hexapeptide (Val-Tyr-Pro-Asp-Gly-Ala) via direct hydrolysis (Figure 5, pathway a). This Asp-hexapeptide, in turn, underwent Asp-Gly amide bond hydrolysis to generate a tetrapeptide (Val-Tyr-Pro-Asp) (Figure 5, pathway a). The formation of the cyclic imide (Asu-hexapeptide; Figure 5, pathway b) was also detected at acidic pH, although its appearance was much slower than the direct hydrolysis reaction, constituting only 10% of the total product. The cyclic imide remained stable in acidic medium and, thus, did not break down further to form the Asp and isoAsp peptide products. From pH 5 to 12, it was shown that deamidation of the Asn-hexapeptide involved the formation of a cyclic imide intermediate followed by its subsequent rapid hydrolysis to form the isoAsp and Asp hexapeptides in a ratio of 4 to 1 (Figure 5, pathway b). Buffer catalysis was observed in the pH range 7–11, but little or no catalysis was observed from pH 5.0 to 6.5. The kinetics of deamidation of the Asn-hexapeptide in solution obeyed the Arrhenius relationship within the temperature range studied (50°–90°C, pH 5.0; 25°C–70°C, pH 7.5).

The pH dependence for the rate of deamidation of the Asn-hexapeptide is illustrated in Figure 6. Overall, the rate constant of deamidation was slower in acid than it was in the neutral to alkaline region. The Asn-hexapeptide experienced maximum chemical stability at pH 3.0 to 4.0. The pH-rate profile showed a unit negative slope on the acidic side (pH 1–2), indicating specific acid catalysis. From pH 5 to 12, the rate was catalyzed by hydroxide ions.

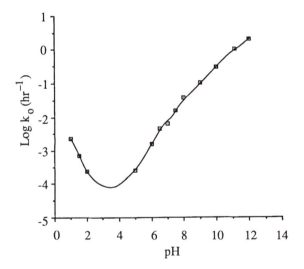

Figure 6. pH-rate profile for the deamidation of the Asn-hexapeptide in aqueous solution (37°C, 0.5 M). (Adapted from ref. 20).

In the Solid State. While extensive research efforts have been devoted to understanding deamidation in aqueous medium, very little is known about this intramolecular chemical event in the solid state. Reports about proteins undergoing deamidation in freeze-dried formulations are limited *(26)*. Often this chemical reaction is further complicated by the concomitant existence of other chemical or physical instability problems in a given system. Consequently, the kinetics and mechanisms of degradation in such a system become extremely complex and, thus, it is not feasible to examine deamidation in a more thorough fashion.

A more complete understanding of the influence of exogenous factors on the deamidation process was made possible through a study of the solid state instability of the model Asn-hexapeptide, Val-Tyr-Pro-Asn-Gly-Ala, whose kinetics and mechanism of degradation in aqueous solution are already well understood (Oliyai, C.; Patel, J. P.; Borchardt, R. T., University of Kansas, unpublished data). The objective of this study was to evaluate the chemical stability of the Asn-hexapeptide in lyophilized formulations as a function of pH of the starting solution, temperature, and residual moisture level.

The disappearance of the Asn-hexapeptide appeared to fit empirically to the pseudo-first-order reversible kinetic model under all experimental conditions. Representative plots of the results along with the calculated lines from this model at different pH values and at 40°C and low moisture level are shown in Figure 7. The theoretical lines were generated using equation 1

$$A = Ao[(k_1/k_1 + k_2)e^{-(k_1+k_2)^t} + (k_2/k_1 + k_2)] \tag{1}$$

where k_1 and k_2 are forward and reverse rate constants, respectively, A is the concentration of peptide remaining at any time t, and Ao is the initial peptide concentration. Analysis of variance calculations indicated that the effects of formulation variables on the reverse rate constant k_2 were not statistically significant irrespective of experimental conditions.

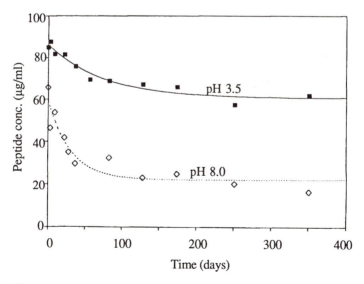

Figure 7. Representative curve-fitted plots for the disappearance of the Asn-hexapeptide at various pH values (40°C) and low moisture level (0.3 % ± 0.09% water).

At pH 3.5, the Asn-hexapeptide deamidated via direct hydrolysis to produce the Asp-hexapeptide which, in turn, was hydrolyzed at the Asp-Gly amide bond to generate a small quantity of tetrapeptide (Figure 8). Additionally, the parent hexapeptide cyclized to form the cyclic imide product in a quantity equivalent to the Asp-hexapeptide (Figure 8). However, the cyclic imide formation was favored over the

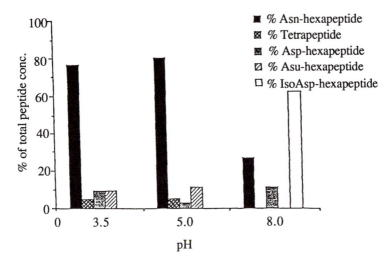

Figure 8. Peptide distribution vs. pH at 40°C and low moisture level (0.3 % ± 0.09% water) after one year.

direct hydrolysis of the Asn side chain at higher temperatures, moisture levels, and pH. The formation of cyclic imide required not only the appropriate pH environment but also, more importantly, the chain flexibility suitable for adopting the necessary local conformation for cyclization. Increasing either temperature or residual moisture content would enhance chemical reactions by increasing chain flexibility. Thus, the amount of cyclic imide significantly increased at pH 3.5 as a function of temperature or moisture level. While the propensity of the Asn-hexapeptide to cyclize increased slightly when going from pH 3.5 to 5.0, the rate of hydrolysis of the Asn side chain decreased (Figure 8). The Asn-hexapeptide was most unstable at pH 8.0, such that about 73% of the Asn-hexapeptide deamidated after 12 months at 40°C, generating isoAsp-hexapeptide and Asp-hexapeptide as the major and minor products, respectively (Figure 8). The overall rate of decomposition of the Asn-hexapeptide was minimal at pH 5.0 where the gradual transition from direct hydrolysis of the Asn side chain to deamidation via cyclic imide intermediate occurred.

Conclusion

Using small model peptides, we have illustrated that the chemical reactivities of Asn and Asp residues in solution and in the solid state are significantly affected by the exogenous environment. Degradation pathways and reaction mechanism(s) of selected chemical transformations can be elucidated by these studies. Undoubtedly, in an intricate system such as proteins, other factors (e.g. local conformation) would also influence the chemical instability of potentially labile Asp and Asn residues.

This type of information on the influence of exogenous and endogenous factors on chemical degradation of proteins may lead to the option of site-directed mutagenesis in which the labile sites or adjacent amino acids which catalyze the instability are being replaced by more inert residues. Obviously, in a large complex protein, this type of stabilization strategy will need to avoid both disrupting the conformational state of the protein needed for biological activity and invoking an unwanted immune response. Often, more traditional formulation methods, which lead to the use of the optimal exogenous factors, will have to be employed to stabilize proteins.

Acknowledgments

C.O. acknowledges the financial support provided by the Parenteral Drug Association Pre-Doctoral Fellowship, Abbott Laboratories, and The National Institute of General Medical Sciences Biotechnology Traineeship. The authors like to thank Professors Richard Schowen (The University of Kansas) and Jeffrey Fox (The University of Utah) for their invaluable comments and discussions. We would also like to acknowledge the following individuals from Abbott Laboratories and the University of Kansas: Dr. Jitendra Patel, Dr. Madhup Dhaon, Linda Carr, Dr. Steve Krill, Dr. John Quick, Jerry Sutherland, Dr. Alex Puko, Christopher Smith, and Brian Moon.

Literature Cited

1. Oliyai, C.; Schöneich, C.; Wilson, G. S.; Borchardt, R. T. In Topics *in Pharmaceutical Sciences* ; Crommelin, D. J. A.; Midha, K. K., Eds.; Medpharm Scientific Publisher: Stuttgart, 1992; pp. 23–46.
2. Manning, M. C.; Patel, K.; Borchardt, R. T. *Pharm. Res.* **1989**, *6*, 903–918.
3. *Stability of Protein Pharmaceuticals Part A: Chemical and Physical Pathways of Protein Degradation;* Ahern, T. J.; Manning, M. C., Eds.; Pharmaceutical Biotechnology; Plenum Press: New York, 1992; Vol. 2.
4. *Stability of Protein Pharmaceuticals Part B: In Vivo Pathways of Degradation and Strategies for Protein Stabilization;* Ahern, T. J.; Manning, M. C., Eds.; Pharmaceutical Biotechnology; Plenum Press: New York, 1992; Vol. 3.
5. Ota, I. M.; Clarke, S. *Biochemistry* **1989**, *28*, 4020–4027.
6. George-Nascimento, C.; Lowenson, J.; Borissenko, M.; Calderon, M.; Medina-Selby, A.; Kuo, J.; Clarke, S.; Randolph, A. *Biochemistry* **1990**, *29*, 9584–9591.
7. Marcus, F. *Int. J. Peptide Protein Res.* **1985**, *25*, 542–546.
8. Clarke, S. *Annu. Rev. Biochem.* **1985**, *54*, 479–506.
9. Tsuda, T.; Uchiyama, M.; Sato, T.; Yoshino, H.; Tsuchiya, Y.; Ishikawa, S.; Ohmae, M.; Watanabe, S.; Miyake, Y. *J. Pharm. Sci.* **1990**, *79*, 223–227.
10. Inglis, A. S. Methods Enzymol. **1983**, *91*, 324–332.
11. Kirsch, L. E.; Molloy, R. M.; Debono, M.; Baker, P.; Farid, K. Z. *Pharm. Res.* **1989**, *6*, 387–393.
12. Johnson, B. A.; Shirokawa, J. M.; Hancock, W. S.; Spellman, M. W.; Basa, L. J.; Aswad, D. W. *J. Biol. Chem.* **1989**, *264*, 14262–14271.
13. Ondetti, M. A.; Deer, A.; Sheehan, J. T.; Kocy, O. *Biochemistry* **1968**, *7*, 4069–4075.
14. Perseo, G.; Forino, R.; Galantino, M.; Gioia, B.; Malatesta, V.; DeCastiglione, R. *Int. J. Peptide Protein Res.* **1986**, *27*, 51–60.
15. Schultz, J. *Methods Enzym.* **1967**, *11*, 255–263.
16. Oliyai, C.; Borchardt, R. T. *Pharm. Res.* **1993**, *10*, 95–102.
17. Schön, I.; Kisfaludy, L. *Int. J. Peptide Protein Res.* **1979**, *14*, 485–494.
18. Bodanszky, M.; Kwei, J. Z. *Int. J. Peptide Protein Res.* **1978**, *12*, 69–74.
19. Blake, J. *Int. J. Peptide Protein Res.* **1979**, *13*, 418–425.
20. Patel, K.; Borchardt, R. T. *Pharm. Res.* **1990**, *7*, 703–711.

21. Humphrey, M. J.; Kingrose, P. S. *Drug Metab. Rev.* **1986**, *17*, 283–310.
22. Mendenhall, D. W. *Drug Dev. Ind. Pharm.* **1984**, *10* , 1297–1342.
23. Hora, M. S.; Rana, R. K.; Wilcox, C. L.; Katre, N. V.; Hirtzer, P.; Wolfe, S. N.; Thomson, J. W. *Dev. Biol. Stand.* **1992**, *74*, 295–303.
24. Hora, M. S.; Rana, R. K.; Smith, F. W. *Pharm. Res.* **1992**, *9*, 33–36.
25. Hora, M. S.; Simpkins, J. W.; Bodor, N. *Pharm. Res.* **1991**, *8*, 792–795.
26. Pikal, M. J.; Dellerman, K. M.; Roy, M. L.; Riggin, R. M. *Pharm. Res.* **1991**, *8*, 427–436.
27. Roy, M. L; Pikal, M. J.; Rickard, E. C. Maloney, A. M. *Dev. Biol. Stand.* **1992**, *74,* 323–339.
28. Prestrelski, S. J.; Arakawa, T; Carpenter, J. F. *Arch. Biochem . Biophys.* **1993**, *303*, 465–473.
29. Carpenter, J. F.; Prestrelski, S. J.; Arakawa, T. *Arch. Biochem. Biophys.* **1993**, *303*, 456–464.
30. Carpenter, J. F; Arakawa, T.; Crowe, J. H. *Dev. Biol. Stand.* **1992**, *74*, 225–238.
31. Arakawa, T.; Kita, Y; Carpenter, J. F. *Pharm. Res.* **1991**, *8*, 285–291.
32. Kovalcik, T. R.; Guillory, J. K. *J. Parenter. Sci. Technol.* **1988**, *42*, 29–37.
33. Strickley, R. G.; Visor, G. C.; Lin, L.; Gu, L. *Pharm. Res.* **1989**, *6*, 971–975.
34. Pearlman, R.; Nguyen, T. *J. Pharm. Pharmacol.* **1992**, *44*, 178–185.
35. Franks, F. *Cryo-Letters* **1990**, *11*, 93–110.
36. Violand, B. N.; Schlittler, M. R.; Toren, P. C.; Siegel, N. R. *J. Protein Chem.* **1990**, *9*, 109–117.
37. Daumy, G. O.; Wilder, C. L.; Merenda, J. M.; McColl, A. S.; Geoghegan, K. F.; Otterness, I. G. *FEBS Lett.* **1991**, *278*, 98–102.
38. Maeda, H.; Kuromizu, K. *J. Biochem.* **1977**, *81*, 25–35.
39. Friedman, A. R.; Ichhpurani, A. K.; Brown, D. M.; Hillman, R. M.; Krabill, L. F.; Martin, R. A.; Zurcher-Neely, H. A.; Guido, D. M. *Int. J. Peptide Protein Res.* **1991**, *37*, 14–20.
40. Wearne, S. J.; Creighton, T. E. *Proteins: Struct. Funct. Genet.* **1989**, *5*, 8–12.
41. Bhatt, N. P.; Patel, K.; Borchardt, R. T. *Pharm. Res.* **1990**, *7*, 593–599.
42. Gleed, J. H.; Hendy, G. N.; Kimura, T.; Sakakibara, S.; O'Riordan, J. L. H. *Bone Miner.* **1987**, *2*, 375–382.
43. Yuksel, K. U.; Gracy, R. W. *Arch. Biochem. Biophys.* **1986**, *248*, 452–459.
44. Stephenson, R. C.; Clarke, C. *J. Biol. Chem.* **1989**, *264*, 6164–6170.
45. Patel, K.; Borchardt, R. T. *Pharm. Res.* **1990**, *7*, 787–793.
46. Clarke, S. *Int. J. Peptide Protein Res.* **1987**, *30*, 808–821.
47. Kossiakoff, A. A. *Science* **1988**, *240*, 191–194.
48. Geiger, T.; Clarke, S. *J. Biol. Chem.* **1987**, *262*, 785–794.

RECEIVED June 8, 1994

Chapter 4

Oxidation Degradation of Protein Pharmaceuticals

Tue H. Nguyen

Pharmaceutical Research and Development, Genentech, Inc., South San Francisco, CA 94080

Oxidation of proteins in pharmaceutical formulations often occurs through the conversion of methionine to methionine sulfoxide and cysteine to cystine. The degradation proceeds slowly, catalyzed by trace amounts of impurities which are derived from many and sometimes unexpected sources. An understanding of the basic mechanism of the reactions involved provides a rational approach to stabilize the products. Solution pH, ionic strength, co-solvent, metal chelating agents, inert gas and antioxidants are some of the parameters that directly affect oxidation rate. Formulation components have to be tested for impurities and tight specifications must be established. Phase separation upon freezing of the solution can create pockets of concentrated protein and buffer components where inter-chain disulfide bridge formation is favored. Thus, excipients which ensure the homogeneity of the frozen solution or the lyophilized matrix should play an important role in assuring the stability of the protein in these dosage forms.

Oxidation is one of the major degradation pathways for proteins in pharmaceutical products. Recent reviews have thoroughly summarized current biochemical literature on the subject (1-4). These reviews provide an important background for understanding the reaction but they are not always directly applicable to the various protein formulations. The experiments reported were usually carried out under a specific set of conditions such that the nature of the oxidants and catalysts were well defined. The concentration of the reactants were calibrated to drive the reactions to completion within a reasonable amount of time.

The side chains in proteins oxidize under very different conditions. Energy input in the form of photon or high-energy particle irradiation, the presence of sensitizing agents, radical initiators and transition metal ions are necessary for the oxidation of the aromatic side chains of histidine, tyrosine, and tryptophan. The reaction invariably leads to the opening of the aromatic ring to generate multiple by-products (5-10). Methionine can be oxidized to methionine sulfoxide and, in the presence of peracid, to its sulfone derivative (11-15). The thiol side-chain in cysteine undergoes oxidation to cystine in the presence of oxygen and trace metal ions. Under more drastic conditions it can be converted directly to sulfenic acid, sulfinic acid then to sulfonic acid with successive oxidation steps (16).

0097–6156/94/0567–0059$08.00/0

Protein pharmaceuticals are usually formulated within a range between 4.0 and 7.5. Solutions are manufactured and stored with all the precautions necessary to insure the integrity of the molecule. Low temperature storage, protection from excessive exposure to light, air and agitation are common practice. Lyophilized preparations are stoppered under vacuum or inert gas head space. Under these conditions, oxidative degradation of proteins is often limited to the conversion of methionine to methionine sulfoxide and cysteine to cystine. The reaction occurs over a period of months catalyzed by trace amounts of contaminants. The concentration and the nature of the reactive species are seldom definitively identified. For these reasons, oxidation in pharmaceutical formulations is often referred to as spontaneous oxidation, auto-oxidation or non-catalyzed oxidation.

Oxidation of Cysteine

Oxidation of cysteine is a major challenge in the formulation of interferon -b (INF-b), interleukin-1b (IL-1b) and interleukin-2 (IL-2) *(17-21)*, since it usually leads to significant changes in tertiary structure, dimerization and aggregation with complete loss of bioactivity. In the presence of air and metal ions, typically Cu^{++} and Fe^{++}, the thiol side-chain reacts with another cysteine residue to form a disulfide bridge *(22)*. This reaction was studied in detail by Misra *(23)* using dithiotreitol (DTT) as a model compound. A sequence of steps involving generation of super oxide radicals, hydroxyl radicals and hydrogen peroxide was proposed

$$RS^- + Me^{n+} \rightarrow RS\cdot + Me^{(n-1)+}$$

$$RSH + RS\cdot + O_2 \rightarrow RSSR + H^+ + O_2^-$$

$$Me^{(n-1)+} + O_2 \rightarrow Me^{n+} + O_2^-$$

$$O_2^- + O_2^- + 2H^+ \rightarrow H_2O_2 + O_2$$

$$O_2^- + H_2O_2 \rightarrow OH\cdot + OH^- + O_2$$

$$RS^- + HO\cdot \rightarrow RS\cdot + OH^-$$

$$HO\cdot + O_2^- \rightarrow OH^- + O_2$$

$$RSH + RS\cdot + H_2O_2 \rightarrow RSSR + OH^- + OH\cdot$$

Scheme I

In this study, the rate of oxidation in carbonate buffer at pH 10.2 increased five folds when the solution was purged with oxygen. Removal of metal ions by chelation with sodium ethylene diamine tetraacetate (EDTA) resulted in a marked decrease in the rate of cyclization of DTT. Superoxide dismutase and radical scavengers such as mannitol and formate also reduced the rate of oxidation. Catalase on the other hand did not affect the reaction velocity. This result was explained by the low steady state concentration of H_2O_2 and the low affinity of the enzyme toward its substrate (large Km).

A similar phenomenon of oxidation and dimer formation was observed by Tomazic and Klibanov *(24)*. They investigated the inactivation of bacillus a-amylase at high temperature and demonstrated that at pH 8.0, loss of activity was mainly due to dimerization of the enzyme with a concomitant disappearance of the single free thiol moiety in the molecule. The formation of intermolecular disulfide bridge was greatly accelerated by the addition of Cu^{++} and conversely, replacing atmospheric air with argon significantly reduced the formation of dimer.

As shown in Scheme I, auto-oxidation of cysteine is initiated by the abstraction of an electron from thiolate anion RS⁻ by metal ion generating an RS· radical. Thus, the oxidation rate is directly related to the ionization constant of the thiol side chain and to the pH of the solution. Simple aliphatic thiols have a pK_a between 7.5 and 10.5 *(25)*. The ionization constant of the cysteine thiol side-chain in proteins generally falls within the same range *(25)*. This explains the increased propensity for thiol oxidation in solution buffered at neutral to basic pH. However, more reactive thiols with a much lower pK_a have been observed. Ionization of neighboring residues was invoked to explain the unusually low pK_a (4.1) of the active-site Cys-25 in papain *(26)* and dipole moment of a nearby a-helix was thought to depress the pK_a of Cys-35 in thioredoxin to around 6.7 *(27)*. As a general rule, strain in the disulfide bridge *(28)*, and the presence of positively charged residues which enhances the ionization potential of thiol, result in a highly reactive functional group. Conversely, an electro-negative environment induced by neighboring groups tends to stabilize the neutral species RSH with the concurrent reduction in oxidation potential *(29-31)*.

Storage at low temperature is a common approach to minimize degradation of pharmaceuticals. However, the activation enthalpy of sulfur oxidation is relatively small, ranging between 12 and 18 kcal mole⁻¹. As expected, the activation entropy of the reaction is highly negative ranging between -15 to -30 cal mole⁻¹ deg⁻¹ *(25,32,33)*. Consequently, the reduction in oxidation rate with low temperature storage is not large. This result is sometimes compounded by the fact that oxygen solubility in aqueous solution increases as temperature is lowered. The solubility of oxygen in water at 0°C and 25°C is 4.9 cm³/100 mL and 3.2 cm³/100 mL respectively *(34)*. Increasing the buffer concentration or adding salt to the formulation should have a stabilizing effect since the aqueous solubility of oxygen is inversely related to solution ionic strength. Storage at -70°C has often been considered a standard procedure for protein. However, there are examples indicating that the stability of the molecule is not assured under these conditions. Recombinant INF-b loses 60% of its activity following 3 months storage at -70°C due to oxidation of Cys-17 *(21)*. In this situation, disulfide-linked aggregates are formed mainly due to the increase in protein concentration in the solution which is excluded from the forming ice matrix *(36)*. This concentration effect can also be observed in the lyophilized state if similar phase separation occurs during the freezing cycle. Subsequent removal of water would result in a heterogeneous lyophilized cake in which regions of high protein concentration are interspersed with areas of pure excipient. Recombinant tissue plasminogen activator is formulated as a lyophilized powder. As shown in Table I, storage of a formulation at 50°C for approximately one year generates up to 15% dimers. Most of these dimers are dissociable by SDS. A small percentage of SDS non-dissociable dimers is also present. The amount of these covalently-linked dimers increases with time and storage temperature. Incubation of the sample in DTT prior to analysis by size exclusion-HPLC dissociates these aggregate suggesting that they are linked by an intermolecular disulfide bridge, presumably through the single free Cys-83. In solution, a sharp increase in disulfide-linked aggregate formation is observed at high temperature *(19,24)*. This aggregation is attributed to the unfolding of the molecule which exposes reactive sites that otherwise are buried. Aggregation of denatured protein also enhances the formation of intermolecular disulfide bonds due to the close proximity of the molecules. Similarly, scrambling of intramolecular disulfide bridges accelerates as the result of increase in the polypeptide chain flexibility and better accessibility to reactive sites at high temperature *(36)*.

Thiol-disulfide exchange, by definition is not a pure oxidation process because the overall oxidation state of the molecule is not altered. But it is a redox system that is closely identified with protein degradation, and bears resemblance to cysteine

Table I: Aggregate formation in lyophilized formulation of recombinant tissue plasminogen activator. The reconstituted solution of rt-PA contains 1.0 mg/mL protein.

DIMER AS PERCENTAGE OF TOTAL PROTEIN determined by SEC-HPLC in the presence of SDS				
Time (days)	Storage Temperature			
	5°C	30°C	40°C	50°C	60°C
Initial	1.2				
90	3.4	1.6	1.7	3.3	9.9
180	1.2	1.9	1.9	4.0	13.5
270			2.0	3.6	

DIMER AS PERCENTAGE OF TOTAL PROTEIN determined by SEC-HPLC in the presence of SDS and reducing agent				
Initial	0.0				
30	0.0	0.0	0.0	0.8	2.1
90	0.0	0.0	0.0	1.3	6.0
180	0.0	0.0	0.0	2.1	8.3
285	0.0	0.0	1.1	2.2	

oxidation. The reaction is often initiated by an SN_2 nucleophilic attack of a thiolate anion on a disulfide bridge along the S-S axis *(33)* (Scheme II):

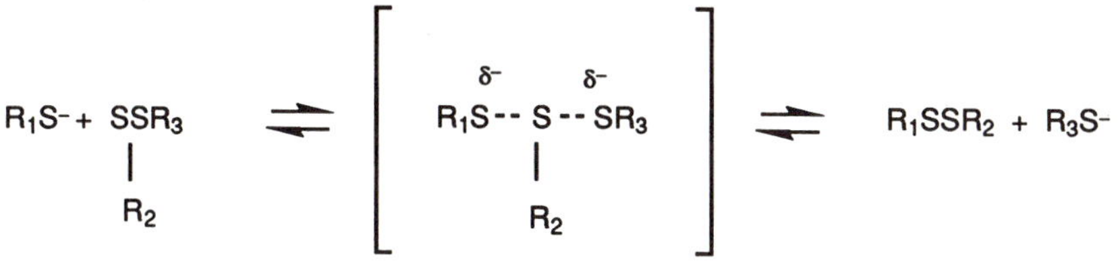

$$R_1S^- + \underset{\underset{R_2}{|}}{S}SR_3 \; \rightleftharpoons \; \left[R_1S\overset{\delta^-}{--}\underset{\underset{R_2}{|}}{S}\overset{\delta^-}{--}SR_3 \right] \; \rightleftharpoons \; R_1SSR_2 + R_3S^-$$

Scheme II

Quantitative structure activity relationship studies with model compounds indicate that the reaction follows the Bronsted relationship. Singh and Whiteside *(32)* described the correlation between the rate of thiolate-disulfide reaction and the pKa of the attacking nucleophile (nuc), the pKa of the central sulfur atom (c), and the pKa of the leaving group (lg) by the relationship:

$$\text{Log } k = 6.3 + 0.59 \, pKa^{nuc} - 0.40 \, pKa^{c} - 0.59 \, pKa^{lg} \tag{1}$$

Thus, the kinetics and thermodynamics of the reaction depend on the pKa of the reactants and products. High ionic strength and high dielectric constant stabilize the charge on the reactant and the transition state. However, the localized negative charge on RS^- is more favored than the diffuse charge of the transition state; as a result the reaction rate is reduced in high salt solution *(33)*.

Disulfide-linked aggregates and protein with scrambled disulfide bridges usually can be reduced and refolded back to the native conformation by incubation with reducing agents such as cysteine, DTT or b-mercaptoethanol *(37,38)*. However, inclusion of thiol compounds in the formulation is usually not a preferred stabilization strategy since it may generate disulfide-linked derivatives of the active protein. Spontaneous oxidation of these materials also produces a variety of unwanted active oxygen species. Sodium thiosulfate has been reported to reduce the formation of disulfide-linked high molecular weight species of OKT3 antibody in solution *(39)*. Purging the solution with inert gas, formulation at low pH and addition of a metal chelating agents are well known approaches to minimize oxidative degradation. However, metal chelating agents such as EDTA may be themselves a source of metal catalysts because traces of metal are often found in these materials. Obviously, replacement of free cysteine by alanine or serine would eliminate the problem altogether if the residue is not essential for the bio-activity of the molecule and the change does not create new immunogenic epitopes or altered conformations *(17,21)*.

Oxidation of methionine

At low pH, protein oxidation occurs primarily at methionine residues generating methionine sulfoxide. Among the many reported oxidative pathways for methionine *(1-3)*, oxidation catalyzed by metal ions and oxidation induced by alkylhydroperoxides are of interest to pharmaceutical scientists.

Slow autooxidation of protein pharmaceuticals has often been attributed to the presence of trace amounts of metal ion. Transition metal ions such as Cu^{++}, Fe^{++} or Fe^{+++} can form redox systems with molecular oxygen or prooxidants such as cysteine, dithiotreitol, glutathione and ascorbic acid *(9,40,41)* to generate a variety of reactive oxygen species. These oxidants react with many amino acid side chains including the free thiol side chain of cysteine, the imidazole ring of histidine and the thio-ether side chain of methionine. In their studies on metal catalyzed oxidation of model peptides, Li et al *(42)* and Schoneich et al *(43)* reported that formation of methionine sulfoxide is favored at low pH. The reaction was not affected by the addition of superoxide dismutase, catalase, methanol and EDTA, implying that free superoxide, hydrogen peroxide and hydroxyl radicals may not be the reactive species, and metal bound oxidant complexes were invoked to explain the experimental results *(43)*. The peptide primary structure also played a significant role in determining the reactivity of methionine. Histidine appeared to facilitate the oxidation of neighboring methionine, and C-terminal methionine oxidized more readily than N-terminal or intra-chain residues *(42,43)*. However, these studies do not provide information on the influence of secondary and tertiary structures on the reaction kinetics, an important consideration for proteins.

The mechanism of methionine oxidation by akylhydroperoxide is better understood. The characteristics of the reaction have been summarized in several reviews *(11,44)*. The reaction is first order with respect to sulfide and peroxy compounds. It is acid-catalyzed, but oxygen pressure, catalysts and inhibitors of free radical reactions do not affect the reaction kinetics. A one-step mechanism involving a nucleophilic attack of sulfide on a peroxide-solvent complex followed by a series of concerted electronic displacements leads to the transfer of oxygen to the sulfur atom.

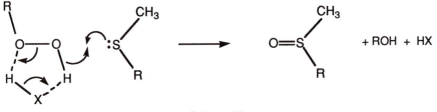

Scheme III

This reaction scheme implies that the peroxidic oxygen is electrophilic and readily released. Evidence supporting this mechanism comes from structure activity relationship studies. Electron-attracting groups on peroxide accelerate the reaction by increasing the electrophilicity of oxygen and electron-releasing groups in the vicinity of sulfur increases its nucleophilicity with a corresponding increase in oxidation rate *(45,46)*. The concerted transfer of atoms described in this mechanism is also consistent with the negative entropy observed for this type of reaction *(45)*. Although the mechanism predicts a specific acid catalysis component, the reaction rate is constant from pH 1 and 5 *(47)*. Thus, if the reaction has an acid catalyzed component, it must be fairly weak and in aqueous solution of moderate acidity it must be masked by a water catalyzed reaction (Scheme III, X=OH). Figure 1 illustrates the effect of water activity on the oxidation of recombinant human relaxin by hydrogen peroxide. The reaction was carried out in mixtures of water and ethanol. Far U.V. circular dichroism (C.D.) data shown in Figure 2 indicate a slight trend of increase in C.D. signals with increasing ethanol concentration. These changes are small, and it appears that the secondary structure of relaxin in water : ethanol mixtures remains essentially the same as that in acetate buffer, pH 5.0. Furthermore, the methionine residues involved in the reaction are fully exposed to solvent, and small perturbations to the secondary structure of the molecule would not significantly influence their accessibilty. Consequently, the observed reduction in oxidation rate of relaxin in water : ethanol mixtures can be attributed to the water concentration in the mixtures supporting the mechanism depicted in Scheme I. In agreement with the general reaction mechanism

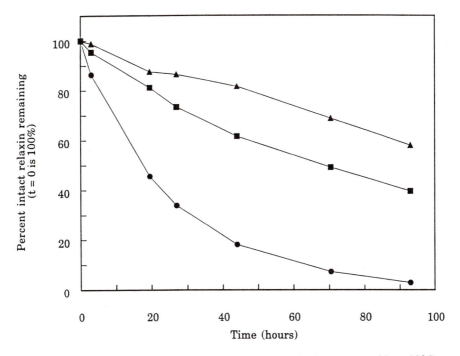

Figure 1: Oxidation of recombinant human relaxin by hydrogen peroxide at 20°C.
Relaxin: 1.7×10^{-5} M; hydrogen peroxide: 8.4×10^{-4} M;

- ■ aqueous solution pH 5.0 : ethanol 3:1 (mole:mole)
- ▲ aqueous solution pH 5.0 : ethanol 1:1 (mole:mole)
- ● aqueous solution pH 5.0

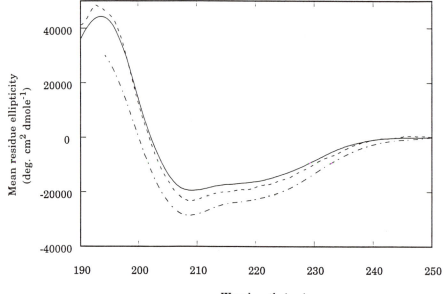

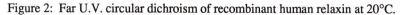

Figure 2: Far U.V. circular dichroism of recombinant human relaxin at 20°C.

—   in aqueous solution at pH 5.0

- -   aqueous solution pH 5.0 : ethanol 3:1 (mole:mole)

- - - aqueous solution pH 5.0 : ethanol 1:1 (mole:mole)

(44), the rate of methionine oxidation by hydrogen peroxide in this study is independent of the solution ionic strength and buffer concentration *(48)*.

While there is no evidence that amino acids adjacent to methionine on the polypeptide chain play any role in determining the reactivity of a particular methionine to peroxide *(48)*, the difference in the rates of oxidation of different methionine residues in many proteins appears to reflect their accessibility to solvent; methionine buried inside the hydrophobic core of globular proteins are normally not susceptible to oxidation *(30)*. For example, human growth hormone has three methionine residues, Met-14, Met-125 and Met-170. Met-14 is readily oxidized by hydrogen peroxide under native conditions. Met-125 reacts more slowly. Unfolding the molecule in 8 M urea is required to induce oxidation at Met-170 *(49-51)*. The reactivity of these three methionine residues correlates directly to their solvent-accessible surface areas of 29.1 $Å^2$, 23.5 $Å^2$ and 0.8 $Å^2$ respectively (Bart De Vos, personal communications). Oxidation of methionine does not always lead to loss in bioactivity *(48,50,54-59)*. Human growth hormone with Met-14, Met-125 oxidized exhibits a 25% drop in bioactivity based on the rat tibia test *(49)* but retains its full activity on the hypophysectomized rat model *(60)*. When all three methionine were oxidized, most of the bioactivity is lost *(49)*. It is interesting to note that although they were not involved in the electrostatic interaction with the receptor themselves, Met-14 resides inside the binding site I, and Met-170 in the binding site II *(52,53)*. Circular dichroism studies show that Met-14, Met-125 di-sulfoxide hGH retains most of the native secondary and tertiary structure while hGH with all three methionine oxidized exhibits a more loosened, opened structure *(49)*. These data and others *(49, 59)* suggest that in many cases oxidation of buried methionine affects the overall structure of the molecule with the expected reduced affinity to the receptor, and oxidation of surface methionine does not lead to loss in bioactivity unless the residue involved is an integral part of the active site or binding site of the molecule *(61-63)*.

Studies on protein oxidation by hydrogen peroxide have direct application to pharmaceutical formulation since peroxide is found in many pharmaceutical excipients. The least conspicuous source of peroxide comes from the surfactants used to prevent surface denaturation of the protein *(64)*. Polysorbate 80 and polysorbate 20 have a long safety record in pharmaceutical products and are widely used in protein solutions. Both food grade and pharmaceutical grade polysorbates often contain trace amounts of alkylhydroperoxide and hydrogen peroxide as a result of a bleaching step in the manufacturing process *(66)*. Reduction of peroxide by sodium borohydride has been used to reduce the level of oxidant *(65)*. During storage, polysorbates degrade generating peroxides. The degradation is accelerated by exposure to air, light, elevated temperature and traces of metal catalysts *(65,66)*. Dividing the material into small aliquots for storage in cool dry places under inert gas head space should minimize oxidative degradation. Polyethylene glycol is another commonly used pharmaceutical excipient which has an oxidative degradation profile similar to polysorbate *(67,68)*, and the same cautions should also be applied to this compound. Another potential source of peroxide is leachable materials from the container closure system. For example, gray silicon rubber stoppers contain 2, 4, dichlorobenzoyl peroxide as a curing agent *(69)*. Silicon emulsions are applied to rubber stoppers to prevent stopper adhesion and to facilitate automatic stoppering operation. A commercially available silicon emulsion (General Electric Silicone Division) is formulated with 2% polysorbate 80 and 0.5% polysorbate 20. The containers in which the emulsion is supplied are often rinsed with 2% hydrogen peroxide for disinfection, contributing to the potential sources of peroxide.

Cellulose-based gelling agents also contain low levels of alkylhydroperoxide degradation by-products. Recombinant human relaxin was formulated in a methylcellulose gel for topical application. Oxidation of Met-4 and Met-25 on the B-chain is the primary route of relaxin degradation in solution *(48,55)*. Figure 3 is a plot of the stability of recombinant human relaxin in solution and 3% methylcellulose gel. Relaxin degraded much faster in the gel formulation than in a solution formulated in

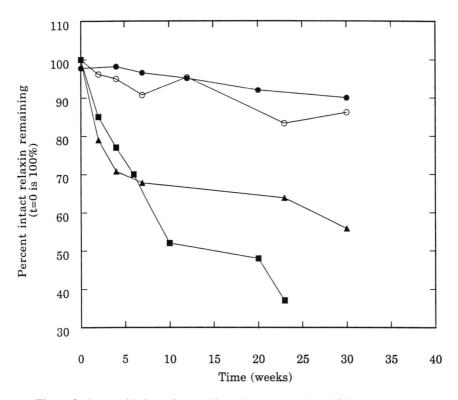

Figure 3: Auto-oxidation of recombinant human relaxin at 5°C.

●   Aqueous solution pH 5.0; relaxin: 1.7×10^{-5} M

■   3% methyl cellulose gel pH 5.0; relaxin: 1.7×10^{-4} M; lot A

▲   3% methyl cellulose gel pH 5.0; relaxin: 1.7×10^{-4} M; lot B

○   3% methyl cellulose gel pH 5.0 formulated with 10% glycerol and 0.5% methionine; relaxin: 1.7×10^{-4} M

the same buffer. Inert gas head space and chelating agents had no significant effect on the stability of the gel.

Antioxidants are often added to the formulation to prevent oxidative degradation. Well-known antioxidants such as butylated hydroxy toluene, butylated hydroxy anisole, propyl gallate and vitamin E are not sufficiently soluble in water to be considered. The selection of water soluble antioxidants is fairly limited. Ascorbic acid, reducing agents containing thiol such as glutathione, thioglycerol and cysteine may provide some short term protection. But, as discussed above, these agents may induce degradation in the long run. In a recent publication, ascorbic acid 2-phosphate was reported to be an effective, non-toxic and more stable antioxidant than ascorbic acid. Its usefulness still has to be demonstrated with long term stability data (70). Sodium sulfite and sodium bisulfite are strong nucleophiles which react with protein. Addition of 0.2% sodium bi-sulfite to relaxin gel induced complete loss of native protein in just a few days (Tue Nguyen, unpublished data). Methionine appears to be the agent of choice in this situation. As seen in Figure 3, when used in large excess, methionine effectively stabilized relaxin gel formulation, probably acting as a peroxide scavenger.

Conclusion:

Oxidation is an ubiquitous process. It is to be expected when the protein contains unpaired cysteine or methionine residues. Because pharmaceutical proteins are manufactured and formulated with several precautions to avoid physicochemical alterations, oxidative degradation of these products is usually a slow process catalyzed by trace amounts of impurities derived from many, sometimes unexpected sources. An understanding of the basic mechanism of the reactions provides a rational for a general stabilization strategy. Storage at low but above freezing temperature, inert gas head space, protection from excessive exposure to light and agitation are good preventive measures. The optimum solution pH and ionic strength, the use of co-solvent, metal chelating agents and antioxidants should be selected based on the properties of each protein. Examples available suggest that a homogenous frozen solution or lyophilized cake is also advantageous, as it would isolate and immobilize the protein molecule in an inert matrix. Thus, the selection of an appropriate cryoprotectant or bulking agents is critical. On a per weight basis, inert excipients comprise the majority of the formulation components. These materials have to be tested carefully for impurities, and tight specifications must be established for the storage of these materials. The container closure system is part of the formulation. Stability testing in the final packaging configuration often prevents annoying and time-consuming surprises down the line.

Acknowledgments: The author thanks Dr. Jim Kou, Ms. Rita Wong, Ms Xanthe Lam for their contribution to the relaxin project, and Ms. Jessica Burdman for her editorial assistance.*

* *Current address: Institute of Pharmaceutical Sciences, Syntex Research, Palo Alto, CA 94304*

Literature Cited

1   Oeswein, J. Q. and Shire, S. J. In *Peptide and Protein Drug Delivery*; Lee, V. H. L., Ed.; Advances in Parenteral Sciences 4; Marcel Dekker Inc., New York, New York,1990; pp 167-202.
2   Manning, M. C.; Patel, K. and Borchardt, R. T.; *J. Pharm. Res.* **1989**, *6*, pp 903-918.
3   Cleland, J. L.; Powell, M. F. and Shire, S. J.; In *Critical Reviews In Therapeutics Drug Carrier Systems*, Bruck S., Ed. CRC Pres Inc., Boca Raton, Florida in press.
4   Kosen, P. A.. In *Stability of Protein Pharmaceuticals, Part A: Chemical and Physical Pathways of Protein Degradation*, Ahern, T. J. and Manning, M. C., Eds.; Plenum Press, New York, New York, 1992, pp 3166.
5   Foote, C. S.; *Science*, **1968**, *162*, pp 963-969.
6   Holt, L. A.; Milligan, B.; Rivett, D. E. and Frederick, H. C.; *Biochim. et Biophys. Acta*, **1977**, *499*, pp 131-138.
7   Bertolotti, S. G. and Garcia, N. A.; *J. Photochem. Photobiol. B: Biol.*; **1991**, *10*, pp 57-70.
8   Prutz, W. A.; Butler, J. and Land, E. J.; *Int. J. Radiat. Biol.*; **1983**, *44*, pp 183-196.
9   Farber, J. M. and Levine, R. L.; *J. Biol. Chem.*, **1986**, *261*, pp 4574-4578.
10   Huggins, T G.; Wells-Knecht, M. C.; Detories, N. A.; Baynes, J. W. and Thorpet, S. R.; *J. Biol. Chem.*, **1993**, 268, pp 12341-12347.

11   Behrman, E. J. and Edwards, J. O. In *Progress in Physical Organic Chemistry*, Streitwieser Jr., A. and Taft, R. W., Eds., Interscience Publishers, New York, New York, 1967; 4, pp 93-123.
12   Bruice, T. C.; *J. Chem. Soc., Chem. Commun.*, **1983**, pp 14-15.
13   Curci, R. and Modena, G.; *Tetrahedron*, **1966**, *22*, pp 1227-1233.
14   Chen, S. H.; Kenley, R. A. and Winterle, J. S.; *Int. J. Pharmceutics*, **1991**, *72*, pp 89-96.
15   Modena, G. and Todesco, P. E.; *J. Chem. Soc.*, **1962**, pp 4920-4926.
16   Little, C. and O'Brien, P. J.; Eur. J. Biochem., 1969, 10, pp 533-540.
17   Kenney, W. C.; Watson, E.; Bartley, T., Boone, T. and Altrock, B. W.; *Lymphokine Res.*, **1986**, *5*, pp S23-S27.
18   Geigert, J.; Ziegler, D. L.; Panschar, B. M., Creasey, A. A. and Vitt, C. R.; *J. Interferon Res.*, **1987**, *7*, pp 203-211.
19   Gu, L. C.; Erdos, E. A.; Chiang, H. S.; Calderwood, T.; Tsai, K., Visor, G. C.; Duffy, J., Hsu, W. C. and Foster, L. C.; *Pharm. Res.*, **1991**, *8*, pp 485-489.
20   Browning, J. L.; Mattaliano, R. J.; Chow, E. P.; Liang, S. M.; Allet, B.; Rosa, J. and Smart, J. E.; *Anal. Biochem.*, **1986**, *155*, pp 123-128.
21   Mark, D. F.; Lu, S. D.; Yammamoto, R. and Lin, L. S.; *Proc. Natl. Acad. Sci.*, **1984**, *81*, pp 5662-5666.
22   Barron, E. S. G.; Miller, Z. M. and Kalnitsky G.; *Biochem. J.*, **1947**,*41*, pp 62-68.
23   Misra, H. P.; *J. Biol. Chem.*, **1974**, *249*, pp 2151-2155.
24   Tomazic, S. J. and Klibanov, A. M.; *J. Biol. Chem.*, **1988**, *263*, pp 3086-3091.
25   Jocelyn, P. C. In Biochemistry of the SH group, Academic Press, New York, New York, 1972, pp 47-62.
26   Shaked, Z.; Szajewski, R. P. and Whitesides, G. M.; *Biochemistry*, **1980**, *19*, pp 4156-4166.
27   Katti, S. K.; Lemaster, D. M. and Eklund, H.; *J. Mol. Biol.*, **1990**, *212*, pp 167-184.
28   Kuwajima, K.; Ikeguchi, M.; Sugawara, T.; Hiraoka, Y. and Suagai, S.; *Biochemistry*, **1990**, *29*, pp 8240-8249.
29   Snyder, G. H.; Cennerazzo, M. J.; Karalis, A. and Field, D.; *Biochemistry*, **1981**, *20*, pp 6511-6519.
30   Parente, A.; Merrifield, B.; Geraci, G. and D'Alesio, G.; *Biochemistry*, **1985**, *24*, pp 1098-1104.
31   Pearlman, R. and Bewley, T. A. In Stability and Characterization of Protein and Peptide Drugs - Cases Histories, Wang,Y. J. and Pearlman, R., Eds., Plenum Press, New York, New York, 1993, pp 1-58.
32   Singh, R. and Whiteside, G. M.; *J. Am. Chem. Soc.*, **1990**, *112*, pp 1190-1197.
33   Whiteside, G. M.,; Houk, J. and Patterson, M. A. K.; *J. Org. Chem.*, **1983**, *48*, pp 112-115.
34   CRC Handbook of Chemistry and Physics, Weast, R. C. and Astle. M. J., Eds., CRC Press, Boca Raton, Florida, 1981.
35   Volkin, D. B. and Klibanov, A. N. In *Protein Function: A practical Approach*, Creighton, T. E., Ed.; IRL Press, Oxford University, England, 1989; pp 1-24.
36   Creighton, T. E.; *Bioessays*, **1988**, *8*, pp 57-63.
37   Anfinsen, C. B.; Sela, M. and Cooke, J. P.; *J. Biol. Chem.*, **1962**,*237*, pp 1825-1831.
38   Galat, A.; Creighton, T. E.; Lord, R. C. and Blout, E. R.; *Biochemistry*, **1981**, *20*, 594-601.
39   Rao, P. and Kroon D. J. In Stability and Characterization of Protein and Peptide Drug - Case Histories, Wang, Y. J. and Pearlman, R., Eds, Plenum Press, New York, New York, 1993, pp 135-158.
40   Barron, E. S. G.; Miller, Z. B. and Kalinitsky, G.; Biochem. J., 1947, 41, pp 62-68
41   Stadman, E.; *Free Radical Biology & Medicine*, **1990**, 9, pp 315-325.

42   Li, S.; Schoneich, C., Wilson, G. S. and Borchard, R. T.; *Pharm. Res.*, **1993**, 10, pp 1572-1578.
43   Schoneich, C.; Zhao, F.; Wilson, G. S. and Borchart, R. T.; *Biochim. Biophys. Acta*, in press.
44   Barnard, D.; Bateman, L. and Cuneen, J. L. In The Chemistry of Oganic Sulfur Compounds, Kharasch, N. and Meyers, C. Y., Eds.; Pergamon Press, New York, 1961, pp 229-247.
45   Caldwell, K. A. and Tappel, A. L.; *Biochemistry*, **1964**, 3, pp 1643-1647.
46   Yamamoto, H.; Miura, M.; Nojima, M and Kusabayashi, S.; *J. Chem. Soc. Perkin Trans. I*, **1986**, pp 173-182.
47   Tennies, C. and Callan, T. P.; *J. Biol. Chem.*, **1939**, pp 481-489.
48   Tue, N. H., Burnier, J. and Meng, W., *Pharm. Res.*, **1993**, 11, pp 1563-1571.
49   Houghten, R. A.; Glaser, C. B. and Li, C. H.; *Arch. Biochem. and Biophys.*, **1977**, *178*, pp 350-355.
50   Teh, L. C.; Murphy, L. J.; Huq, N. L.; Surust, A.; Friesen, H. G.; Lazarust, L. and Chapman, G. E.; *J. Biol. Chem.*, **1987**, 262, pp 6472-6477.
51   Teshima, G. and Canova-Davis, E.; *J. of Chrom.*, **1992**, *625*, pp 207-215.
52   Cunnigham, B. C. and Wells, J. A.; *Proc. Natl. Acad. Sci.*, **1991**, 88, pp 3407-3411.
53   De Vos, A. M.; Ultsch, M. and Kossiakoff, A. A., *Science*, **1992**, 255, 306312.
54   Heath, W. F. and Merrifield, R. B., *Proc. Natl. Acad. Sci.*, **1986**, *83*, pp 6367-6371.
55   Cipolla, D. C. and Shire, S. J. In Techniques in Protein Chemistry II, Academic Press, New York, New York, 1991, pp 543-555.
56   Vale, W.; Speiss, J.; Rivier, C. and Rivier, J.; *Science*, **1981**, *213*, pp 13941397.
57   Glaser, C. B. and Li, C. H.; *Biochemistry*, **1974**, *13*, pp 1044-1047.
58   Cascone, O.; Biscoglio, M. J.; Bonino, J. and Santome, J. A.; *Int. J. Pep. Protein Res.*, **1980**, *16*, pp 299-
59   Frelinger III, A. L. and Zull, J. E., *Arch. Biochem. Biophys.*, **1986**, 244, pp 641-649.
60   Becker, G. W.; tackitt, P. M.; Bromer, W. W.; Lefeber, D. S. and Riggin, R. M.; *Biotech. and Appl. Biochem.*, **1988**, *10*, pp 326-337.
61   Caldwell, P.; Luk, D. C.; Weissbach, H. and Brot, N.; *Proc. Natl. Acad. Sci.*, **1978**, 75, pp 5349-5352.
62   Holeysovsky, V. and Lazdunski, M.; *Biochim. Biophys. Acta*, **1968**, 154, pp 457-467.
63   Stauffer, C. E. and Etson, D.; *J. Biol. Chem.*, **1969**, *244*, pp 5333-5338.
64   Hora, M. S.; Rana, R. K.; Wilcox, C. L.; Katre, N. V.; Hirtzer, P.; Wolfe, S. N. and Thomson, J. W. In International Symposium on Biological Product Freeze-Drying and Formulation, Dev. Biol. Std., 74, pp 295-306.
65   Hamburger, R., Azaz, E. and Donbrow, M.; *Pharmaceutica Acta Helvetica*, **1975**, 50, pp 10-17.
66   Donbrow, M.; Azaz, E. and Pillersdorf, A; J. Pharm. SCi., 1978, 67, pp 1676-1681.
67   McGinity, J. W.; Patel, T. R.; Naqvi, A. H. and Hill, J. A.; *Drug Dev. Commun.*, **1976**, 2, pp 505-519.
68   Decker, C; *J. Polym. Sci.,, Polym. Chem. Ed.,,* **1977**, 15, 781-798.
69   Smith, E. J. and Nash, R. J. In Elastomeric Closures for Parenterals, Marcel Decker, Inc., New York, 1986, pp 178-179.
70   Hata, R. I. and Senoo, H.; *J. Cell. Physiol.*, **1989**, *138*, pp 8-16.

RECEIVED April 19, 1994

Chapter 5

Origin of the Photosensitivity of a Monoclonal Immunoglobulin G

Henryk Mach, Carl J. Burke, Gautam Sanyal, Pei-Kuo Tsai, David B. Volkin, and C. Russell Middaugh

Department of Pharmaceutical Research, WP78–302, Merck Research Laboratories, West Point, PA 19486

An IgG monoclonal antibody has been encountered which undergoes significant decreases (i.e., 70%) in its intrinsic fluorescence emission intensity upon UV-irradiation. Despite these large changes, second derivative UV-absorbance spectroscopy suggests that only one or two tryptophan residues out of a total of 28 undergo photooxidation. It is shown that these light sensitive sidechains are probably located on the surface of the protein in heavy chain hypervariable regions. The surprisingly large light-induced decrease in the immunoglobulin's fluorescence emission spectrum appears to result from the production of a small number of N-formylkynurenine residues which quench the bulk of the protein's remaining indole sidechains fluorescence presumably by a fluorescence resonance energy transfer mechanism. The relevance of these observations to the formulation of immunoglobulins as pharmaceutical and diagnostic agents is briefly discussed.

Monoclonal antibodies constitute a structurally homogeneous but functionally diverse class of proteins of unique pharmaceutical importance. Fortunately, immunoglobulins are generally stable proteins facilitating their potential use as clinical and diagnostic agents. In the process of examining a wide variety of such proteins by intrinsic fluorescence spectroscopy, however, we and others have found that immunoglobulins occasionally display what appears to be a marked photosensitivity to UV light. This effect is most often manifested in large, time dependent changes in their fluorescence emission spectra when light of 260-300 nm is used for excitation.

It has long been recognized that the aromatic amino acids in proteins can be modified by the absorption of UV light, especially in the presence of sensitizing agents such as molecular oxygen or ozone (1-4), proximate disulfide bonds (5,6), riboflavin (7), porphyrins (8), and various dyes (3,4,7). The most frequently observed photooxidation products of tryptophan residues are kynurenine (Kyn) and N-formylkynurenine (NFK), while tyrosine is primarily converted to 3,4-dihydroxyphenylalanine (DOPA), 3-(4-hydroxyphenyl)lactic acid and dityrosine (4,9). In this work, we examine a particularly dramatic case of photosensitivity in a monoclonal IgG$_1$ (mAb1) in which photooxidation of, on the average, only one or two Trp residues produces large spectral changes. This observation can be explained by a nonradiative energy transfer mechanism in which limited photoproduct generation leads to extensive quenching of the overall fluorescence of the protein.

MATERIALS AND METHODS

Monoclonal Antibody. The photosensitive immunoglobulin employed in these studies was expressed in mammalian cell culture and purified to homogeneity by methods described previously (10,11). Other IgGs used for comparative purposes were similarly obtained. Samples were irradiated in a Hitachi F-2000 spectrofluorometer equipped with a 150 W xenon lamp employing a 20-nm excitation bandwidth and 280-nm radiation. Solutions of mAb1 were prepared by diluting an 8.5 mg/ml stock solution into 2 ml of 6 mM sodium phosphate, 150 mM sodium chloride, pH 7.0 to obtain a final concentration of 0.2 mg/ml. Both during irradiation and measurement of the fluorescence spectra, samples were stirred continuously with a Precision Controller stirring device (Spectracell).

Fluorescence. Steady-state fluorescence measurements were performed with a Hitachi F-4500 spectrofluorometer by using 1- and 10-nm excitation and emission bandwidths, respectively. To minimize photodegradation, spectra were rapidly scanned at a rate of 1200 nm/min. Individual spectra were, therefore, obtained in less than 5-10 sec, a period over which photodegradation was minimal. Sample temperature was maintained at 20°C by a thermostatically controlled cell holder. Tryptophan fluorescence was monitored as the integral of fluorescence intensity between 320 and 380 nm with excitation at 300 nm. NFK-type fluorescence was measured by excitation at 320 nm and emission at 410 nm, while Kyn-type fluorescence was detected with excitation at 380 nm and emission at 490 nm (12,13). Three dimensional spectra generated by excitation from 250 to 400 nm every 5 nm and emission monitoring from 300 to 600 nm every 10 nm were routinely collected to test for the presence of fluorescence spectral components other than the three aforementioned components (Trp, NFK, Kyn), but none were detected. The degree of solvent exposure of Trp residues in mAb1 was assessed by titration of 2 ml of a 0.2 mg/ml protein solution with 5 μl additions of 3 M KI solution up to a final concentration of 120 mM iodide. The iodide stock solution was supplemented with 50 mM sodium bisulfite to slow the formation of molecular iodine. An N-acetyl-L-tryptophanamide (NATA) solution of the same absorbance was titrated in a similar manner to estimate the response expected from fully solvent exposed Trp residues. For comparison between immunoglobulin preparations, an SLM 8000 was employed to obtain fluorescence spectra using 295 nm excitation with emission monitored at 340 nm. In these studies, all mAb solutions (10 μg/ml) were equilibrated at 10°C and bandwidths of 4 and 16 nm were used for excitation and emission, respectively. Fluorescence lifetime experiments were performed with an ISS GREG200 multiple frequency phase fluorometer. Samples were excited at 292 nm and monitored after the emitted light passed through a 345 nm cut-off filter. Modulation frequencies ranged from 20 MHz to 120 MHz. As a reference, *p*-terphenyl (τ = 1.05 ns) was employed. Phase shifts and demodulation were analyzed by a multiexponential least-squares analysis (14).

UV Absorbance Spectroscopy. Near-UV spectra were acquired with a Hewlett-Packard 8450A diode-array, dual beam spectrophotometer. This instrument acquires one spectrum in approximately half a second. Thus, the 30 second data acquisition times employed in this work provided an average of 60 measurements. No significant spectral changes were observed over this time period. Spectra were transferred to an IBM personal computer and second derivatives were calculated with a 5-point, quadratic/cubic formula (15) by using a Lotus 1-2-3 spreadsheet program. For assessment of the relative amplitudes of the second derivative peaks, the absolute values of the second derivative spectral values between 280 and 300 nm were integrated yielding a cumulative signal from tryptophan and tyrosine residues. The degree of solvent exposure of the tryptophan and tyrosine sidechains was assessed by fitting the second derivative spectrum in the 270-300 nm region to a model consisting of sets of systematically shifted Trp and Tyr second derivative spectra. Individual spectra in these sets simulated the spectra of these aromatic amino acids over the range of polar (H_2O) to nonpolar (protein interior) environments. Experimental perturbation of protein samples with a nonpolar solvent (viz., ethylene glycol) resulted in a redistribution of Trp and Tyr component spectra and an apparent redshift of the peaks. The extent of such shifts, quantitated by changes in the positions of the intersection with the abscissa of the fitted second derivative spectra near 288 nm for Trp and 285 for Tyr, were assumed to be proportional to the average degree of solvent exposure of these residues. NATA and N-acetyl-L-tyrosinamide were again used to estimate the shifts expected from fully solvent exposed residues (H. Mach, unpublished results).

Circular Dichroism (CD). CD spectra were measured with a Jasco J-720 spectropolarimeter calibrated with d-10-camphorsulfonic acid. For far-UV studies, 10 mM sodium phosphate, pH 7, was used as the solvent. A quartz cuvette with a pathlength of 0.1 cm was employed and a temperature of 20°C was maintained by a circulating bath and a thermostatic cell holder during all measurements. A spectral bandwidth of 5 nm was used and an average of 5 scans from 250 to 200 nm collected at 20 nm/min was employed to generate spectra. Photodegradation under these spectral conditions was judged to be minimal by complementary fluorescence experiments.

Light Scattering and Chemical Crosslinking. The size of any mAb1 aggregates induced by photochemical events was assessed with a Malvern 4700 light scattering system (Malvern, England) equipped with a 5W 488 nm argon laser (Spectra Physics) as previously described (28). The intensity of scattered light was measured in parallel with the optical density at 320 nm. Under conditions of limited aggregation, scattering intensity is approximately proportional to the weight-average molecular weight (29). In addition to monitoring the static scattering, simultaneous dynamic light scattering analyses were performed employing a Malvern autocorrelator and data analyzed by the method of cumulants (30). In order to estimate the potential optical density produced by scattering from mAb1 oligomers, glutaraldehyde crosslinking was employed to generate protein aggregates. Thus, the optical density at 320 nm due to photodegradation products could be evaluated by subtraction of optical densities expected for similarly sized aggregates in the absence of chromophores absorbing above 300 nm. No evidence was found that chemical crosslinking with 0.2% glutaraldehyde over the time scale employed (up to 30 min at 20°C) produced any chromophoric absorption that might interfere with these measurements.

RESULTS

Photo-induced Spectral Changes in mAb1. The fluorescence spectrum of mAb1, a monoclonal IgG$_1$, is characterized by a wavelength of maximum emission (λ_{max}) of 346 nm when light of 300 nm is employed to primarily excite Trp residues (Fig. 1A). Prolonged exposure of mAb1 to intense UV light (280 nm) decreases the emission intensity to about 10-20% of its original value and shifts the λ_{max} to 336 nm. This shift of the emission maximum to lower wavelength after irradiation suggests that the photochemically altered Trp residues possess a higher wavelength emission maximum characteristic of more solvent exposed residues. The loss of Trp emission is accompanied by the appearance of fluorescence at 410 nm when the sample is excited at 320 nm (Fig. 1B). Both observations are consistent with UV-induced Trp photooxidation resulting in the formation of N-

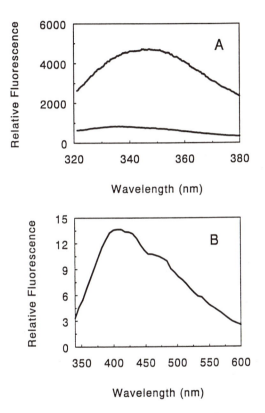

Figure 1. **A:** Steady-state fluorescence emission spectra (λ_{ex} = 300 nm) of intact mAb1 before (upper spectrum) and after UV-irradiation at 280 nm for 80 minutes (lower spectrum). **B:** Weak fluorescence (approximately 1% of initial mAb1 signal) observed upon excitation at 320 nm after 20 minutes of irradiation.

formylkynurenine (NFK) as previously observed in an immunoglobulin light chain (13) as well as other proteins (16-19). However, the presence of an ozonide (20), however, another product of Trp oxidation that possesses spectral properties similar to NFK, cannot be rigorously excluded. A very weak third emission peak was detected at 490 nm when the protein was excited at 380 nm (not illustrated) which is consistent with the formation of kynurenine (Kyn), another previously observed product of Trp photooxidation (12,16).

Although prolonged irradiation produces a large change in the fluorescence spectrum of mAb1, a much smaller change is seen in the protein's UV absorption spectrum (Fig. 2A). In general, the irradiated sample manifests increased optical density over the entire UV region including significantly enhanced optical density above 310 nm. Both NFK and Kyn absorb light above 300 nm and probably contribute to the observed optical density between 300 and 350 nm. In addition, mAb1 was found to increase in average molecular weight approximately 3-fold after UV exposure as monitored by static and dynamic light scattering with the mean hydrodynamic radius of the protein growing from 9 to 11 nm (not illustrated). In order to evaluate the relative contribution of light scattering from aggregated mAb1 to the total UV spectrum, glutaraldehyde crosslinking was employed to chemically aggregate the immunoglobulin. When the aggregation state reached that of irradiated mAb1 as determined by light scattering, UV spectra were recorded. The glutaraldehyde-treated samples produced scattering contributions that account for only about 15% of the baseline increase above 300 nm that is observed in samples that were exposed to 280-nm irradiation. It was therefore tentatively concluded that most of the optical density increase in this region arises from absorbance by Trp photooxidation products such as NFK and Kyn.

Another method that can be used to deconvolute the relative contributions of light absorption and scattering is examination of the second derivative of the UV spectrum (21). Unlike Trp and Tyr, the new spectral components and the contribution from light scattering lack obvious fine structure. Thus, the presence of broad unstructured signals does not generally interfere with the second derivative spectra of sharper absorbance peaks (21-24). Furthermore, the amplitudes of second derivative peaks obey the Beer-Lambert Law and can therefore be used for concentration determination (25,26). As shown in Figure 2B, this method finds that irradiation decreases the second derivative UV absorbance of mAb1 by less than 5%, while fluorescence emission is decreased by approximately one-half under the same conditions (Fig. 3). The amplitudes of the three major second derivative UV peaks (negative signals at 284 and 292 nm and a positive peak at 288 nm) are decreased by a similar amount (< 5%) after irradiation (Fig. 2B). This suggests that Tyr residues are also altered by the incident light since the second derivative peak at 284 nm might be expected to display less change relative to the other peaks in the absence of Tyr modification (22,23). In spite of the fluorescence and UV spectral changes, the overall secondary structure does not appear to be significantly altered as monitored by far-UV CD measurements (not illustrated).

Kinetics of Trp Loss in mAb1. The decay of Trp fluorescence observed during continuous UV irradiation is biphasic in nature (Fig. 3). An initial fast phase accounts for greater than 70% of the signal decrease and is followed by a much more slowly decaying second phase. As shown in the inset to Figure 3, this decrease in emission intensity parallels the observed shift in the λ_{max}. Since the blueshift in the λ_{max} suggests that solvent exposed Trp residues may be being preferentially quenched or modified, the addition of KI, a selective quencher of fluorescent residues on the surface of proteins, was employed to monitor the effect of irradiation on mAb1 after iodine-induced quenching of solvent exposed indole sidechains. In the presence of 1 M KI, the fast UV-induced decay phase of fluorescence is accelerated while the extent of change and the magnitude of the λ_{max} shift

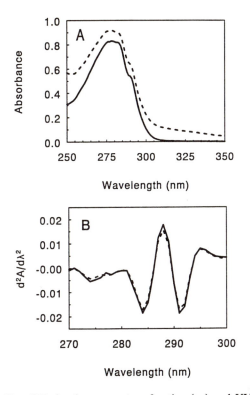

Figure 2. A: Near-UV absorbance spectra of native (—) and UV-irradiated (---) mAb1 samples. B: Second derivative calculated from the spectra shown in A.

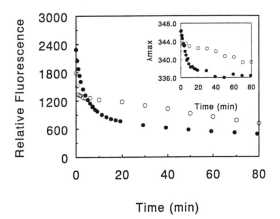

Figure 3. Kinetics of the fluorescence emission decay of mAb1 upon UV irradiation at 280 nm. Inset: Kinetics of shift in the fluorescence emission peak maxima. ●: mAb1 only; O: mAb1 in the presence of 1 M KI.

is decreased (Fig. 3). The latter two observations suggest that KI may either inhibit the photo-induced modification of surface Trp residues or prevent energy transfer to other quenching moieties (e.g., NFK, see discussion). High ionic strength alone does not alter the decay kinetics since no effect was observed in the presence of 1 M KCl compared to results observed in lower ionic strength buffers.

The photodegradation of Trp is known to be accelerated in aerated solutions. Therefore the effect of dissolved oxygen on the rate of the fluorescence decay was examined. The fluorescence of mAb1 was monitored under several different conditions: untreated; purged with argon; and argon-purge in the presence of an O_2-scavenging glucose/glucose oxidase/catalase system. Each sample manifested a decreasing rate of fluorescence loss with the latter sample demonstrating about half the decrease in intensity that was observed in the untreated sample (not illustrated). This partial dependence of Trp reactivity on dissolved oxygen has been previously observed and is consistent with photolysis of Trp (4).

Formation of NFK. Since a fluorescence emission peak near 410 nm is detected after the photo-induced decay of Trp emission and is consistent with NFK formation (Fig. 1B), the kinetics of the formation of this signal were examined by exciting the sample at 320 nm and observing the appearence of fluorescence at 410 nm as a function of irradiation time. Similar to the time dependence of Trp fluorescence decay, the kinetic profile of putative NFK formation was characterized by an initial fast phase which was complete after approximately 10 min. (Fig. 4A). This rapid growth in fluorescence intensity followed first-order kinetics (Fig. 4B) suggesting the appearance of NFK results from the alteration of a well defined population of Trp residues. As shown in Fig. 4C, a linear relationship between Trp decay and NFK formation is observed which suggests that these processes are linked.

Solvent Exposure of Trp Residues. To further probe the extent of solvent exposure of Trp residues, the formation of NFK sidechains as monitored in the presence of the quenching agent, KI (Fig. 4A). To ensure that any quenching observed was not due to high ionic strength itself, control experiments showed that no significant change was observed in the presence of 1 M KCl. The observed fluoresence decrease in the presence of KI could result from either the direct quenching of NFK fluorescence or inhibition of NFK formation. The contribution of the latter can be assessed by examining changes in the Trp second derivative peaks of the near UV absorbance spectrum. After the fast phase (ca. 10 min.), Trp absorbance is decreased by less than 5% indicating that, on the average, modification of only 1-2 of the 28 Trp residues in mAb1 are responsible for the majority of the initial spectral changes. The kinetics of the loss of the Trp second derivative UV peak are shown in Fig. 5. Additional residues appear to be altered after prolonged irradiation, but these changes cannot account for the dramatic fluorescence change. The rate of Trp absorbance loss is also slowed in the presence of KI suggesting that the kinetics of Trp disappearance are controlled, at least in part, by the susceptibility of these residues (Fig. 5). The decreased rate in the presence of KI is most simply explained by the ability of KI to quench or prevent the formation of the excited state of Trp which is postulated to be an intermediate in NFK formation (5).

To further support the observation that the large fluorescence spectral changes are the result of modification of only very few residues, the solvent exposure of Trp was also probed by ethylene glycol and KI-induced UV absorbance and fluorescence wavelength shifts, respectively (Fig. 6). When compared to Trp in solution (i.e., NATA), the Trp residues in mAb1 manifest only a slight UV absorbance spectral shift upon addition of up to 50% ethylene glycol. Specifically, the intersection with the wavelength axis of the deconvoluted Trp second derivative spectrum of NATA shifted from 286.2 to 287.2 nm and

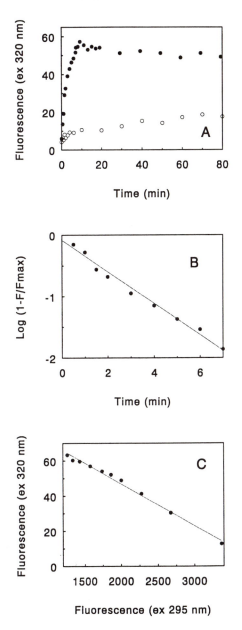

Figure 4. **A:** Fluorescence emission intensity at 410 nm with 320 nm excitation of mAb1 as a function of irradiation time. ●: mAb1 only; O: mAb1 in the presence of 1 M KI. **B:** Plot of the natural logarithm of the fraction of the Trp remaining as judged from the fluorescence at 410 nm versus irradiation time. **C:** Correlation between the formation of the NFK-like product (fluorescence emission at 410 nm with excitation at 320 nm) and the decay of the intrinsic Trp fluorescence (excitation at 295 nm and emission at 340 nm).

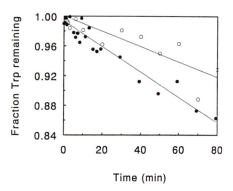

Figure 5. The kinetics of decrease of the amplitude of the second derivative UV-absorption peaks upon UV irradiation. ●: mAb1 only; O: mAb1 in the presence of 1M KI. The lines represent linear fits to the data.

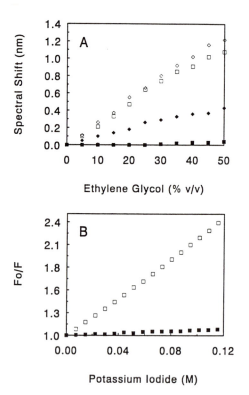

Figure 6.A: Apparent second derivative band shifts induced by perturbation of mAb1 with ethylene glycol. The second derivative spectra of perturbed samples were deconvoluted to yield Trp (■) and Tyr (◆) components. NATA (□) and N-acetyl-L-tyrosinamide (◊) were used to determine the shifts expected for fully exposed side chains. B: Stern-Volmer plot of the KI-induced quenching of the fluorescence emission at 320-380 nm with 295 nm excitation of mAb1 (■) and NATA (□).

from only 289.22 to 289.26 nm in the case of mAb1. This small spectral change observed upon shifting to a more apolar environment suggests that most of the Trp residues in mAb1 are inaccessible to the solvent. Assuming that the spectral peak positions are directly proportional to solvent exposure, 1-2 of the indole sidechains are again predicted to be on the surface of the molecule although a larger number of marginally exposed Trp residues could also account for this observation. The larger shift observed in the zero-order spectrum is found to arise from the Tyr residues which are more exposed to the solvent as indicated by the larger shift of the deconvoluted Tyr peak (Fig. 6A).

Fluorescence Excited State Lifetimes of the Trp Residues in mAb1. The average tryptophan fluorescence lifetimes of mAb1 were measured before and after UV radiation in order to characterize the relative contributions of various populations of these sidechains to the overall emission signal. Although this protein contains 28 Trp residues, the phase shift and demodulation data of a sample prior to irradiation could be satisfactorily fit to a 2-component exponential model ($\chi^2 = 3.73$). The major component, contributing 86% of the total intensity, had a lifetime of 3.5 ns while the minor component exhibited a lifetime too short to be resolved by these measurements. After irradiation, the lifetimes exhibited a marked increase in heterogeneity and the data could no longer be satisfactorily fit to a single or multicomponent exponential model ($\chi^2 = 17.2$ for a two-component fit). At each modulation frequency between 20 and 100 MHz, the phase angle was smaller by 8-15%, but the extent of demodulation of the emitted light was higher (5-20%) for the irradiated compared to the native sample. These data suggest that a different and/or a more heterogeneous population of tryptophans dominate the fluorescence decay kinetics after the photomodification process.

Location of the Photosensitive Trp Population. Examination of the loss of fluorescence intensity with time in several other mAbs finds an apparently related phenomena (Fig. 7). To this end, three additional monoclonal IgG molecules were compared: mAb2, an IgG_1 with lower affinity for the same antigenic peptide (P1) as mAb1; mAb3, an IgG_3 possessing the same complimentary determining regions (CDRs) and binding affinity as mAb1; and mAb4, an IgG_4 unrelated to the other three mAbs. Although all four of the immunoglobulins manifest detectable photosensitivity, the spectral changes seen in mAb1 and 3 are clearly much more dramatic than those seen in 2 and 4, which are on the order of only 20% those seen in the two antibodies with the same CDRs (Fig. 7A). If the data are normalized relative to the total observed change in each immunoglobulin, all of the mAbs display an initial rapid phase followed by a slower decay (Fig. 7B). The similarity in behavior of mAb1 and mAb3 suggests a photolabile Trp residue(s) may be located in one of the common CDRs of the two immunoglobulins. In fact, examination of the amino acid sequence of mAb1 shows that one-half of the Trp residues are in or immediately adjacent to the CDRs. Titrating mAb1 with its antigenic peptide, P1, results in a large quenching of the fluorescence signal with maximum attenuation achieved after two moles of P1 are added to one mole of mAb1 (Fig. 7C). The efficient quenching of mAb1 by P1 may at least partially arise from a disulfide bond present in the antigenic peptide since cystine residues are known quenchers of indole fluorescence (5). Interestingly, the presence of bound peptide results in enhanced production of NFK and Kyn fluorescence upon irradiation compared to the unliganded protein (not illustrated).

DISCUSSION

What is the origin of the dramatic decrease in fluorescence seen when mAb1 is irradiated with UV light? Although the large spectral changes might seem to suggest a correspondingly large structural alteration of the molecule, this does not appear to be the case. Far-UV CD spectra do not indicate a change in the secondary structural content of

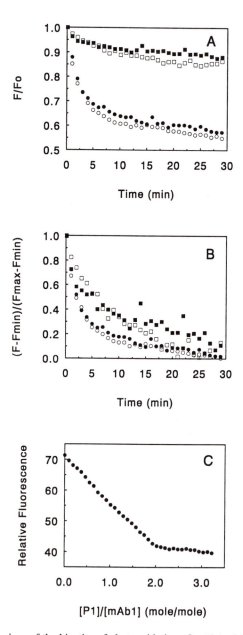

Figure 7. Comparison of the kinetics of photooxidation of mAb1 with other IgGs and effect of interaction with P1. **A:** Time-dependent fluorescence change at 340 nm (295 nm ex.): mAb1 (●); mAb2 (□); mAb3 (○); and mAb4 (■). **B:** Kinetic data from A normalized to the overall change in fluorescence. **C:** Titration of mAb1 with the P1 peptide.

the antibody and UV absorbance spectroscopy reveals that only one or two of the protein's 28 tryptophan residues are modified. This modification appears to be due to photooxidation of indole moieties since spectral features characteristic of Kyn and NFK are produced concomitantly with the large decrease in tryptophan fluorescence and smaller changes in the Trp absorption peaks. The most probable explanation for these seemingly contradictory observations is that the production of the NFK or NFK-like photooxidation product leads to an extensive overall quenching of the tryptophan fluorescence through a resonance energy transfer mechanism. This hypothesis is supported by several observations. NFK is known to be an efficient energy transfer acceptor for indole since its absorption peak extensively overlaps the emission peak of tryptophan. It has, in fact, previously been shown that the incorporation of a single molecule of NFK into aggregates of NATA containing an average of 36 indole groups quenched 95% of the tryptophan's fluorescence (16). An energy transfer mechanism for the observed fluorescence quenching is also supported by the increased heterogeneity of the fluorescence lifetime distribution upon irradiation.

A second question is why such dramatic photo-induced fluorescence changes are seen in mAb1 but not in all IgG molecules, a series of highly homologous proteins. The fact that almost identical alterations are also seen in mAb3 which contains the same CDRs as mAb1 immediately suggests that one or more indole groups in the hypervariable regions are involved in this phenomenon. mAb1 does in fact contain a tryptophan residue in the heavy chain CDR3 as well as single residues at the interface between the hypervariable and framework regions of the heavy chain CDRs 1 and 3. Such residues are expected to be exposed to solvent. The ability to partially quench the fluorescence of mAb1 with a peptide that binds to the CDRs is entirely consistent with the presence of indoles in the antigen binding site of this immunoglobulin. Further support for the location of the photosensitive indoles is the observation that they are exposed to solvent as demonstrated by the ability to quench their fluorescence with KI.

All of this is consistent with the hypothesis that solvent exposed tryptophan residues may be more photolabile than their more buried counterparts (6). For example, it has been shown that the rate of photooxidation of Trp in mellitin is strongly inhibited when these residues were occluded by conformational changes (12). It has, in certain cases, also been possible to correlate the photooxidation of specific residues in proteins with their exposure to solvent (27). In the case of mAb1, the presence of a conserved disulfide bond in the first domain of the heavy chain and its potential proximity to CDR localized indoles may contribute to the latter's photosensitivity. Photooxidation probably requires the formation of an excited state indole (5). This idea is supported by the ability of iodide ion to suppress the formation of photoproducts observed in this work. Disulfide bonds have been proposed to accept ejected electrons from such excited states enhancing the early stages of photoproduct generation. This process may also explain the surprisingly large fluorescence changes induced by binding of the disulfide-containing antigenic peptide P1 to mAb1 since this process may also enhance photooxidation (and therefore more global quenching) as described earlier.

Finally, we note that these photodegradative reactions are simply prevented by the storage of this protein in UV opaque containers constructed of commonly employed materials such as glass or plastic. Furthermore, simple minimization of exposure of mAb1 to intense UV light is sufficient to inhibit significant photooxidation. Perhaps surprisingly, the structural modifications of the antigen binding region produced by photooxidation have no detectable effect on the affinity of the immunoglobulin for the peptide suggesting that the altered indole moieties still enable mAb1 to accurately accommodate the antigen.

ACKNOWLEDGMENTS

We thank Dr. Anthony Conley for providing the immunoglobulins and Dr. Richard Tolman for the P1 peptide.

REFERENCES

1. Foote, C.S. (1968) *Science 162*,963-970.
2. Foote, C.S. (1976) *Free Radicals Biol. 2*, 85-133.
3. Gollnick, K. (1968) *Adv. Photochem. 6*, 1-122.
4. Creed, D. (1984) *Photochem. Photobiol. 4*, 537-562.
5. Bent, D.V. & Hayon, E. (1975) *J. Am. Chem. Soc. 87*, 2612-2619.
6. Grossweiner, L.I., Blum, A. & Brendzel, A.M. (1982) In *Trends in Photobiology* (Helene, C., Charlier, M., Montenay-Garestier, T. & Laustriat, G., Eds.) p. 67. Plenum Press, New York.
7. Veno, N., Sebag, J., Hirokawa, H. & Chakrabarti, B. (1987) *Exp. Eye Res. 44*, 863-870.
8. Cozzani, I., Fori, G., Bertoloni, G., Milanesi, C., Carlini, P., Sicuro, T. & Ruschi, A. (1985) *Chem. Biol. Interact. 53*, 131-143.
9. Creed, D. (1984) *Photochem. Photobiol. 4*, 563-575.
10. Daugherty, B.L., DeMartino, J.A., Law, M.-F., Kawka, D.W., Singer, I.I. & Mark, G.E. (1991) *Nucleic Acids Res. 18*, 2471-2476.
11. DeMartino, J.A., Daugherty, B.L., Law, M.-F., Cuco, C., Alves, K., Silberklang, M. & Mark, G.E. (1991) *Antibody Immunoconj. & Radiopharm. 4*, 828-835.
12. Pigault, C. & Gerard, D. (1984) *Photochem. Photobiol. 40*, 291-296.
13. Okajima, T., Kawata, Y. & Hamaguchi, K. (1990) *Biochemistry 29*, 9168-9175.
14. Lakowicz, J.R., Lacsko, G., Cherek, H., Gratton, E. & Limkeman, M. (1984) *Biophys. J. 46*, 463-477.
15. Savitzky, A. & Golay, J.E. (1964) *Anal. Chem. 36*, 1627-1639.
16. Spodheim-Maurizot, M. Charlier, M. & Helene, C. (1985) *Photochem. Photobiol. 42*, 353-359.
17. Pileni, M.-P., Walrant, P. & Santus, R. (1976) *J. Phys. Chem. 80*, 1804-1809.
18. Vorkert, N.A. & Ghiron, C.A. (1973) *Photochem. Photobiol. 17*, 9-16.
19. Pirie, A. (1971) *Biochem. J. 125*, 203-208.
20. Pryor, W.A. & Uppu, R.M. (1993) *J. Biol. Chem. 268*, 3120-3126.
21. Fell, A.F. (1983) *Trends Anal. Chem. 48*, 312-318.
22. Levine, R.L. & Federici, M.M. (1982) *Biochemistry 21*, 2600-2606.
23. Balestrieri, C., Colonna, G., Giovanne, A., Irace, G. & Servillo, L. (1978) *Eur. J. Biochem. 90*, 433-440.
24. Ichikawa, T. & Terada, H. (1977) *Biochim. Biophys. Acta 494*, 267-270.
25. Mach, H., Thomson, J.A. & Middaugh, C.R. (1989) *Anal. Biochem. 181*, 79-85.
26. Mach, H. & Middaugh, C.R. (1993) *BioTechniques 15*, 240-241.
27. Sellers, D.R. & Ghiron, C.A. (1973) *Photochem. Photobiol. 18*, 393-402.
28. Mach, H., Volkin, D.B., Burke, C.J., Linhardt, R.J., Fromm, J.R., Loganathan, D., Mattsson, L. & Middaugh, C.R. (1993) *Biochemistry 32*, 5480-5489.
29. Herskovits, T.T., Russell, M.W. & Carberry, S.E. (1984) *Biochemistry 23*, 1875-1881.
30. Koppel, D.E. (1972) *J. Chem. Phys. 57*, 4814-4820.
31. Englander, S.W., Calhoun, D.B., and Englander, J.J. (1987) *Anal. Biochem. 161*, 300-306.

RECEIVED April 19, 1994

Chapter 6

Disulfide-Linked Oligomerization of Basic Fibroblast Growth Factor

Effect of Sulfated Compounds

Z. Shahrokh[1], V. Sluzky, P. R. Stratton, G. A. Eberlein, and Y. J. Wang

Pharmaceutical Research and Development, Scios Nova, Inc., 2450 Bayshore Parkway, Mountain View, CA 94043

Heparin, sulfate ion, and a number of polysulfated saccharides enhance the stability of bFGF against thermal denaturation. To assess the potential enhancement of bFGF shelf life, solution formulations containing these excipients were incubated at 35°C and analyzed by heparin affinity HPLC for soluble protein content as well as monomer/multimer distribution. Loss of soluble protein due to precipitation was seen in all formulations. Fluorescence spectroscopy was used to show that the precipitates without added ligands were made up of unfolded protein, consistent with the requirement for chaotropes to solubilize them. In the presence of sulfated compounds, however, the precipitates dissociated by 1 M NaCl and were enriched in disulfide-linked multimers with native tertiary structure. Disulfide-linked multimers also appeared in solution, at ~2-fold greater amount in the presence than the absence of sulfated compounds. In buffer alone, multimers dissociated to unfolded monomers via spontaneous thiol-disulfide exchange, resulting in aggregation and precipitation. Heparin which increased structural stability of bFGF, prevented disulfide interchange and indirectly promoted multimerization; beyond a threshold multimer concentration in solution, the multimers precipitated with the sulfated compounds. Thus, despite stabilization of native bFGF structure by sulfated compounds, loss of soluble monomer was unabated.

Basic fibroblast growth factor (bFGF) is a globular protein with mitogenic and angiogenic properties which is currently under investigation for its ability to accelerate wound healing (1-2). bFGF belongs to a family of several structurally related heparin binding growth factors including acidic FGF (aFGF) with ~55 % sequence homology and superimposing folded structures (3-6). The solution stability of FGFs under physiological conditions ranges from a few hours for aFGF (7-9) to weeks for bFGF (10). Hence, the successful solution formulation of FGFs with extended shelf lives requires evaluation

---

[1]Current address: Genentech, Inc., South San Francisco, CA 94080

of their interaction with stabilizers.   Several *in vitro* studies on aFGF have shown that heparin protects against heat and acid denaturation and protease digestion *(7, 11-12)*, thus enhancing bioactivity and solution stability *(7-9)*. There is also *in vivo* precedence for protection of FGFs against degradative processes when bound to polyanionic glycosaminoglycans (GAGs) at the subendothelial basement membrane *(1)*. Studies on the effect of heparin on bFGF stability, on the other hand, are limited. Early reports show similar protection from protease and acid digestion by heparin *(11-14)*. Thus, we investigated the effect of heparin and a number of polysulfated compounds on the solution stability and the physico-chemical characteristics of bFGF. These findings provided evidence for predominant pathways of bFGF aggregation.

Effect of Sulfated Excipients on Thermal Stability

Recombinant human bFGF was produced from an E. *coli* expression system and purified to homogeneity by HPLC methods *(10,15)*. The effect of sulfated compounds on thermal denaturation of bFGF was assessed by differential scanning calorimetry. bFGF formulated at 1 mg/ml in 100 mM phosphate, 1 mM EDTA buffer at pH 6.5 (Buffer A) showed a narrow exothermic melting profile with a peak at 60 °C which is the melting temperature, T_m. Heparins, polysulfated hexoses, and sodium sulfate increased T_m by 10 to 30 °C (Table I). The minimum weight ratio of excipient to bFGF which yielded the maximum T_m and did not compromise bFGF solubility was used in subsequent formulations.

Table I. Effect of Sulfated Compounds on Heat Denaturation of bFGF Determined by Differential Scanning Calorimetry

Excipient *(excipient:FGF weight ratio[a])*	T_m *(°C)*
none	60
heparin (0.3)	90
LMW heparin (0.3)	90
sucrose octasulfate (1)	80-85[b]
inositol hexasulfate (0.5)	82
sodium sulfate (44)	68

[a] Denaturation profiles were obtained for a range of excipient:bFGF weight ratios which did not compromise bFGF solubility. Results for the minimum ratio that yielded the maximum T_m are shown here.
[b] SOS caused a broad multipeak denaturation profile.

Effect of Sulfated Excipients on bFGF Solution Stability

The effect of sulfated compounds on stability of bFGF was investigated by incubating 1 mg/ml bFGF and various excipients in Buffer A at 35 °C. Heparin affinity HPLC was used to determine multimer distribution and measure the amount of soluble protein after removal of precipitates by filtration. Freshly prepared bFGF was >98% monomeric whereas air oxidized bFGF contained dimers, trimers, and higher oligomers (Figure 1) as determined by size exclusion HPLC (SEC) of the purified components *(16)*.

By 10 days, formulations without heparin contained precipitates. The decrease in soluble protein content as well as the monomer concentration (Figure 2A) reflected such precipitation. In the presence of heparin, an initial lag phase was observed during which soluble protein concentration decreased more slowly than in buffer alone (Figure 2B). Though this observation was consistent with UV turbidimetric analyses of bFGF stability (Eberlein et al, unpublished results), eventual precipitation was not prevented. Interestingly, the rate of monomer loss was greater than in buffer alone. This was initially due to conversion to multimers and eventually due to precipitation (Figure 2B). Note that from these data alone we can not determine whether precipitation proceeded via monomer conversion to multimer, or independent of monomer conversion to multimer. As will be discussed later, we believe that the first scheme is the predominate pathway.

Since heparin has 10-14 potential adjacent binding sites for FGF *(17)*, we investigated whether the heparin-induced increase in dimer formation might have resulted from close contact of bound protein molecules. This hypothesis was not supported by the results which indicated enhanced dimerization by a 3-fold lower molecular weight heparin (Figure 3A) as well as by inositol hexasulfate, sucrose octasulfate, and even sodium sulfate (Figure 3B).

Disulfide Nature of Soluble bFGF Multimers

To determine whether multimers promoted by sulfated compounds were covalent or non-covalent, their dissociation by reducing agents and chaotropes was examined. By heparin affinity HPLC, 90-98% of multimers formed in the presence or absence of sulfated compounds were reduced to monomers following treatment with 20-30 mM dithiothreitol (DTT), suggesting that disulfide association was the main mechanism of bFGF multimerization (Figure 4).

Interestingly, however, 4 M guanidine hydrochloride (GnHCl) also dissociated >90% of the multimers to monomers (Figure 4). A similar dissociation of multimers to monomers in the absence of reducing agents was observed during SDS-PAGE *(16)*. Alkylation of the exposed cysteines during unfolding prevented such dissociation (Figure 5; *(16)*). This phenomenon in bFGF can be explained by the fundamental concept of thiol-disulfide exchange, i.e. rapid exchange of intermolecular for intramolecular disulfides as a result of exposure of a buried free thiol upon unfolding by chaotropes (see model in Figure 5). It should be pointed out that the well-known concept of disulfide redox reactions has previously been shown to explain pathways and rates of protein folding *(18)*, whereas here, we use the concept to explain a pathway for multimer dissociation following protein unfolding.

Composition of bFGF Precipitates

The nature of precipitates was investigated by monitoring the extent of resolubilization with either 1 M NaCl or 4 M GnHCl after removal of any insoluble protein by filtration.

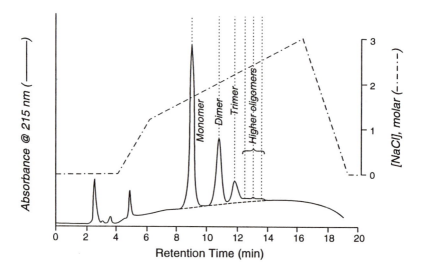

Figure 1: Chromatogram of air-oxidized bFGF obtained by heparin affinity HPLC. Chromatography was conducted on a Toso Haas heparin TSK-gel column using Buffer A as the mobile phase. Monomeric, dimeric, and trimeric species eluted with increasing [NaCl]. The oligomeric state of the protein was determined by SEC of the isolated species. Protein concentration of each component was determined from peak areas using a response factor of 240 mV.s/µg at 1 ml/min. The amount of soluble protein was measured from the total peak area since any precipitates were removed by filtration prior to chromatography.

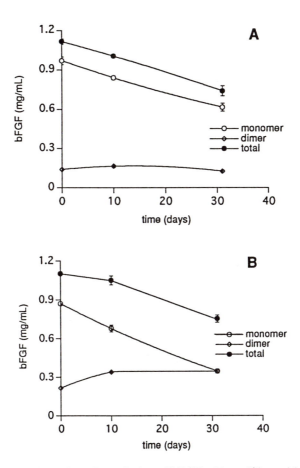

Figure 2: **Time-dependent degradation of bFGF** without (**A**) or with (**B**) heparin (0.3:1 weight ratio). Error bars for measurements on duplicate samples are shown.

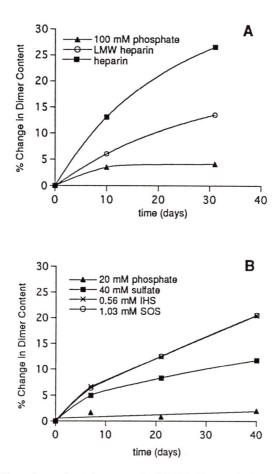

Figure 3: Time-dependent increase in bFGF dimers induced by sulfated compounds. Samples were incubated at 35 °C in the presence of the following excipients: (**A**) Heparin and low M.W. heparin (0.3:1 weight ratio); (**B**) Small sulfated compounds such as sucrose octasulfate (1:1 weight ratio), inositol hexasulfate (0.5:1 weight ratio), and 45 mM sodium sulfate.

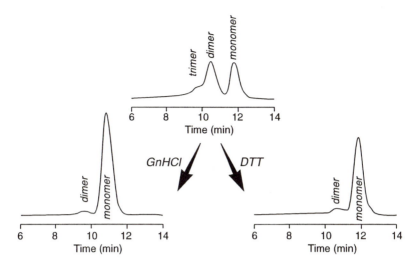

Figure 4: Nature of bFGF multimers. SEC analysis of a heparin-containing bFGF formulation (0.3:1 wt ratio, Buffer A, 35°C, 108 days) showed a high multimer content. Incubation (4°C, 20 h) with either 30 mM DTT or 4 M GnHCl dissociated the majority of these multimers. SEC was performed on a Biosil TSK-125 column using Buffer A + 1 M NaCl + 2 M GnHCl in the mobile phase.

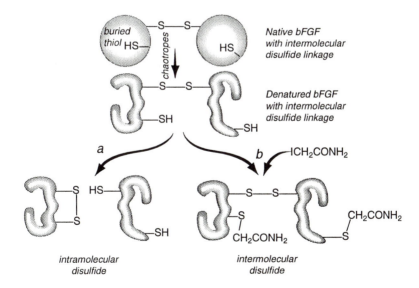

Figure 5: Multimer dissociation via disulfide gymnastics. Exposure of a buried free thiol upon unfolding of a disulfide-linked multimer by chaotropes allows for exchange of inter- for intra- molecular disulfides, and results in dissociation of the native multimers to unfolded monomers in the absence of reducing agents. Such dissociation is prevented by alkylation (e.g., by iodoacetamide) of the free thiols that become exposed during unfolding *(16)*.

The precipitates formed in heparin-containing formulations were completely resolubilized with 1 M NaCl as determined by UV spectroscopy of the filtrates (Table II). These filtrates were enriched in disulfide-linked multimers (71% of soluble protein) compared to the filtrates without salt (58%), as determined by heparin affinity HPLC. These results suggested that heparin precipitated the multimers which it had originally promoted in solution. Evidence for this pathway was also provided by the observation that addition of heparin to a multimer-enriched bFGF solution resulted in immediate precipitation of the majority of multimers (not shown).

In contrast, salt had no effect on the precipitates without heparin (Table II). Only chaotropes (e.g., 4 M GnHCl) completely resolubilized these precipitates, suggesting that the precipitates without heparin were denatured protein that hydrophobically associated. In addition, by SEC, >50% of the resolubilized precipitates were large species (>200 kDa) which dissociated to monomers after treatment with 25 mM DTT (not shown), suggesting that the precipitates without heparin were denatured protein which was associated not only hydrophobically, but also through disulfide linkages.

Table II. Solubilization and Composition of bFGF Precipitates [a]

Samples	Buffer alone		Buffer + 1 M NaCl	
	Protein Recovery[b]	Multimer Content[c]	Protein Recovery[b]	Multimer Content[c]
bFGF alone	66	16	66	16
bFGF + heparin	69	58	100	71

[a] 1 mg/ml bFGF with or without 0.3 mg/ml heparin was incubated at 35°C for 31 days. After 2-fold dilution with either phosphate buffer A or buffer plus 2 M NaCl, any remaining precipitate was removed by filtration and the recovery of soluble protein was determined by UV spectroscopy.
[b] Recovery of soluble protein expressed as % of protein that was placed in the formulation at t_0 ($\pm 5\%$).
[c] Multimer content (measured by heparin affinity HPLC) expressed as % of recovered soluble protein that is shown in the preceding column.

Structure of bFGF within Precipitates

Information about the tertiary structure of precipitated bFGF was obtained by fluorescence spectroscopy of the resuspended particulates. The emission spectrum of soluble bFGF monomer originated from its tyrosines (maximum at 308 nm) with quenched emission of the single tryptophan (Figure 6A). Denaturation of the protein led to increased tryptophan emission (maximum at 350 nm), such that the ratio of emission at 350 nm to that at 308 nm (F350/F308) increased from 0.21 (native) to as high as 1.9 (denatured with 5.5 M GnHCl; Figure 6A).

The F350/F308 ratio for non-solubilized precipitates formed in the presence of heparin was identical to that of the native protein (0.21; Figure 6B). This observation was consistent with enrichment of these precipitates in the multimer forms which also had

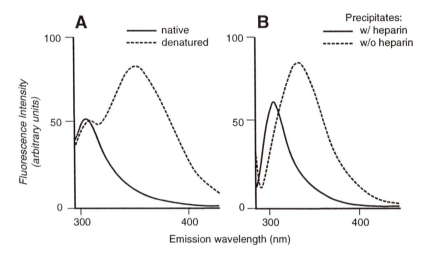

Figure 6: Fluorescence emission spectra of bFGF precipitates. (A) Emission spectra of native soluble bFGF and denatured bFGF in 5.5 M GnHCl (λ_{ex} 277 nm) after dilution to 0.1 mg/ml. (B) Emission spectra of resuspended precipitates with or without heparin (at ~0.1 mg/ml). Note the narrower spectral width of precipitates in the presence of heparin compared to that of soluble monomer without heparin. Emission spectra were obtained on a Shimadzu model RF540 fluorimeter scanned at 400 nm/min. 5 nm excitation and emission slits and 4 mm x 10 mm excitation and emission pathlengths were used.

native-like structures (F350/F308 ratio of 0.25-0.27). In contrast, F350/F308 ratio was markedly greater for the precipitates formed in buffer alone (1.57; Figure 6B), consistent with resolubilization only by chaotropes.

Heparin Binding Site and Cysteine Exposure

The heparin-induced increase in bFGF multimers was contrary to the findings reported for aFGF. To rationalize this dissimilarity in behavior between the two related proteins, their folded structure, heparin binding region, and cysteine localization was examined using computer-aided molecular graphics. Human aFGF has three cysteines, two buried at positions 16 and 47, and one exposed at position 117 (9). Binding of heparin to aFGF occurs in proximity to the sole exposed cysteine (Figure 7A), sterically protecting it from oxidation. bFGF has four cysteines, two at positions 34 and 101 which are buried, and two at positions 78 and 96 which are solvent-accessible (3,4). Note that the residue numbers shown here are for the 154 amino acid bFGF which is 9 residues longer than the truncated forms used for crystallography and documented in the Brookhaven Protein Data Bank reference number 2fgf (3) and 3fgf (4). In contrast to aFGF, the heparin binding domain of bFGF is in a region that is on the opposite side of the two exposed cysteines (Figure 7B), thus having no protective effect against oxidation.

Discussion

bFGF and aFGF both bind to heparin (17). The smallest binding glycan fragment consists of four (9,17) to seven (19) hexose units, the degree of sulfation being crucial to binding and subsequent activity enhancement. Residues 36 and 134 to 144 of bFGF (3,4), encompassing residues 105 to 128 of aFGF (5), make up the lysine/arginine-rich heparin binding domain, with dimensions which could accommodate a tetrasaccharide unit; such an interaction presumably decreases the unfavorably high electrostatic potential created by charge clustering in both FGFs.

Binding of heparin to bFGF dramatically increased its stability against heat denaturation, similar to reports made for aFGF (7). A number of polysulfated hexoses and even sulfate ion showed similar stabilization. The appearance of sulfate ions in bFGF crystals bound to the heparin binding domain (3,4) corroborates a similar site of action and mechanism of stabilization by these sulfated compounds. Structural stabilization was also confirmed by fluorescence spectroscopy, showing that after 108 days at 35°C, both soluble bFGF and the protein within the heparin-bFGF precipitates maintained native-like structure. These results were significantly different from the extensive unfolding of bFGF in precipitates that formed in the absence of heparin. Thus, bFGF without heparin followed the classical denaturation-precipitation process observed in many proteins.

The striking difference in interaction of heparins with bFGF versus aFGF was that heparin promoted disulfide-linked multimers in bFGF which proceeded onto precipitation. It has been shown that the solubility of heparin-bFGF complexes is dependent on the molar ratio of the two components (13). We have now provided evidence that bFGF precipitates in solutions containing heparin dissociated to multimers by salt, suggesting that these precipitates were ionic complexes of bFGF multimers and heparin. Similarly,

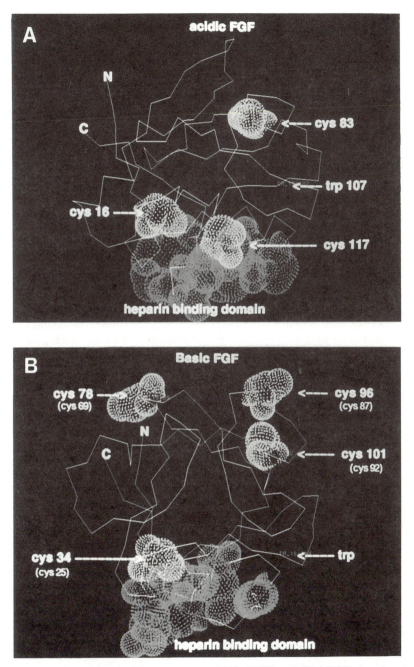

Figure 7: **Crystallographic structures of (A) aFGF and (B) bFGF** showing the spatial relationship of the heparin binding domain (dark dots) and the exposed cysteines (light dots). The single tryptophan in each protein is shown on the right.

the precipitates in solutions with small sulfated compounds also solubilized by salt, suggesting that heparin and other sulfated compounds reduced the electrostatic repulsion between bFGF multimers and rendered them insoluble.

How does one explain a heparin-induced increase in disulfide-linked multimers of bFGF, an observation which is not made for aFGF? Such an effect can not be due to a direct influence of heparin on the reactivity of bFGF cysteines because the two exposed cysteines are in a region which is on the opposite side of the heparin binding domain (Figure 7). Moreover, it is hard to conceive that structural information would translate across the molecule to affect cysteine chemistry, considering that only minimal local structural changes were caused by heparin binding as observed by FTIR (20) or fluorescence spectroscopy (Figure 6). Because heparin has multiple binding sites for FGF, it might bring several bFGFs into close contact to enhance multimerization. However, we ruled out this possibility since multimer formation was enhanced by small sulfated hexoses, even sulfate ions.

We propose the following scheme for the fate of bFGF in solution. Disulfide-linked multimers appear to be the major products of bFGF aging (Figure 8, a). In the absence of excipients, bFGF multimers are inherently unstable such that spontaneous unfolding-refolding of the protein in solution would lead to "disulfide gymnastics". Such disulfide gymnastics would dissociate the native multimers to unfolded monomers with intramolecular disulfides (Figure 8, b). Though direct evidence for an intramolecular disulfide in denatured bFGF awaits future peptide mapping studies, our finding of self-reduction of bFGF multimers by chaotropes suggests that such a reaction is possible as unfolding occurs spontaneously without chaotropes. Moreover, our observation that denatured bFGF very rapidly refolds to its native state (τ of 22 sec; (21)) has led us to postulate that an intramolecular disulfide which is not compatible with protein's native structure (3,4) should irreversibly lock the protein in the unfolded state, push the equilibrium towards further denaturation, and lead to aggregation and eventual precipitation. Then the free cysteines which become exposed upon unfolding oxidize to produce disulfide-linkages in the already hydrophobically associated protein precipitates (Figure 8, c). In contrast, complexation with heparin or small sulfated compounds apparently rigidifies the native structure, minimizes spontaneous unfolding, and preserves the original *inter*molecular disulfide linkages of the native multimers (Figure 8, d). This process indirectly raises the multimer concentration in solution, and with increasing multimer content, larger ionic complexes with heparin are produced which also precipitate (Figure 8, e).

Supporting evidence for the above scheme came from the observation that the cysteine 78- and 96-to-serine double mutant which remained monomeric showed no precipitation with heparin for up to ~3 months at 35 °C and was an order of magnitude more stable than wild type bFGF (degradation rates of 0.0015 d^{-1} versus 0.0132 d^{-1}, respectively). Moreover, because of the close proximity of the buried cysteine 101 to the exposed cysteine 96, we predicted an increased probability of disulfide exchange between thiol 101 and an intermolecular-linkage involving cysteine 96. Indeed, cysteine 96 to serine mutant (which makes only Cys78-Cys78 dimer) showed a 2.5-fold greater stability than cysteine 78 to serine mutant (which makes only Cys96-Cys96 dimer) with degradation rates of 0.0025 d^{-1} and 0.0062 d^{-1}, respectively. Finally, stimulation of disulfide exchange by oxidized glutathione dramatically destabilized the protein and led

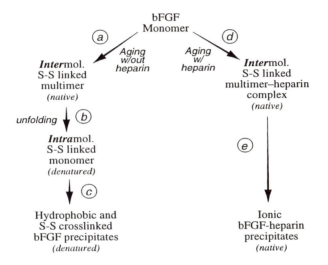

Figure 8. Proposed pathway of bFGF aggregation in solution and the effect of heparin. Formation of structurally native disulfide-linked multimers is the initial step in bFGF aging. Spontaneous unfolding-refolding of the protein in the absence of heparin results in thiol-disulfide exchange, dissociating multimers to irreversibly-denatured monomers which then precipitate. Heparin which stabilizes and rigidifies the native folded structure prevents thiol-disulfide exchange and increases multimer content in solution. Multimers form ionic complexes with heparin, and as their size and amount increases, the complexes precipitate. These precipitates, however, contain native-like protein.

to severe aggregation, whereas alkylation of free thiols delayed precipitation and enhanced stability (data not shown).

In light of the thiol-disulfide exchange phenomenon in bFGF and its prevention by heparin, the physiological state of bFGF and the role of heparin needs to be refined. Low affinity heparin-like molecules in the extracellular matrix facilitate delivery of bFGF to the high affinity cell surface receptor via events that include receptor dimerization *(22)*. Currently, there is no evidence for the oligomeric state of bFGF *in vivo*, though it is thought to be monomeric. Our data suggest that GAG-induced bFGF multimerization might occur. The multimeric form should then facilitate receptor dimerization more so than the monomeric form, simplistically because it is expected to bring more receptors in close proximity. Indeed, our cell proliferation assays show greater mitogenic activity by multimers than by monomers (Eberlein et al, unpublished results) which is consistent with the above notion. Furthermore, release of multimers from GAGs may result in disulfide exchange and formation of monomeric bFGF with intramolecular disulfides. Currently, the *in vivo* state of bFGF thiols is not known (crystallographic structures are obtained on *in vitro* reduced form of bFGF). Our data suggest that at least three potential species of bFGF might exist *in vivo*: the monomeric form with reduced thiols, the multimeric forms, and the monomeric form with an intramolecular disulfide. Considering lack of signal sequence in bFGF *(15)*, the latter species might be the intracellular form which is transported across the cell membrane. The transported protein may refold immediately upon exposure to extracellular GAGs since heparin has been shown to accelerate refolding of aFGF *(23)*. Clearly, the level of each species is expected to be regulated in major part by the physiological redox potential.

In conclusion, bFGF has unique biochemical and biophysical properties which provide clues to pathways of instability in solution. Subtle structural changes among the members of FGF family (e.g., cysteine localization) resulted in marked differences in their interaction with potential stabilizers whose utility could not be predicted from assessment of thermal or structural stabilization alone. Thus, despite structural stabilization of bFGF by heparin, thiol chemistry was not prevented such that for the purpose of preserving soluble monomeric protein, heparin and heparin-like molecules were not useful excipients in solution formulations of bFGF. We found that thiol-containing excipients (which themselves oxidize) were not a long term solution to this problem. Lowering the formulation pH to minimize cysteine reactivity, or modifying the reactive cysteines, should be useful measures to stabilize bFGF.

Acknowledgements: We wish to thank Irena Beylin and Sriram Vemuri for conducting differential scanning calorimetry. We are also grateful to Dr. Stewart Thompson's stimulating discussions on bFGF cysteine chemistry.

Literature Cited

1. Burgess W.H.; Macaig T. *Annu. Rev. Biochem.* **1989**, *58*, 575-606.
2. Hebda P.A., Klingbeil C.K., Abraham J.A., and Fiddes J.C. *J. Invest. Dermat.* **1990**, *95*, 626-631.
3. Eriksson A.E.; Cousens L.S.; Weaver L.H.; and Mathews B.W. *Proc. Natl. Acad. Sci. USA* **1991**, *88*, 3441-3445.

4. Zhang J.; Cousens L.S.; Barr P.J.; Sprang S.R. *Proc. Natl. Acad. Sci. USA* **1991**, *88*, 3446-3450.

5. Zhu X.; Komiya H.; Chirino A.; Faham S.; Fox G.M.; Arakawa T.; Hsu B.T.; Rees D.C., *Science* **1991**, *251*, 90-93.

6. Ago H.; Kitagawa Y.; Fujishima A.; Matsuura Y.; Katsube Y. *J. Biochem.* **1991**, *110*, 360-363.

7. Copeland R.A.; Ji H.; Halfpenny A.J.; Williams R.W.; Thompson K.C.; Herber W.K.; Thomas K.A.; Bruner M.W.; Ryan J.A.; Marquis-Omer D.; Sanyal G.; Sitrin R.D.; Yamazaki S.; Middaugh R., *Arch. Biochem. Biophys.* **1991**, *289*,53-61.

8. Tsai P.K.; Volkin D.B.; Debora J.M.; Thompson K.C.; Bruner M.W.; Gress J.O.; Matuszewska B.; Keogan M., Bondi J.V.; Middaugh R.C., *Pharm. Research* **1993**, *10*, 649-659.

9. Volkin D.B.; Ysai P.K.; Debora J.M.; Gress J.O.; Burke C.J.; Lindhart R.J.; Middaugh R.C. *Arch. Biochem. Biophys.* **1993**, *300* (1), 30-41.

10. Foster L.C.; Thompson S.A.; Tarnowski S.J. International Patent Application no. WO 91/15509, **1991**.

11. Gospodarowicz D.; Cheng J. *J. Cell Physiol.* **1986**, *128*, 475-484.

12. Saksela O.; Moscatelli D.; Sommer A., Rifkin D.B. *J. Cell Biol.* **1988**, 107, 743-751.

13. Sommer A.; Rifkin D.B. *J. Cell Physiol.* **1989**, *138*, 215-220.

14. Folkman J.; Szabo S.; Shing Y. *J. Cell Biol.* **1990**, *223*a, 111.

15. Thompson S.A.; Protter A.A.; Bitting L.; Fiddes J.; Abraham J.A. in *Methods in Enzymology*; Bames D.; Mather J.P.; G.H. Sato, Eds.; Academic Press Inc.:NY, NY; **1991**; Vol 198; pp. 96-116.

16. Thompson S.A.; Fiddes J.C., in *Annals N.Y. Acad. Sci.*; Baird A.; M. Klagsbrun, Eds; Academy of Sciences: NY, NY, **1991**; Vol. 638; pp. 78-88.

17. Mach H.; Burke C.J.; Volkin D.B.; Debora J.M.; Sanyal G.; Middaugh R.C. in *Harnessing Biotechnology for the 21st century*; Ladisch, M.R. and Bose A., Eds., American Chemical Society, Washington D.C., **1993**, pp. 290-293.

18. Gilbert H.F. in *Advances in Enzymology and Related Areas of Molecular Biology*; Alton Meister, Ed.; John Wiley and Sons:NY,NY; **1990**, Vol 63, pp. 69-172.

19. Turnbull J.E.; Fernig D.G., Ke Y.; Wilkinson M.C.; Gallagher J.T. *J. Biol. Chem.* **1992**, *267*, 10337-10341.

20. Prestalski S.J.; Fox G.M.; Arakawa T. *Arch. Biochem. Biophys.*, **1992**, *292*, 314-319.

21. Sluzky V.; Shahrokh Z.; Stratton P.; Eberlein G.; Wang Y.J. *Prot Science*, **1993**, *2(suppl 1)*,139.

22. Moscatelli D. *J. Biol. Chem.*, **1992**, *267*, 25803-25809.

23. Debora J.M.; Sanyal G.; Middaugh R. *J. Biol. Chem.*, **1991**, *266*, 23637-23640

RECEIVED April 26, 1994

Chapter 7

Peptide Stability in Aqueous Parenteral Formulations

Prediction of Chemical Stability Based on Primary Sequence

Michael F. Powell

Pharmaceutical Research and Development, Genentech, Inc., South San Francisco, CA 94080

Certain peptides are chemically unstable in aqueous parenteral formulations at pH 5-7 and room temperature, largely because of reactive peptide sequences. The major degradation pathways and rate constants for peptide breakdown are reviewed herein, where the reactive peptide sequences (such as Asp-Pro or Asn-Gly) are identified, and an estimate made as to whether or not such a sequence permits a two-year shelf-life when formulated in aqueous solution at 25°C. Corroboration is made between the degradation rates for model and therapeutic peptides, showing that an accurate prediction of unstable peptide primary sequences is possible, and should aid in peptide drug and formulation design.

General Concerns

The formulation of peptides for parenteral use is becoming increasingly common with the rising number of peptides and peptide mimetics under development. In some ways, the formulation development of peptides is remarkably similar to the formulation development for small molecules, particularly with respect to chemical stability. Peptides are subject to the usual degradation pathways often found for small molecules: hydrolysis, deamidation, oxidation, racemization, photodegradation, and the like. In other ways, however, the pharmaceutical development of peptides presents some additional challenges not often found for small molecule therapeutics, including physical stability considerations and the propensity for peptides to undergo *in vivo* enzymatic degradation (see below). This review covers the stability of peptides in aqueous solution, both model compounds and known therapeutics, with particular emphasis on delineating some of the reaction rates and pathways that may hamper the development of a parenteral peptide formulation(*1*). Several excellent reviews relevant to peptide parenteral stability have been published: general stability of pharmaceuticals (*2,3*), effect of formulation stabilizers on peptide stability (*4*), protein/peptide degradation pathways (*5*), analytical methods to study polypeptide degradation (*6*) and peptide adsorption (*7*).

Although not the main focus of this review, it is worthwhile mentioning that peptides are often susceptible to enzymatic degradation by peptidases, which are found

0097–6156/94/0567–0100$08.00/0

in the blood and other organs, and so peptide design to circumvent this is tantamount. It is unlikely that unstable peptides, which do not circulate in the blood for more than a few minutes, will be useful as therapeutic agents. The estimation of peptide stability *in vivo* has been carried out by determining *in vitro* peptide stability in plasma or serum. Peptide stability in serum or plasma is usually carried out by adding peptide to serum or plasma at 37°C, removing aliquots at known time intervals and assaying for remaining peptide by HPLC. Although the kinetic determination of peptide stability is straightforward, there are several factors to consider (*8*). The use of non-human serum or plasma often gives different results than human serum, presumably because different species have different levels of certain peptidases, resulting in disparate *in vitro* peptide stabilities (*9,10,11,12*). Inter-subject variation, age of the blood donor or disease state, may cause a variation in the rates of peptide degradation (*13,14,15,16,17,18*). The use of plasma with added EDTA also affects the rate of peptide degradation for *in vitro* stability measurements because EDTA acts as a peptidase inhibitor for metal ion-dependent peptidases (*10,19,20,21,22*). The method for determining peptide stability must also be 'stability-specific', in that it must be able to distinguish between the substrate and it's degradation product(s). The lack of stability-specificity has resulted in different peptide half-lives (*12,13,23,24,25*). Several general observations have been made. Many of the peptides show extremely short half-lives - some of only a few minutes - which precludes their therapeutic usefulness (*8*). Peptide reactivity in heparinized plasma and serum is strikingly similar, and not dissimilar to those reported in whole blood (*15,26,27*). Peptide modification using amide bond isosteres or D-amino acids often increases peptide stability significantly, and not necessarily in an additive fashion (*28,29,30*).

It is also important to note that physical instability is often an issue, largely due to peptide adsorption to surfaces and peptide aggregation. Surface adsorption of peptides usually occurs when peptides are at low concentration, especially for hydrophobic peptides such as the LHRH (*31*) and its hydrophobic analog nafarelin (*32*) at concentrations below 10-100 ug/mL (*33*). Peptide fragments of secretin bind to plastic surfaces including catheters (*34*). Peptide adsorption often can be prevented by increasing the peptide concentration. Overcoming peptide aggregation, however, has not been nearly so simple. Hydrophobic peptides in concentrated aqueous solution tend to aggregate, forming liquid crystals, gels, and precipitates. It is critical to prevent this physical instability as even small amounts of visible precipitate compromise a parenteral product (*35*). The nature of peptide aggregation has been studied in detail for pentagastrin (*36*), and for a few LHRH analogs. The parent peptide LHRH did not aggregate to form liquid crystals, whereas some of the more hydrophobic analogs formed liquid crystals after a few days (*37*). Peptide aggregation was suppressed by lowering the peptide concentration, by adding organic cosolvents (*38*), or by using hydrophilic salts (such as acetate) as buffering agent instead of phosphate (*38*). Adding cyclodextrins also decreased peptide aggregation, presumably by sequestering the hydrophobic side chains (*39*). Physical instability of peptides is difficult to predict based on the primary amino acid sequence, although it is easy to agree that peptides made entirely of hydrophobic amino acids are likely candidates for aggregation or adsorption. Therefore it is necessary to carry out physical preformulation studies for each new peptide drug entity under development.

The focus of this paper is to review peptide degradation in aqueous solution with particular emphasis on the kinetics and the mechanisms of degradation. Why do some peptides degrade quite rapidly, even at reduced temperatures, and yet others appear to be quite robust, even at high temperatures? This understanding of the relationship between the kinetics of peptide stability and the amino acid sequences that contribute to degradation affords some general rules that are useful for the design of peptides, as well as the design of aqueous parenteral formulations for marginally stable peptides.

Peptide Chemical Stability

General. It is generally believed that peptides are chemically stable molecules, largely because the building blocks used to make peptides - the amino acids - are fairly stable themselves. Often, however, peptides have been reported to degrade at reaction rates faster than predicted based simply on the degradation rates of the component amino acid residues. This more rapid degradation occurs for two reasons. First, the reaction at one site may be catalyzed by another residue, either due to favorable proximity of reaction groups, or due to formal catalysis (such as general acid or base catalysis) by a neighboring group. Second, peptides have many more reactive sites than the 'average' small molecule. The combined effect of these sites can result in appreciable rates of degradation, even though the degradation rate at any individual site may be slow. For example, if each reaction site contributes only 0.5% toward peptide degradation after two years, and there were 20 reactive sites, the combined loss would be 10%, giving the product a marginal shelf-life. A moderately sized peptide such as leutinizing hormone-releasing hormone (LHRH; pyro-Glu-His-Trp-Ser-Tyr-Gly-Leu-Arg-Pro-Gly-$CONH_2$) has nine backbone amide bonds that may be subject to hydrolysis, three photoreactive amino acids (His, Trp and Tyr) with six possible photo-products, several sites of racemization (predominantly Ser, Tyr, Trp), and two oxidation sites (Trp and Tyr). It is easy to see how the additive effect of peptide degradation at several sites can result in faster degradation rates than predicted from individual residues.

Model Peptides. Much of our understanding of peptide reactivity in aqueous solution comes from the study of 'model' peptides rather than therapeutic peptide drugs, of which there are only a few, and they are of such disparate chemical structure that comparative studies are tenuous at best. Further, many of these studies have been conducted under accelerated conditions by using high temperatures or strong acids or bases. Although several of these accelerated studies have been carried out to gain mechanistic information that might be useful for the prediction of long term stability at lower temperatures, some have been done to assess peptide stability at elevated temperatures as found in hydrothermal systems in sedimentary basins (40).

Many of these degradation pathways and their reaction kinetics are well understood, and have resulted in a some general rules and stability guidelines. As these rules are specific to the degradation pathway, the following mechanistic summary places particular emphasis on whether or not each degradation pathway (by itself) is sufficiently rapid to compromise the room temperature shelf-life of a parenteral peptide formulation at pH 5-7 (the typical pH of maximum stability for aqueous peptide formulation).

Hydrolysis. The breakdown of peptides may occur by cleavage at the amide bond, similar to the proteolytic reaction(s) observed in proteins (Scheme 1). Many of the early studies focused on peptides such as the polyglycines $(Gly)_5$ and $(Gly)_6$ in order to differentiate the rates of acid-catalyzed exo and endo hydrolytic cleavage (41), or the peptides Gly-Leu-Leu, Leu-Leu, Gly-Phe-Phe and Phe-Phe to determine the effect of bulky substituents upon the rates of acid-catalyzed hydrolysis (42). In these early studies carried out in 2 M HCl at 55-95°C, the relative position of the Gly-Gly bond in a polyglycine peptide did affect the rate of degradation, and steric effects are important. Using the activation parameters given, the rate of Gly-Leu hydrolysis at room temperature and pH 5 (assuming only specific acid catalysis) is greater than 100 years. Most of the other peptides studied in this series of papers show even slower rates of acid-catalyzed peptide hydrolysis, indicating that 'ordinary' hydrolysis by acid (that is, without intramolecular catalysis) is too slow to be kinetically competent for peptide degradation reactions at pH 5-7. This is in good agreement with estimates of amide hydrolysis at room temperature and neutral pH, for example, as determined for the major degradation pathway of amide hydrolysis in lidocaine (*43*). In fact, others have investigated acidity functions using peptides in concentrated acid solutions. For example, Ac-Gly and Gly-Tyr degrade in HCl, $HClO_4$ and H_2SO_4 at 100°C via a bimolecular A-2 mechanism.

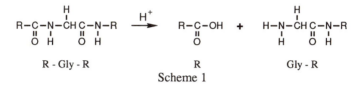

Scheme 1

Peptides containing the Asp residue are much more susceptible to acid-catalyzed hydrolysis than other peptide bonds, as shown by the half-lives of peptide bond cleavage in 0.015 M HCl at 110°C for Asp-Phe (130 min) and Lys-Phe (>1440 min), or Asp-Pro (11 min) and Gly-Pro (>1440 min) (*44*). As shown in Scheme 2, this Asp-catalyzed cleavage reaction proceeds via either the N- or C-terminal bond adjacent to the Asp residue (*45*).

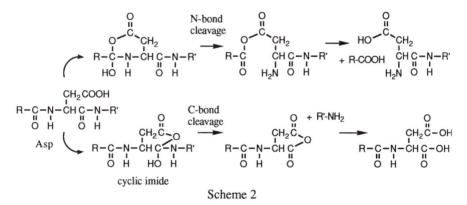

Scheme 2

This cleavage reaction is particularly rapid if Asp-Pro is present (*46*). For example, the reaction half-life of the Asp-X peptide bond in 0.015 N HCl at 110°C is much more rapid for Pro than for other amino acids: X = Pro, 11; Leu, 84; Ser, 108; Phe; 130; Lys, 228 min. This reaction is sufficiently rapid in aqueous mixed media at pH 2.5 and 40°C with a half life of approximately 50 hours (*47*). The enhanced rate of this hydrolytic reaction is due to the increased leaving group ability of the protonated proline in turn due to the higher basicity of the proline nitrogen (Scheme 3). Of concern is whether or not this reaction is sufficiently rapid to compromise an aqueous-based peptide formulation at pH 5-7, the pH of maximum stability for most peptides studied to date (*infra vide*). Correlation of the data above in an Arrhenius plot (and assuming that the rate of Asp-Pro degradation is linearly dependent on the hydronium ion concentration) gives an estimated rate of degradation at pH 5 and 25°C of 7.4×10^{-9} sec^{-1}, corresponding to a shelf life of 0.4 years. Based on this, it is likely that Asp-Pro degradation may limit the shelf life of parenteral peptide formulations between pH 5-7.

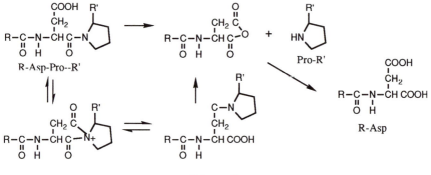

Scheme 3

The Asp-Gly bond is also fairly reactive at neutral pH, giving reversible isomerization between the Asp and iso-Asp forms via the cyclic imide intermediate (Scheme 4). Several Asp-containing peptides also give detectable amounts of this intermediate (48). The higher reactivity of the Asp-Gly bond is observed in the degradation of Val-Tyr-Pro-Asp-X-Ala at pH 7.4 and 37°C. The half-lives for these peptides are: X = Gly, 41; Ser, 168; Ala, 266 days (49). Iso-Asp also forms from Asp when sterically hindered groups are adjacent, such as in glucagon (Asp-Tyr) (50) and calmodulin (Asp-Gln, Asp-Thr) (51). Reaction of Val-Tyr-Pro-Asp-Gly-Ala at pH 7.4 and 37°C has an observed half-life of 53 days. In general, peptides harboring the Asp-Gly linkage may be too reactive to exhibit a 2 yr shelf-life in aqueous formulation at room temperature (52). Oliyai et al (53) determined the effect of pH on the degradation of this peptide. The rate constant for Asp-Gly hydrolysis below pH 3 at 37°C is 7.5×10^{-4} M^{-1}sec^{-1}, corresponding to a shelf-life at pH 5 of approximately 0.5 yrs.

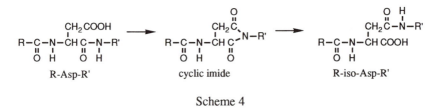

Scheme 4

Serine or threonine amide bonds are more labile than many other amide bonds (54,55). This degradation pathway was found to predominate in the degradation of the LHRH antagonist RS-26306 around pH 5, very near the pH of maximum stability for this peptide (56). Although this peptide degrades approximately 10% after 150-200 hours at 80°C at pH 5, the activation energy for this reaction determined at 50-80°C affords a calculated shelf-life of >10 years at 25°C. Therefore, this reaction pathway alone is unlikely to limit the shelf-life of similar parenteral peptides. The proposed mechanism for this degradation pathway is nucleophilic attack of the serine hydroxyl group on the neighboring amide bond via cyclic tetrahedral intermediate formation (Scheme 5).

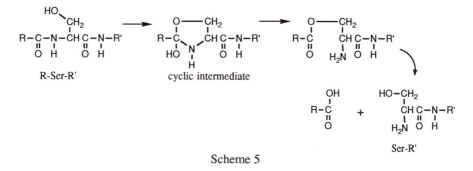

Scheme 5

Deamidation. The most complete review of glutaminyl and aspartyl deamidation is by Robinson and Rudd (*57*). More recent reviews have also been published (*58,59*), including that by Borchardt and Oliyai in this text. In general, deamidation is catalyzed by base, heat and ionic strength, and retarded by the addition of organic solvents (*60*). The deamidation rate for asparagine is usually greater than for glutamine, and is greatest when asparagine or glutamine is adjacent to glycine (*61*). Indeed, the degradation rate of Asn-Gly in 0.1 N HCl is faster than for Asn itself (*62*). The rate of deamidation (as well as the detailed mechanism) is dictated by the pH and the adjacent amino acid(s). The observed peptide half-lives range from days to years (pH 7.4, 37°C). The higher reactivity of the Asn-Gly bond is shown by the degradation of Val-Tyr-Pro-Asn-X-Ala at pH 7.4 and 37°C. The half-lives for these peptides are: X = Gly, 1.1; Ser, 8; Ala, 20; Leu, 70; Pro, 106 days, respectively (*49*). One of the most rapid deamidation reactions is for a peptide with Gly flanking both sides, Val-Tyr-Gly-Asn-Gly-Ala, showing a half-life of only 12 hours at pH 7.5 and 37°C (*63*). Another similar study has reported that the rate constant for degradation of Ac-Gly-Asn-Gly-Gly-NHMe at pH 5.5 and 37°C is 5.0×10^{-4} hr-1 corresponding to a shelf-life of only 8.7 days (*64*). In general, polar amino acids in the position X-Asn or X-Gln accelerate the deamidation rate. Hydrophobic or bulky amino acids in the sequence Asn-X appear to slow the deamidation rate considerably, where X = Gly, Ser, Thr, Ala, or His show the greatest rates. This model peptide (where X = Gly) was found to degrade exclusively via the cyclic imide intermediate from pH 5-12, and via direct hydrolysis of the amide side chain at acidic pH to give the Asp-hexapeptide (*65*). Under similar conditions, the deamidation half-lives for a series of pentapeptides yields values ranging from 6 days (Gly-Ser-Asn-His-Gly) to 3400 days (Gly-Thr-Gln-Ala-Gly) (*57,66,67*). At pH 7.4 and 37°C, the rate of Asn-Gly bond cleavage was found to be 30 to 40-fold faster than for Asp-Gly mentioned above (*49*). A summary mechanism for Asn deamidation is shown in Scheme 6, including direct hydrolysis of the amide side chain and cyclic imide formation. This reaction may also result in racemization forming the D-amino acid analogs.

The pH of maximum stability for most deamidation reactions is around pH 6, where the reaction rate is approximately 5 to 10-fold slower than observed at pH 7.4 (*68*). Even with this correction factor (and assuming that the deamidation rates at room temperature are 2 to 3-fold slower than at 37°C), the half-lives for peptide deamidation (*57*) show that many are not likely to exhibit a shelf-life greater than 2 years at 25 °C.

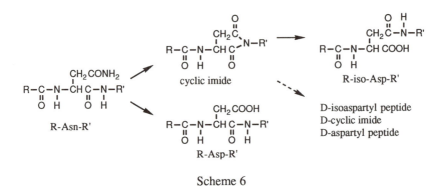

Scheme 6

Racemization. Amino acids (except Gly) are chiral molecules that are subject to base-catalyzed racemization. The mechanism is thought to proceed via α-methine proton removal by base giving the sp³ carbanion. Stabilization of this carbanion transition state by electron-withdrawing side groups (such as Tyr, Ser and Phe) accelerate the racemization reaction (63), as shown for Ser in Scheme 7.

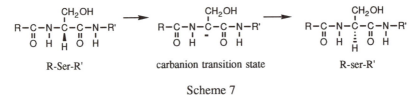

Scheme 7

A summary of factors affecting racemization of amino acids, as well as further mechanistic insight is provided in the study of arylglycines (69). Very fast rates of racemization have been reported for Asp. Indeed, Asp-X may racemize 10^5 times faster than the free amino acid (compared with the 2 to-10 fold rate accelerations seen with other amino acids) (54,70). The rate of base-catalyzed racemization correlates well with the Taft inductive constant for all amino acids studied except for Glu and Asp (70). These amino acids react significantly faster than predicted because they undergo racemization via the cyclic imide intermediate (see Scheme 6). The prediction of racemization rates is not always straightforward, as evidenced by the many factors (inductive and field effects, solvation and steric hindrance) that affected the racemization rates for thirty-seven dipeptides studied at pH 7.5 and 123°C (71). The known activation energies of base-catalyzed amino acid racemization (except for Asp, 20.8 kcal/mol) are fairly large (Phe, 28.7; Glu, 32.5; Ala, 32.4 kcal/mol). These activation energies suggest that even if racemization is detected at higher temperatures, it may still be a minor reaction at room temperature. It is easy to show that using these activation energies and the rate constants determined in 0.1 N NaOH at 65°C (assuming that the rate of racemization is linearly dependent upon hydroxide concentration), the estimated shelf-life at pH 7 exceeds 100 years (neglecting general base catalysis by water). This calculation is supported by the slow rates of racemization for a decapeptide containing several amino acids that are likely to racemize (Ser, Tyr, 3-Cl-Phe and 3-Pyr-Ala). Even though the rate constant for base-catalyzed racemization was 5×10^{-2} $M^{-1}sec^{-1}$ at 80°C, the peptide exhibited a shelf-life longer than 10 years at pH 6 and 25°C (56).

Diketopiperazine and Pyroglutamic Acid Formation. It is well known that dipeptide esters and amides easily cyclize to give diketopiperazines, where the rates are the same order of magnitude for ester hydrolysis (*72*). However it is unknown whether or not the rates of diketopiperazine formation in larger peptides are fast enough to compromise peptide stability at pH 5-7 (*73*). Peptides containing glycine as the third amino acid undergo diketopiperazine formation much more easily than peptides with other amino acids in the third position (*74*). Further, diketopiperazine formation is enhanced by incorporation of Pro or Gly into positions 1 or 2, whereas cyclization is completely prevented by blocking the α-amino group. Unfortunately there is a paucity of data for this reaction, especially at lower temperatures, making the prediction of room temperature stability rather imprecise. Reaction of Pro-Leu-Gly-$CONH_2$ at 100°C and pH 6 results in less than 3% rearrangement after 120 minutes, corresponding to degradation rate constant of 2.5×10^{-4} min^{-1}. By using this rate constant and a realistic E_a of 20 kcal/mol, the calculated shelf-life at 25°C is 260 days. Thus, it is possible that diketopiperazine reactions of this type may be rapid enough at room temperature to compromise shelf-life. The mechanism of diketopiperazine formation involves the nucleophilic attack of the N-terminal nitrogen on the amide carbonyl between the second and third amino acids (Scheme 8).

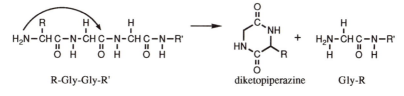

R-Gly-Gly-R' diketopiperazine Gly-R

Scheme 8

The reaction of N-terminal Gln is faster than predicted based on other amino acids, including Asn (*75*). In this case, the Gln-amide undergoes nucleophilic attack by the N-terminal amino group, giving pyroglutamic acid (Scheme 9). Fukawa has shown that Gln-Gly reacts much faster than the other dipeptides studied, including Pro-Gln, Ala-Gln and Leu-Gln (*76*). The tripeptide Gln-Gln-Gly showed an initial rapid rate of degradation at 100°C, and then slowed down when approximately half of the Gln was gone, presumably because of the rapid reaction to form Pyr-Gln-Gly with a slower rate for the internal Gln.

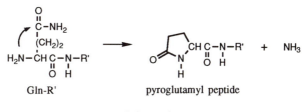

Gln-R' pyroglutamyl peptide

Scheme 9

The few papers reporting on the kinetics of pyroglutamic acid suggest that this cyclization reaction may be of kinetic consequence at pH 5-7 at room temperature. For example, Melville reports that peptides N-terminated with Gln lost ammonia (due to pyroglutamic acid formation) at 37°C and pH 2-8. Peptides terminated with Gln-Lys-Asn show a reaction half-life of approximately 8 hours at 37°C when formulated in ammonium bicarbonate at pH 7.9, and reacted slightly faster in 0.01 M phosphate buffer at pH 8 (*77*). Conversion is also observed under acid conditions (0.1 M acetic acid), where 15% was lost after 7 months. Further, this sequence is not expected to be unusually reactive. It has been shown in the Gln-X-OCH$_2$-resin model system that the fastest rates of pyroglutamic acid formation are for X = Asn, Leu and Gly (*78*). Based on these reports, it is expected that many peptides terminated in Gln will not show a 2 year shelf-life at room temperature and pH 5-7.

Oxidation. Autooxidation is 'uncatalyzed' thermal oxidation, and is studied infrequently because the reaction rate is slow and variable. Uncatalyzed or thermal protein oxidation is actually a misnomer, as the degradation rate is often governed by trace amounts of peroxide, metal ions, light, base and free radical initiators (*79*). Even though a great deal is known about reactive oxygen species, the presence (or absence) of these initiators makes the prediction of autooxidation in parenteral formulations imprecise. For example, free radical oxidation involves the separate effects of initiation, propagation and termination. Further, there are several reactive oxygen species, including: singlet oxygen $^1O_2$, superoxide radical O_2^-, alkyl or hydrogen peroxide ROOH or H_2O_2, hydroxyl radicals (HO• or HOO•), and halide oxygen complexes (ClO$^-$) (*80*), and the order of reactivity is generally HO•, HOO• > O_2^- > ROOH, H_2O_2 > $^1O_2$ ClO$^-$ > O_2.

There is limited published data on the oxidation of peptides in pharmaceutical formulations because only a few peptides have been developed thus far - and only a few of these are susceptible to oxidative degradation. Much of our understanding of polypeptide oxidation comes from the biochemical literature. Although there are several reactive aminoacids that are known to oxidize (methionine, cysteine, cystine, histidine, tryptophan and tyrosine), a review of the literature shows that, under mild oxidative conditions at pH 5-7, methionine and cysteine are the predominant amino acids undergoing oxidation (*58,81*). Methionine oxidizes by both chemical and photochemical pathways to give methionine sulfoxide and, under extremely oxidative conditions, methionine sulfone (Scheme 10).

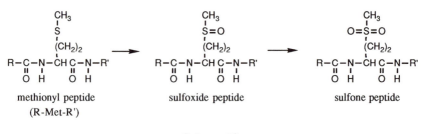

methionyl peptide (R-Met-R')   sulfoxide peptide   sulfone peptide

Scheme 10

Methionine residues in polypeptides show widely varying reactivity, largely because some are protected from oxidation by steric effects or inaccessibility, buried in the hydrophobic core of the molecule (*82*). Methionine can also be photoxidized by a free radical pathway (*83*), via a singlet oxygen intermediate (*84*), or by thermal reaction via nucleophilic substitution (*85*). This mechanism of nucleophilic displacement is preferred over one in which there is an acid-base equilibrium with the solvent generating the species HOOH$_2^+$, based lack of salt effects (*86*), negative entropy of activation (*85*), and leaving group substituent effects (*87*). The second order rate constants ($M^{-1}sec^{-1}$) of methionine oxidation by hydrogen peroxide have been determined at room temperature for methionine free amino acid (0.93), Ac-Ser-Trp-Met-Glu-Glu-CONH$_2$ (1.07), Ac-CysNH$_2$-S-S-AcCys-Gly-Met-Ser-Thr-CONH$_2$ (1.0) and the methionines in relaxin B chain at positions 25 and 4 (Met B^{25}, 0.85; Met B^4, 0.34) (*88*). This study shows that the peroxide catalyzed degradation of methionine has little temperature effect (ΔH ~ 10-12 kcal/mol), and may be partially offset by the increased solubility of oxygen in water at lower temperatures (*89*). Further, there was a negligible effect of pH or ionic strength. The amount of peroxide in some excipients such as polyethylene glycols and surfactants varies widely (*90,91*), and so the use of these with methionyl peptides should be done cautiously. Using the data of Nguyen et al. (*88*), it is estimated that 1 nM peroxide in a methionyl peptide formulation would shorten the shelf-life to less than 2 years.

Cysteine is also easily oxidized. For example, cysteine can react with oxygen to yield cystine disulfide (Scheme 11). The rate of cysteine oxidation is usually accelerated at higher pH where the thiol is deprotonated (*92*), although exceptions to this rule are known and are usually attributed to protein unfolding at lower pH with a concomitant increase in the rate of thiol oxidation (*93,94*). The oxidation of cysteine to give cystine is also very dependent on the nature of the neighboring amino acids, where hydrophobic residues significantly retard the cysteine oxidation rates (*95*).

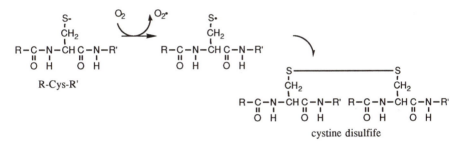

Scheme 11

The pH dependence on polypeptide oxidation may be used to distinguish between cysteine and methionine oxidation. The hydrogen peroxide-catalyzed oxidation of cysteine decreases as the pH is lowered, whereas the rate of methionine oxidation increases slightly from pH 5 to 1 (*96*). Once the cystine disulfide is formed, it is much less susceptible to reaction with mild oxidants (such as H$_2$O$_2$) than either methionine or cysteine, and so is often found as the stable end product in an oxidative reaction. As this reaction is dependent on molecular oxygen (or other oxidants), formulation of cysteine-containing peptides under nitrogen should prolong shelf-life. This reaction is fairly rapid and so it is unlikely that 'ordinary' cysteinyl peptides can be formulated in aqueous solution at pH 5-7 and room temperature for 2 years without appreciable amounts of dimerization.

Formulation Stability of Peptide Drugs

Except for a few isolated reports of unusual peptide reactivity such as selected neuroendocrine peptides (*97*) and tuftsin (Thr-Lys-Pro-Arg) (*98*), the peptide stability guidelines derived from model peptides can be used to predict the stability of peptide therapeutics formulated at pH 5-7 and room temperature. For example, adrenocorticotrophic hormone (ACTH, a 39-amino acid polypeptide which contains Asn-Gly) degrades primarily via deamidation to give iso-Asp-ACTH at neutral to high pH, presumably by cyclic imide formation, and Asp-ACTH by direct deamidation at lower pH (99). The observed shelf-life for this peptide is 140 hours at pH 5, and 55 hours at pH 7. An activation energy of 20 kcal/mol predicts a shelf-life of 21 and 8 days, respectively, significantly less than the two years desired for a parenteral formulation.

Other peptide therapeutics also degrade by reaction to form iso-Asp. The lipopeptide antibiotic, daptomycin, forms iso-Asp by cyclic imide formation from its Asp precursor molecule (100), similar to amphotericin (101). This reaction occurs with a half-life of approximately 1 day at 60°C and pH 5.3, and 0.2 days at pH 7.4, presumably due to the labile Asp-Gly sequence. It is unfortunate that the activation energies for daptomycin are not reported and therefore the rate of degradation at room temperature cannot be calculated.

Other hormones also degrade at Asp to give iso-Asp. Secretin (His-Ser-Asp-Gly-Thr-Phe-Thr-Ser-Glu-Leu-Ser-Arg-Leu-Arg-Asp-Ser-Ala-Arg-Leu-Gln-Arg-Leu-Leu-Gln-Gly-Leu-Val-CONH$_2$) degrades at the Asp-Gly (102) and the Asp-Ser (103) positions to give iso-Asp, where the pH of maximum stability is 6.8-7 (104). Further analysis of the secretin degradation pathway at 60°C reveals that the major pathway at neutral pH is formation of iso-Asp at position 3 (iso-Asp^3), whereas formation of the cyclic imide of Asp^3 predominates at lower pH (105,106). Unfortunately these studies were carried out only at a single temperature making extrapolation to room temperature difficult. Growth hormone releasing factor (GRF$^{1-44}$-CONH$_2$) also degrades slowly at pH 5.5 via cyclic imide formation to give iso-Asp^3-GRF-CONH$_2$ with a shelf life of 52 hours at 55°C (107). Interestingly, the related peptide, Leu^{27}-GRF$^{1-32}$-CONH$_2$ degraded primarily by deamidation of Asn^8 (giving iso-Asp^8, 46%; Asp^8, 16%; iso-asp^8, 5%) with a limited shelf-life of 28 hours at pH 7.4 and 37°C (108). These deamidation reactions are not limited to small peptides, but are often the major degradation pathway in larger polypeptides such as insulin (109,110) or human growth hormone (111).

The degradation pathway of LHRH (112), as well as the LHRH agonist, histerelin (<Glu-His-Trp-Ser-Tyr-X-Leu-Arg-Pro-NHEt, where X = N-imidazole benzyl-his) yields the products expected from model peptide studies (113). The primary cleavage sites for histerelin at pH 5.1-5.4 are cleavage of the <Glu-His and the Trp-Ser bonds, giving pyroglutamic acid (<Glu), <Glu-His-Trp, and the corresponding C-terminal peptide fragments. Reaction of this peptide at 87°C for 18 days at pH 5.4 results in an approximate 50% loss of parent peptide, whereas storage for 6 months at 37°C yields only a few percent loss, indicating that this peptide should exhibit a room temperature shelf-life of greater than two years. This conclusion is supported by the high stability of other, structurally similar LHRH peptides such as nafarelin (114), trp^6-Pro^9-NEt-LHRH (115), and the parent LHRH peptide itself (116). Other peptides not containing any obvious reactive groups such as the neurotensin activity peptide (Me)Arg-Lys-Pro-Trp-tert-Leu-Leu do not degrade at elevated temperatures (but at conditions where the C-terminal methyl ester analog shows an estimated shelf-life of 3.6 years (pH 5.1, 25°C)) (117).

The stability of muramyl dipeptides presents an interesting borderline scenario, where one of two peptides studied has a shelf-life longer than two years, and the other does not (118). This difference may be attributable to increased steric hindrance by the amino acid side groups in the more stable peptide. The more reactive peptide contains a glycine which may be susceptible to hydrolysis catalyzed by the muramic acid hydroxy group in the 2-position. Although both of these MDP compounds have an C-terminal amide, this group does not appear to be the major degradation pathway at pH ~5. The rates of acid catalyzed C-terminal amide hydrolysis have been determined for several peptides at 75-80°C (acetamide, 1×10^{-3} $M^{-1}sec^{-1}$(*119*); nafarelin, 1.2×10^{-3} $M^{-1}sec^{-1}$ (*114*); RS-26306, 0.5×10^{-3} $M^{-1}sec^{-1}$(*56*) and are remarkably similar to the value for MDP (0.9×10^{-4} M^{-1} sec^{-1}) indicating that this reaction, albeit undetected in this study, is also likely.

In addition to hydrolysis, racemization has also been observed in peptide drugs, such as the LHRH analogs (*39,56*). As expected, racemization occurs at high pH. In the first of these two studies on LHRH analogs, 0.1 N NaOH was used at 40°C to effect only partial racemization after two days. The more relevant work in the second study showed that a peptide with many potential sites of racemization did not react sufficiently fast enough to shorten the room temperature shelf-life at pH 5-6. Although not observed directly in ACTH, racemization of Asp in the ACTH[22-27] pentapeptide gave at half-life of <20 hours at pH 7.4 and 37°C, and it is likely that ACTH itself racemizes at a similar rate. This rate of degradation is also in agreement with the rapid rates of Asp racemization observed in model compounds.

Few pharmaceutical peptides degrade by oxidation as the major pathway. For example, solutions of relaxin are easily oxidized by hydrogen peroxide (*88*), as well as photooxidized (*120*). It is difficult to corroborate these rates with the known rates of methionine oxidation in other polypeptides (*58*). Peptide oxidation is also a problem in the production and isolation of peptide hormones such as cholecystokin-33, a polypeptide having three methionine residues (*121*). Stabilization of methionine containing peptides has been solved by replacing methionine with another amino acid that is stable towards oxidation such as the Thr and norLeu replacement analogs of CCK (*122*), or the Leu (*123,124*) and methoxinine (*125*) analogs of gastrin peptides. Oxidation of cysteine also has its precedent in therapeutics. For example, captopril is a peptide mimetic compound with a free thiol showing a fairly rapid dimerization reaction (shelf-life of 2 days at pH 6 and ~50°C) (*126*). Additionally, unusual oxidation reactions occur occasionally, such as the oxidation of the antipsychotic therapeutic Gly-Gly-Phe-Met(O_2)-lys-Phe to give $HONHCH_2CO$-Phe-Met(O_2)-lys as one of the minor degradation products (*127*).

Summary

Many peptides are predicted to be stable in aqueous parenteral formulations at pH 5-7 if certain amino acid combinations, such as -Asn-Gly- or -Asp-Pro-, are absent from the primary peptide sequence. Table I lists the common peptide functional groups that are thought to be reactive in aqueous solution, with an estimation (based on the recent literature data) regarding the stability of these groups at pH 5-7 in an aqueous parenteral formulation. As mentioned above, a single peptide often has several degradation pathways available due to the varied chemical reactivities of the different amino acids, and so peptides are often more reactive than predicted based on the analysis of a single functional site. Further, reaction rates for oxidation and aggregation are difficult to predict because of the unknown amount of catalyst and effect of the neighboring amino acids. As long as these caveats are kept in mind, the Table I may be used to show that certain peptides (containing reactive functional groups) may not demonstrate an adequate shelf-life in aqueous parenteral formulations at neutral pH and room temperature.

Table I. Reactive Peptide Sequences and Formulation Stability

Sequence (a)	Primary Reaction	Major Products	Est. Reactivity (b)
-X-X-	typical amide bond hydrolysis	-X + X-	slow
-Asp-X- (X ≠ Pro)	hydrolysis	-Asp + X- or -isoAsp + X-	slow to moderate
-Asp-Pro-	hydrolysis	-Asp + X- or -isoAsp + X-	**may compromise shelf-life, fast**
-Asn-X- (X ≠ Gly)	deamidation	-Asp-X-, -isoAsp-X- Asp + X-, -isoAsp -asp, -iso-asp L or D cyclic imide	may compromise shelf-life, moderate
-Asn-Gly-	deamidation	-Asp-Gly-, -Asp,- isoAsp-Gly-, -isoAsp -asp, -iso-asp L or D cyclic imide	**may compromise shelf-life, fast**
-X-Ser- or -X-Thr-	hydrolysis	-X + Ser- or -X + Thr-	moderate
-Ser-, -Tyr-,-Phe-, -X- (≠ Gly, Asp)	racemization	-ser-, -thr-, -phe- -x-	slow to moderate
-Asp-	racemization	-asp-, (-iso-asp-)	**may compromise shelf-life, fast**
X-X-Gly-	diketopiperazine formation	cyclo-X-X + Gly-	**may compromise shelf-life, fast**
Gln-X-	pyroglutamic acid formation	<Gln-X-	**may compromise shelf-life, fast**
-Met-	oxidation	-Met(O)-	may compromise shelf-life, moderate
-Cys-	oxidation, disulfide formation	-Cys Cys-	**may compromise shelf-life, fast**
-X- (X = many hydrophobic AA's)	precipitation	aggregates, liquid crystals	may compromise shelf-life, moderate

a) X = any amino acid unless otherwise specified. Amino acids drawn as -X or X- denote the possibility of additional amino acids attached. See notes in Reference 1 regarding nomenclature of amino acids.

b) Reactions which are likely to compromise peptide stability (t90 < 2 yrs at 25°C) at pH 5-7 are shown in bold, where other limiting reactions are adjusted for their propensity to limit shelf life accordingly.

Acknowledgements

Helpful comments by S. Shire, J. Cleland, R. Pearlman and A. Jones, all of Genentech. Technical editing and manuscipt preparation were done by J. Burdman.

References

1   The following descriptors are used: l- aminoacids are represented using the standard capitalized three letter codes; d-aminoacids are in standard three letter codes without capitalization; peptides terminated with 'Ac-' or '-CONH$_2$' represent the N-terminal acetylated or C-terminal amide derivatives, respectively; '<Gln' or 'Pyr' represents pyroglutamic acid; sequences in italics (followed by numbers only in parenthesis) represent the peptide by its general name and peptide number. Amino acid position is denoted using by a superscript notation following the amino acid three letter code, or the common peptide name where applicable. Common peptide names (where the sequence is too long to show) are given in parenthesis after their first notation.

2   Manning, M. C.; Patel, K. and Borchardt, R. T., *Pharm. Res.*, **1989**, *6*, 903.

3   Mollica, J. A.; Ahuja, S. and Cohen, J., *J. Pharm. Sci.*, **1978**, *67*, 443.

4   Wang, Y. -C. J. and Annson, M. A., *J. Parenter. Sci. Tech.*, **1988**, *42*, S2.

5   Chen, T., *Drug Dev. Indus. Pharm.*, **1992**, *18*, 1311.

6.   Jones, A. J. S., *Adv. Drug Del. Rev.*, (in press, **1993**).

7   Andrade, J. D. and Hlady, V., *Adv. Polym. Sci.*, **1986**,*79*, 2.

8   Powell, M.F., *Ann. Rev. Med. Chem.*, (in press, **1993**).

9   Verhoef, J.; van den Wildenberg, H. M. and van Nispin, J. W., *J . Endocrin.*, **1986**, *110,* 557.

10   McDermott, J. R.; Smith, A.I.; Biggins, J. A.; Hardy, J. A.; Dodd, P. R. and Edwardson, J. A., *Reg. Peptides*, **1981**, *2*, 69.

11   Brewster, D. and Waltham, K., *Biochem. Pharmacol.*, **1981**, *30*, 619.

12   Walter, R.; Neidle, A. and Marks, N., *Proc. Fed. Soc. Exper. Bio. Med.*, **1975**, *148*, 98.

13   Iverson, E., *J. Endocrin.*, **1988**, *118*, 511.

14   White, N.; Griffiths, E. C.; Jeffcoate, S. L.; Milner, R. D. G. and Preece, M. A., *J. Endocrin.*, **1980**, *86*, 397.

15   Springer, C. J.; Eberlein, G. A.; Eysselein, V. E.; Schaeffer, M.; Goebell, H. and Calam, J., *Clinica Chimica Acta.* **1991***, 198*, 245.

16   Koop, H., *Clin. Gastroent.*, **1984,** *13*, 739.

17   Johnson, A. R.; Coalson, J. J.; Ashton, J.; Larambide, M., and Erdos, E. G., *Am. Rev. Respir. Dis.*, **1985**, *132,* 1262.

18   Rutenberg, A. M.; Goldbarg, J. A. and Pineda, E. P., *New Eng. J. Med.*, **1968**, *259*, 469.

19   Springer, C. J. and Calam, J., *Gastroenterol.*, **1988**, *95*, 143.

20   Graf, M. V.; Saegesser, B. and Schoenenberger, G. A., *Peptides*, **1987**, *8*, 599.

21   Visser, T. J.; Klootwjk, W.; Docter, R. and Hennemann, G., *Acta Endocrinol.*, **1977**, *86*, 449.

22   Artner-Dworzak, E.; Lindner, H. and Puschendorf, B., *Clin. Chim. Acta.*, **1991**, *203*, 235.

23   Frohman, L. A.; Downs, T. R.; Williams, T. C.; Heimer, E. P.; Pan, Y-C. E., and Felix, A. M., *J. Clin. Invest.*, **1986**, *78,* 906.

24   Benuck, M. and Marks, N., *Life Sci.*, **1976**, *19*, 1271.

25   Walter, R.; Neidle, A. and Marks, N., *Proc. Fed. Soc. Exper. Bio. Med.*, **1975**, *148*, 98.

26   Hichens, M., *Drug Met. Rev.*, **1983**, *14*, 77.
27   Hussain, M. A.; Rowe, S. M.; Shenvi, A. B. and Aungst, B. J., *Drug. Met. Disp.*, **1990**, *18*, 288.
28   Heavner, G. A.; Kroon, D. J.; Audhya, T. and Goldstein, G., *Peptides*, **1986**, *7*, 1015.
29   Su, C.; Jensen, L. R.; Heimer, E. P.; Felix, A. M.; Pan, Y. C. and Mowles, T. F. *Horm. Metab. Res.,* **1991**, *23*, 15.
30   Benovitz, D. E. and Spatola, A. F., *Peptides*, **1985**, *6,* 257.
31   Sandow, J.; Konig, W. and Krauss, B., *Acta Endocrin.,* **1977**, *84*, 32.
32   Anik, S. T. and Hwang, J.-Y., *Int. J. Pharm.*, **1983**, *16,* 181.
33   Mizutani, T. and Mizutani, A., *J. Pharm. Sci.*, **1978**, *67*, 1102.
34   Bitar, K. N.; Zfass, A. M. and Makhlouf, G. M.,*Gastroenterol.*, **1978**, *75*, 1080.
35   Borchert, S. J.; Abe, A.; Aldrich, D. S.; Fox, L. E.; Freeman, J. E. and White, R. D., *J. Parenter. Sci. Tech.*, **1986**, *40*, 212.
36   Attwood, D.; Florence, A. T.; Greig, R. and Smail, G. A. *J. Pharm. Pharmac.*, **1974**, *26*, 847.
37   Powell, M. F.; Sanders, L. M.; Rogerson, A. and Si, V. *Pharm. Res.*, **1991**, *8*, 1258.
38   Fleitman, J. ; Powell, M. F.; Sanders, L. M. and Si, V. *J. Pharm Sci.,* (in press, **1993**).
39   Rossi, F.; Zaccaro, L.; DiBlasio, B.; Pavone, V.; Pedrone, C.; Cucinotta, V. et al. *Conference Proceedings, 22nd European Peptide Symposium,* September 13, **1992**.
40   Schock, E. L. *Geochimica Cosmochimica Acta.*, **1992**,*56,* 3481.
41   Lee, R. G.; Long, D. A. and Truscott, T. G. *Trans. Faraday Soc.*, **1968**, *64*, 503.
42   Long, D. A. and Truscott, T. G. *Trans. Faraday Soc.,* **1968**, *64*, 1866.
43   Powell, M. F. *Pharm. Res.*, **1987**, *4* , 42.
44   Piszkiewicz, D.; Landon, M. and Smith, E.L. *Biochem. Biophys. Res. Commun.*, **1970**, *40*, 1173.
45   Inglis, A. S. *Methods Enzymol.*, **1983**, *91*, 324.
46   Schultz, J. *Methods Enzymol.*, **1967**, *11*, 255.
47   Landon, M. *Methods Enzymol.*, **1977**,*47*, 145.
48   Bodansky, M.; Ondetti, M. A.; Levine, S. D. and Williams, N. J. *J. Amer. Chem. Soc.*, **1967**, *89*, 6753.
49   Stephenson, R. C. and Clarke, S. *J. Biol. Chem.*, **1989**, *264*, 6164.
50   Ota, I. M., Ding, L. and Clarke, S. *J. Biol. Chem.*, **1987**, *262*, 8522.
51   Ota, I. M. and Clarke, S. *J. Biol. Chem.*, **1989**, *264*, 54.
52   Geiger, T. and Clarke, S. *J. Biol. Chem.*, **1987**, *262*, 785.
53   Oliyai , C. and R. T. Borchardt, *Pharm. Res.*, **1993**, *10*, 95.
54   Sanger, F. and Tuppy, H. *Biochem. J.* **1951**, *49*, 463.
55   Harris, J. I.; Cole, R. D. and Pon, N. G. *Biochem. J.*, **1956**, *62*, 154.
56   Strickley, R. G.; Brandl, M.; Chan, K. W.; Straub, K. and Gu, L. *Pharm. Res.*, **1990**, *7*, 530.
57   Robinson, A. B. and Rudd, C. J. In *Current Topics in Cellular Regulations,* Horecker, B. L. and Stadman, E. R. (Eds.), Academic Press, NY, 1974, Vol. 8, pp. 247-295.
58   Cleland, J. L.; Powell, M. F. and Shire, S. J. *CRC Crit. Rev. Biochem. Mol. Biol.,* (in press, **1993**).
59   Wright, H. T. *CRC Crit. Rev. Biochem. Mol. Biol.*, **1991**, *26*, 1.
60   Capasso, S.; Mazzarella, L. and Zagari, A. *Pep. Research*, **1991**, *4*, 234.
61   Robinson, A. B.; Scotchler, J. W. and McKerrow, J. H. *J. Amer. Chem. Soc.*, **1973**, *95*, 8156.

62   Leach, S. J. and Lindley, H. *Trans. Faraday Soc.*, **1953**, *49*, 915.
63   Patel, K. and Borchardt, R. T. *Pharm. Res.*, **1990**, *7*, 787.
64   Capasso, S.; Mazzarella, L.; Sica, F.; Zagari, A.; Baldoni, G. and Salvadori, S. In *Peptides 1992* C. H. Schneider and A. N. Eberle (Eds.), ESCOM Science Publishers, 1993, pp 251.
65   Patel, K. and Borchardt, R. T. *Pharm. Res.*, **1990**, *7*, 703.
66   McKerrow, J. H. and Robinson, A. B. *Science*, **1974**, *183*, 85.
67   Robinson, A. B.; Scotchler, J. W. and McKerrow, J. H. *J. Amer. Chem. Soc.*, **1973**, *95*, 8156.
68   Scotchler, J. W. and Robinson, A. B. *Anal. Biochem.*, **1974**, *59*, 319.
69   Smith, G. G. and Sivakua, T. *J. Org. Chem.*, **1983**, *48*, 627.
70   Friedman, M. and Masters, P. M. *J. Food Sci.*, **1982**, *47*, 760.
71   Smith, G. G. and DeSol, B. S. *Science*, **1980**, *207*, 765.
72   Purdie, J. E. and Benoiton, N. L.*J. Chem. Soc. Perkin Trans II*, **1973**, *13*, 1845.
73   Steinberg, S. M. and Bada, J. L. *J. Org. Chem.*, **1983**, *48*, 2295.
74   Sepetov, N. F.; Krymsky, M. A.; Ovchinnikov, M. V.; Bespalava, Z. D.; Isakova, O. L.; Soucek, M. and Lebl, M. *Pept. Res.*, **1991**, *4*, 308.
75   Blomback, B. *Methods Enzymol.*, **1967**, *11*, 398.
76   Fukawa, H. *J. Chem. Soc. Jpn.*, **1967**, *88*, 459.
77   Khandke, K. M.; Fairwell, T.; Chait, B. T. and Manjula, B. N. *Int. J. Pep. Protein Res.*, **1989**, *34*, 118.
78   Orlowska, A.;Witkowska, E. and Izdebski, J. *Int. J. Pep. Protein Res.*, **1987**, *30*, 141.
79   Johnson, D. M. and Gu, L. C. In *Encyclopedia of Pharmaceutical Technology, Volume 1. Absorption of Drugs to Bioavailability and Bioequivalence,,* Swarbrick, J. and Boylan, J. C., Eds. New York, Marcel Dekker Inc. 1988.
80   Halliwell, B. and Gutteridge, M. C., In *Methods in Enzymology*, Volume 186. Free Radicals and Metal Ions in Human Disease, Academic Press, 1990.
81   Stadtman, E. R., *Free Radical Biol. Med.*, **1990**, *9*, 315.
82   Teh, L. C-.; Murphy, L. J.; Huq, N. L.; Surus, A. S.; Friesen,H. G.; Lazarus, L. and Chapman G. E., *J. Biol. Chem.*, **1987**, *262*, 6472.
83   Weil, L. *Arch. Biochem. Biophys.,* **1965**, 110, 57.
84   Sysak, P. K.; Foote, C. S. and Ching, T.-Y. *Photochem. Photobiol.*, **1977**, *26*, 19.
85   Behram, E. J., and Edwards, J. O., *Prog. Phys. Org. Chem.*, **1967**, *4*, 93.
86   Barnard, D.; Bateman, L. and Cunneen, J. I. In *Organic Sulfur Compounds*, N. Kharasch, Ed. Pergamon, London, 1961, Volume 1, 229-247.
87   Bruice, T. C.*J. Chem. Soc. Chem. Commun.,* **1983**, 14.
88   Nguyen, T. H.; Burnier, J. and Meng, W. *Pharm. Res.*, (in press, **1993**).
89   Stewart, P. J. and Tucker, I. G. *Austr. J. Hosp. Pharm.*, **1985**, *15*, 111.
90   Hamburger, R.; Azaz, E. and Donbrow, M. *Pharm. Acta Helv.*, **1975**,*50*, 10.
91   McGinty, J. W.; Hill, J. A. and La Via, A. L., *J. Pharm. Sci.*, **1975**, *64*, 356.
92   Philipson, L. *Biochim. Biophys. Acta.,* **1962**, *56*, 375.
93   Jaques, L. B., and Bell, H. J. *Can. J. Research*, **1946**, *24E*: 79.
94   Neumann, N. P.; Moore, S.; and Stein, W. H. *Biochemistry*, **1962**, *1*, 68.
95   Davies, D. E.; Alexander, P.; Higginbotham, C. L.; Sharma, R. P. and Richter, A. In *Peptides 1992,* Schneider, C. H. and Eberle, A. N. (Eds.), ESCOM Science Publishers, 1993, pp 543.
96   Means, G. E. and Feeney, R. E. In *Chemical Modification of Proteins*, Holden Day Inc., San Francisco, 1971, 162-65.
97   Feldman, J. A.; Cohn, M. L. and Blair, D. *J. Liquid. Chromat.*, **1978**, *1*, 833.

98   Nishioka, K.; Dessens, S. E. and Rodriguez, Jr., T. J. *Peptide Res.,* **1990**, *4*, 230.
99   Bhatt, N. P.; Patel, K. and Borchardt, R. T. *Pharm. Res.*, **1990**, *7*, 593.
100   Kirsch, L. E.; Molloy, R. M.; Bebono, M.; Baker, P. and Fariad, K. Z. *Pharm. Res.*, **1989**, *6*, 387.
101   Bodansky, M.; Sigler, G. F.and Bodansky, A. *J. Amer. Chem. Soc.*, **1973**, *95*, 2352.
102   Beyerman, H. C.; Grossman, M. I; Solomon, T. E. and Voskamp, D. *Life Sci.*, **1981**, *29*, 885.
103   Wunsch, E., *Biopolymers*, **1983**, *22*, 493.
104   Jaeger, E.; Knof, S.; Scharf, R.; Lehnert, P.; Schultz, I. and Wunsch, E. *Scand. J. Gastroenterol.*, **1978**, *13*, 93.
105   Tsuda, T.; Uchiyama, M.; Sato, T.; Yoshino, H.; Tsuchiya, Y.; Ishikawa, S.; Ohmae, M.; Watanabe, S. and Miyake, Y. *J. Pharm. Sci.*, **1990**, *79*, 53.
106   Tsuda, T.; Uchiyama, M.; Sato, T.; Yoshino, H.; Tsuchiya, Y.; Ishikawa, S.; Ohmae, M.; Watanabe, S. and Miyake, Y. *J. Pharm. Sci.*, **1990**, *79*, 223.
107   Felix, A. M.; Lambros, T.; Heimer, E. P.; Cohen, H.; Pan, Y. C-; Campbell, R. and Bongers, J. Proceedings of the 21st European Peptide Symposium, 1990, 732-733.
108   Friedman, A. R.; Ichhpurani, A. K.; Brown, D. M.; Hillman, R. M.; Krabill, L. F.; Martin, R. A.; Zurcher-Heely, H. A. and Guido, D. M. *Int. J. Peptide Protein Res.*, **1991**, *37*, 14.
109   Brange, J. *Acta. Pharm. Nord.*, **1992**, *4*, 209.
110   Brange, J.; Langkjaer, L.; Havelund, S. and Volund, A. *Pharm. Res.,* **1992**, *9*, 715.
111   Pearlman, R. and Bewley, T. A. In *Stability and Characterization of Peptide and Protein Drugs: Case Histories.* Y. C. Wang and R. Pearlman (Eds.) Plenum Press, NY (1993).
112   Motto, M. G.; Hamburg, P. F.; Graden, D. A.; Shaw, C. J. and Cotter, M. L. *J. Pharm. Sci.*, **1991**, *80*, 419.
113   Oyler, A. R.; Naldi, R. E.; Lloyd, J. R.; Graden, D. A.; Shaw, C. J. and Cotter, M. L. *J. Pharm. Sci.*, **1991**, *80*, 271.
114   Johnson, D. M.; Pritchard, R. A.; Taylor, W. F.; Conley, D.; Zuniga, G. and McGreevy, K. G. *Int. J. Pharm.*, **1986**, *31*, 125.
115   Winterer, J.; Chatterji, D.; Comitem, F.; Decker, M. H.; Loriaux, D. L.; Gallelli, J. F. and Cutler, G. B. *Contraception*, **1983**, *27*, 195.
116   Powell, M. F.; Sanders, L. M.; Rogerson, A. and Si, V. *Pharm. Res.*, **1991**, *8*, 1258.
117   Tokumura, T.; Tanaka, T.; Sasaki, A.; Tsuchiya, Y.; Abe, K. and Machida, R. *Chem. Pharm. Bull.*, **1990**, *38*, 3094.
118   Powell, M. F.; Foster, L. C.; Becker , A. R. and Lee, W. *Pharm. Res.*, **1988**, *5*, 528.
119   Bolton, P. D. *Aust. J. Chem.*, **1966**, *19*, 1013.
120   Cippolla, D. C. and Shire, S. J. *Tech. Protein Chem.*, **1991**, *II*.
121   Bacarese-Hamilton, A. J.; Adrian, T. E.; Chohan, P.; Antony, T. and Bloom, S. R. *Peptides*, **1985**, *6*, 17.
122   Moroder, L.; Wilschowitz, L.; Gemeiner, M.; Goehring, W.; Knof, S.; Scharf, R.; Thamm, P.; Gardner, J. D.; Solomon, T. E. and Wunsch, E. Z. *Physiol. Chem.*, **1981**, *362*, 929.

123 Kenner, G. W.; Mendive, Y.J. and Sheppard, R. C. *J. Chem. Soc. C.*, **1968**, 761.
124 Wunsch, E.; Jaeger, E.; Deffner, M. and Scharf, R. *Z. Physiol. Chem.*, **1972**, *353*, 1716.
125 Wunsch, E.; Moroder, E.; Gillessen, D.; Soerensen, U. B. and Bali, J. P. *Z. Physiol., Chem.*, **1982**, *363*, 665.
126 Timmins, P.; Jackson, I. M. and Wang, Y. J. *Int. J. Pharm.*, **1982**, *11*, 329.
127 van den Oetelaar, P. J. M.; Jansen, P. S. L.; Melgers, P. A. T. A.; Wagenaars, G. N. and ten Kortenaar, P. B. W. *J. Control. Rel.*, **1992**, *21*, 11.

RECEIVED April 19, 1994

LYOPHILIZATION

Chapter 8

Freeze-Drying of Proteins

Process, Formulation, and Stability

Michael J. Pikal

Lilly Research Laboratories, Eli Lilly and Company, Indianapolis, IN 46285

Although freeze drying is often the process of choice for production of therapeutic proteins, protein degradation may occur in-process and/or during storage. This review begins with a presentation of physical and engineering principles critical to process control and concludes with a discussion of the relationship between formulation and stability. In-process degradation is minimized by rapid freezing and drying below the relevant glass transition temperature to the optimum residual moisture, which is normally less than 1%. Formulation has a major impact on both process design and stability, but "mechanisms" of stability enhancement are controversial. It is tentatively concluded that: (a) stabilization during freezing is largely due to the increase in thermodynamic stability of the protein caused by solutes that are "excluded" from the protein/aqueous surface; (b) stability of the dried product is enhanced by immobilization in a glassy (chemically inert) matrix; and (c) stability enhancement during drying may involve aspects of both the immobilization and the "water substitute" mechanisms.

Freeze drying, also termed lyophilization, is a desiccation process where the solvent (usually water) is first frozen and then is removed by sublimation in a "vacuum" environment *(1)*. Typically, an aqueous solution of the protein is filled into glass vials, and the vials are loaded onto temperature controlled shelves in a large drying chamber. The shelf temperature is lowered, and the solution is frozen. Next, the system is evacuated by vacuum pumps, and the shelf temperature is increased to supply heat for sublimation of ice. Chamber pressure is normally controlled by a regulated nitrogen leak into the drying chamber. The stage of the process where ice is sublimed is termed "primary drying". After all the ice has been removed, the product typically contains large amounts of water "dissolved" in the amorphous solute phase. This "dissolved" water, conventionally denoted "unfrozen water", is removed by desorption during the "secondary drying" stage of the process, usually by using elevated product temperature.

Since freeze drying is a low temperature process, relative to spray drying, it is normally considered less destructive to protein products. However, proteins may suffer loss of activity during freeze drying, and a freeze dried product may not have adequate storage stability. Careful attention to process details and proper selection of formulation components are often required to minimize stability problems.

0097–6156/94/0567–0120$08.00/0

The Freeze Drying Process

Freezing.   The general objective in the freezing process is to totally convert the solution to a solid so that most water is removed by sublimation. Thus, the product is left in the form of a porous "cake" that has good rehydration properties and acceptable appearance. Most of the water is converted to ice, and all solutes are converted to either crystalline solids or amorphous solids below their glass transition temperature. For a solute that crystallizes, the system temperature must be reduced well below the eutectic temperature to insure crystallization of the eutectic mixture, as eutectic systems often supercool. Although some formulation components, such as NaCl and buffer salts, may crystallize, proteins do not crystallize during freezing. Rather, the protein remains amorphous and is transformed into a rigid solid (i.e., a glass) at the glass transition temperature of the <u>maximally concentrated solute</u>, T_g' *(2)*. Thus, the temperature of an amorphous system must be reduced below its T_g' during freezing to insure a "solid" system. Note that while a glass transition temperature is denoted, T_g, the notation becomes, T_g' when the system is the "maximally concentrated solute". Eutectic temperature and T_g' are conventionally measured by using differential scanning calorimetry (DSC). A eutectic melt is detected as a small endotherm while a glass transition is observed as an abrupt change in the baseline, reflecting an increase in heat capacity as the system passes from a glass to a fluid. Eutectic temperatures close to the equilibrium freezing point usually are obscured by the ice melt endotherm. Also, in some systems, particularly in pure proteins, T_g' cannot be detected by DSC. The lack of an observable glass transition may simply mean that the glass transition is obscured by the ice melt endotherm. However, it is also possible that the glass transition does not produce a significant change in heat capacity over a narrow temperature range. Freeze drying microscopy can help decide between the alternatives. Microscopic observation of a sample during primary drying may be used to determine the collapse temperature, T_c, a parameter closely related to T_g'*(3)*. The collapse temperature is the temperature above which the "dried" region adjacent to the ice:vapor interface attains sufficient fluidity to flow during the course of the experiment and destroy the cake structure. Microscopic observation is employed to note the temperature at which the cake structure is lost. Since "attainment of fluidity" corresponds to attainment of a temperature above T_g', the collapse temperature and T_g' are closely related. Generally, the collapse temperature is several degrees higher than T_g'*(3)*.

In principle, the "degree of supercooling" and the "ice growth rate", are independent freezing variables. However, freezing methodology that produces rapid growth often also produces a high degree of supercooling. Since the concentrated aqueous system formed during freezing is potentially a hostile environment for a protein, the time between onset of ice formation and completion of freezing (i.e., reaching T_g') should be minimized. This requirement means a rapid ice growth rate. On the other hand, a very high degree of supercooling is generally undesirable. The degree of supercooling determines the number and size of ice crystals and therefore determines both the surface area of ice crystals and the pore size of the dried cake. A very high degree of supercooling will lead to a high surface area of the ice:solution interface and very small pores. Small pores means slower primary drying, and some evidence suggests that a high surface area of ice may accentuate protein aggregation *(4)*. Conversely, a very low degree of supercooling may produce both compositional heterogeneities and extended secondary drying times, the latter arising from low specific surface area. Thus, the ideal freezing methodology produces moderate supercooling and rapid ice growth.

As water is converted into ice, all solutes are concentrated. Thus, while a normal saline solution (0.9% NaCl) may be optimal for protein stability, the formulator must realize that "freeze concentration" will produce a NaCl solution of about 5 molar before eutectic crystallization removes NaCl from the protein phase *(5)*. Buffer salts are also concentrated and may crystallize, producing significant shifts in pH during freezing *(5)*. Crystallization of disodium phosphate monohydrate during freezing will shift the pH from around 7 to as low as 3.5. However, depending on cooling rate and initial buffer concentration, crystallization of a given buffer component may be incomplete or may not occur at all *(6)*. We find that buffer crystallization normally does not occur in formulations in which the buffer is a minor component relative to the other solutes.

Proteins and typical lyoprotectants are also concentrated during freezing but do not crystallize. Freeze concentration of a non-crystallizable component is illustrated by Figure 1 where product temperature and percentage of "unfrozen water" are shown as a function of time during a typical freezing process. After supercooling by about 10°C (at 10 minutes), ice nucleates and freezing begins. A small amount of ice immediately forms, and the percentage of unfrozen water sharply decreases, liberating heat and driving the product temperature up to about -5°C. As additional heat is removed, more water crystallizes until the glass transition temperature of the maximally concentrated solute, T_g', is reached. At temperatures below T_g'(-24°C), the system becomes a rigid glass, and further crystallization of ice is minimal. Note that the freeze concentrated glass is about 24% water.

Primary Drying. Primary drying is normally the longest stage of the freeze drying process. Here, the objective of process design is to minimize process time yet maintain product quality. Since the time required for primary drying decreases by about a factor of 2 for a 5°C increase in product temperature, the product temperature should be maintained as high as possible. However, to avoid collapse and possible product degradation, the product temperature needs to be carefully controlled a safe margin (2-5°C) below the collapse temperature, or in the case of very unstable proteins, maintained a safe margin below the T_g'. Thus, product temperature control is critical. Note that, due to self-cooling, the product remains much colder than the shelf during primary drying.

Except for short periods of time following a large change in shelf temperature or chamber pressure, product temperature is relatively constant, and primary drying proceeds in a pseudo "steady state" process where the rate of heat removal by sublimation, $\Delta H_s \cdot (dm/dt)$, is balanced by the rate of heat supplied by the shelves, dQ/dt. Here, ΔH_s is the heat of sublimation, 660 cal/g, and dm/dt is the sublimation rate in g/sec. Thus, there exists a coupling between heat and mass transfer given by the equation,

$$dQ/dt = \Delta H_s \cdot (dm/dt) \tag{1}$$

where the sublimation rate, dm/dt is given by,

$$dm/dt = (P_0 - P_c)/(R_p + R_s), \tag{2}$$

and the rate of heat transfer from the shelves to the product is,

$$dQ/dt = A_v K_v (T_s - T_p), \tag{3}$$

where A_v is the vial cross-sectional area. The "driving force for sublimation", (P_0-P_c), is the difference between the vapor pressure of ice in the frozen product, P_0, and the total pressure in the drying chamber, P_c (mostly water vapor)*(7)*. Since the vapor pressure of ice increases exponentially with temperature, the driving force, and therefore, the sublimation rate, increases sharply with an increase in product temperature. The mass transfer barriers are the resistance of the product, R_p, and the resistance of the stopper in the semi-stoppered position, R_s (i.e., allowing small gaps for escape of water vapor). As long as the desired chamber pressure can be maintained, condenser efficiency has no impact on the process. The product resistance is normally more than 80% of the total resistance. The product resistance depends on the nature of the product, is inversely proportional to the cross sectional area of the frozen product, and increases with the thickness of the dried cake. Heat input from the shelf depends on the temperature difference between the shelf surface, T_s, and the product temperature, T_p, through the vial heat transfer coefficient, K_v. Since much of the heat is conducted via collisions between gas molecules and the two surfaces (i.e., shelf surface and vial bottom), K_v increases with increasing chamber pressure. Combining Equations 1-3 yields,

$$\Delta H_s \left\{ (P_0 - P_c)/(R_p + R_s) \right\} = A_v K_v (T_s - T_p) \qquad (4)$$

Since the vapor pressure of ice is a known function of temperature, specification of the process control variables, P_c and T_s, then allows Equation 4 to be solved for the remaining variable, the product temperature. The resistances and the vial heat transfer coefficient must also be known. Thus, while the product temperature does depend on the process control variables, product temperature is not uniquely defined unless the nature of the product and the nature of the container system (i.e., vial and stopper) are also defined.

The process control objective is to maintain the product temperature constant at a target temperature close to but safely below the collapse temperature throughout all of primary drying. Since the dried layer increases in thickness as primary drying proceeds, the product resistance, R_p, increases with time, thereby decreasing the ability of the product to "self-cool" by sublimation. Thus, to avoid an increase in product temperature, heat input from the shelves must decrease with time. Heat input from the shelf is decreased by a decrease in shelf temperature and/or a decrease in chamber pressure (i.e., a decrease in K_v). Mathematically, Equation 4 states that when R_p increases as drying proceeds, either P_0 increases (i.e., product temperature, T_p, increases) or the right hand side of Equation 4, $K_v(T_s - T_p)$, must decrease. Note that a decrease in chamber pressure decreases K_v, and a decrease of shelf temperature decreases $(T_s - T_p)$. After freezing, the shelf temperature should be increased quickly to bring the product temperature to the "target" within a few hours. Thereafter, one or two step-wise reductions in shelf temperature are sufficient to maintain approximate constant product temperature. In cases where a few degrees rise in product temperature is not a problem, the shelf temperature may often be held constant for the duration of primary drying. However, the practice of gradually increasing the shelf temperature throughout primary drying is without scientific foundation. This procedure produces little drying during the first half of the process and risks collapse of the cake during the final stages of primary drying because the highest heat input occurs at the time when the cooling rate provided by sublimation is lowest.

Secondary Drying. Since secondary drying is often very slow at low product temperature, the shelf temperature is normally increased significantly after all ice has been removed to increase the product temperature to about 25°C-35°C. Clearly, to avoid collapse, the product temperature must not be increased before all ice (and some unfrozen water) has been removed. While product temperature monitoring has traditionally been used to determine the end of primary drying, techniques which monitor the composition of the gas in the drying chamber provide a far better alternative *(8)*. These methods depend upon the observation that as long as water evolution is rapid (i.e., ice and/or very high levels of unfrozen water remain in the product), the gas in the drying chamber is mostly water. However, when it is "safe" to increase the shelf temperature for secondary drying, the composition of gas in the chamber shifts sharply to mostly nitrogen. For example, in a typical process where chamber pressure is controlled at 0.15 Torr, the partial pressure of water is 0.15 Torr during primary drying and roughly 0.02 Torr during secondary drying.

During secondary drying, water is removed from the amorphous freeze concentrate. Drying proceeds rapidly during the initial part of secondary drying when the concentration of water is high, but slows greatly as drying proceeds. The decreasing drying rate is <u>not</u> a reflection of water "strongly bound" thermodynamically, but rather is a direct result of kinetics; motion in the solid and evaporation at the solid surface become very slow as the freeze concentrate is dried*(9)*. While one would expect the drying rate to decrease as the concentration of water in the solid decreases, the plausible assumption that drying rate is proportional to residual water concentration is <u>not</u> valid. The drying rate decreases much faster than such a "first order" rate model would predict *(9)*. Achieving low levels of residual water may require the use of high temperatures. However, to avoid loss of product quality, the product temperature should not exceed its glass transition temperature during secondary drying. Since drying sharply increases the glass transition temperature, maintaining the product in a glassy state is normally not a serious problem.

Secondary drying is allowed to proceed until a target moisture content is reached. Clearly, the optimum moisture content must be established by stability studies. If a protein is subject to damage by overdrying, it must be recognized that the moisture content is not necessarily uniform within the vial (Figure 2). As expected, the residual moisture near the top of the dried cake is somewhat lower than at the bottom, but the driest section of the product is that near the walls of the container.

Prevention of overdrying is difficult unless the freeze dryer can provide controlled condenser temperatures as high as about - 35°C. With this capability, a combination of low product temperature and high condenser temperature can provide a high relative humidity throughout the process, thereby preventing overdrying in <u>any</u> of the product *(10)*. Alternately, use of low product temperature and abbreviated drying times may limit the degree of overdrying to tolerable levels *(9)*.

Experimentally, it is found that the rate of secondary drying is independent of chamber pressure in the pressure ranges normally employed in freeze drying*(9)*. Since the rate determining step in secondary drying involves diffusion in the solid and/or evaporation at the solid/vapor interface, this result is expected. Since drying at very low pressures can lead to problems, secondary drying should be carried out at moderate pressure (i.e., $\approx$ 0.2 Torr) *(9)*.

Residual Moisture and Storage Stability

While it is a common assumption that proteins exhibit optimum stability at an "intermediate" water content, experimental evidence for this "damage by overdrying" is limited. Influenza virus is clearly less stable when the moisture content deviates from the optimum 1.7% *(11)*, but the extrapolation of this observation to protein formulations is uncertain. Aggregation of excipient-free tissue plasminogen activator

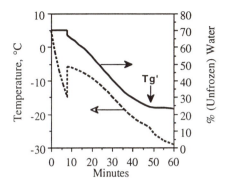

Figure 1. Freeze Concentration in an Amorphous System: Moxalactam Disodium Freezing in Glass Vials.Shelf temperature of -30°C. The product temperature is shown as the broken line while the percentage of unfrozen H_2O is given as a solid line.

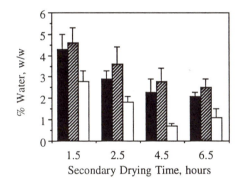

Figure 2. Intravial Distribution of Moisture During Secondary Drying: Human Serum Albumin (HSA) Drying at 13°C. The solution was 5% (w/w) HSA at a fill depth of 1 cm in 10 cc tubing vials. Primary drying was carried out at a shelf temperature of 13°C and a chamber pressure of 0.10 Torr (13.3 Pa) giving a product temperature of -35°C to -32°C during primary drying. A Virtis SRC-25X freeze dryer equipped with a sample thief was used. A core sample of the freeze dried cake was taken and sectioned into two parts, "top" and "bottom", where bottom means the section closest to the vial bottom. The portion of the cake next to the vial wall remained in the vial and is denoted "outer" section. Key: solid bar = "top", shaded bar = "bottom", open bar = "outer". *(M. Pikal and S. Shah, unpublished observations.)*

(tPA) at high temperature is faster at low water content *(12)*, and aggregation of excipient-free hGH at 25°C is faster for material that had been highly dried*(10)*. However, numerous experimental observations demonstrate that storage stability normally improves as the residual water content decreases. This generalization is valid for both small molecules and a wide range of protein systems. For example, the rate of hemoglobin oxidation at room temperature doubles as the residual water content increases from 1% to 4%*(13)*, the rate of HSA aggregation increases linearly with increasing moisture content*(14)*, and the rate of chemical degradation in formulated hGH at 25°C increases by nearly a factor of 10, in highly non-linear fashion, as the moisture content varies from about 1% to 2.5%*(10,15)*. While the optimum moisture content for stability is normally zero, the stability gained by decreasing the residual water content much below 1% is often minimal. Thus, a "practical" definition of "dry" is often "less than 1% water".

The detrimental effect of moisture on storage stability is often interpreted in terms of mobility in the solid and therefore, reactivity, of the protein. One concept states *(16)* that above monolayer levels of water, the protein has increased conformational flexibility, and the additional water has the ability to mobilize potential reactants in the amorphous phase, both effects increasing the rate of protein degradation. Alternately, since water plasticizes the amorphous phase, thereby lowering its glass transition temperature, T_g, a protein at high water content could well be above its glass transition temperature during storage. Therefore, the protein would exist in a liquid state rather than molecularly dispersed in an inert glassy matrix. The increased molecular motion characteristic of a liquid would be expected to facilitate most degradation reactions*(2)*. The "monolayer" argument and the "glass transition" concept are similar but are not identical. Both attribute increased reactivity by addition of water to increased molecular motion, but a system above monolayer coverage is not necessarily above the glass transition temperature. Further, while "monolayer" coverage is essentially independent of temperature, the difference between a sample's temperature and its glass transition temperature obviously depends upon both water content and temperature. Thus, using the "monolayer" argument, one would predict stability trends with increasing water content that are qualitatively independent of temperature. Conversely, the "glass transition" concept predicts that stability trends with increasing moisture depend greatly upon sample temperature. Here, the key variable is "T_g -T".

Formulation and Stability

Formulation details have a major impact on both in-process and storage stability. Commonly, formulation components termed "lyoprotectants" are added to maintain stability during freeze drying. Lyoprotectants must maintain stability during both freezing and drying. Lyoprotectants may also enhance stability during storage.

A number of excipients stabilize proteins extremely well during freezing and thawing (i.e, are "cryoprotectants") but provide little or no protection during the drying stage(s) of freeze drying. For example, proline and trimethylamine N-oxide are effective "cryoprotectants" for phosphofructokinase (PFK), preserving 65% of initial activity, but allow almost total loss of activity during freeze drying *(17)*. Conversely, several disaccharides preserve 65% of the initial activity during both freeze/thaw and freeze drying. Glucose is relatively ineffective. Pure catalase retains 80% of its initial activity during freeze/thaw, but is only 30% active after freeze drying. The addition of glucose to the catalase solution allows retention of nearly 90% of initial activity during both freeze/thawing and freeze drying. Mannitol, maltose, and maltotriose also preserve about 90% of the initial activity during freeze drying, and maltohexose preserves about 80% of initial activity *(18)*. There is no correlation between lyoprotection of catalase and glass forming characteristics of the pure lyoprotectant.

Pure mannitol crystallizes and does not form glasses during normal freeze drying while the glass transition temperatures, T_g, of glucose, maltose, maltotriose, and maltohexose are, respectively, 39°C, 95°C, 135°C and 175°C. However, it should be noted that both the tendency to crystallize and the glass transition temperature may be significantly altered by the presence of the protein and buffers. Glass transition temperatures of the formulation are more relevant to stability, but unfortunately such data are usually not available.

Lyoprotectant-free L-asparaginase retains activity after a freeze/thaw cycle, but loses about 80% of its initial activity during freeze drying *(19)*. Glucose, tetramethylglucose[TMG], mannose, sucrose, and poly(vinylpyrrolidone)[PVP] are all extremely effective lyoprotectants, preserving 100% of the initial activity. Mannitol preserves only about 50% of the initial activity. Stabilization does not correlate with residual moisture *(19)*. Here, as with catalase, a monosaccharide is as effective as a disaccharide. Further, the effectiveness of TMG and PVP demonstrate that "disaccharide-type" hydrogen-bonding to the protein is not essential for stability.

Typically, therapeutic protein candidates may be freeze dried with little or no loss in activity. Instability of the dried product during storage is the more common problem. Human growth hormone, hGH, is an example*(15)*. Dimer formation in hGH during storage is rapid and formulation dependent (Figure 3). The "initial" level of dimer refers to the dimer level in the formulation immediately after freeze drying, which for formulation D is also the dimer level in the starting solution. Note that excipient free hGH freeze dries without significant increase in dimer content, but dimer formation occurs rapidly upon storage at 25°C. Formulation D, the glycine:mannitol combination, is clearly the most stable formulation. The greater stability relative to either the "mannitol only" or "glycine only" formulations is attributed to the fact that the glycine remains amorphous in the combination formulation*(15)*. Thus, the excipient is able to dilute and immobilize *(2)* both protein and reactants in the solid state, thereby decreasing the rate of bimolecular reactions (i.e., diffusion in a solid is slow) and allowing interaction with the protein, perhaps promoting stability via this interaction. A simple physical mixture of the protein phase with a crystalline excipient phase would not be expected to alter stability. While the comparison of formulations E and F with formulation A suggests that some of the excipient remained amorphous in these formulations, formulation D, with all the glycine remaining amorphous, would be expected to provide a greater amount of amorphous phase in which the protein is "dissolved", thereby providing greater stability. However, the dextran formulation is 100% amorphous, but is quite unstable. Storage stability is essentially the same as that observed with formulation A (no excipient), and "in-process" dimer formation actually appears greater than with formulation A. Thus, the requirement that the stabilizer must remain amorphous can only be a necessary condition for stabilization, not a sufficient condition. The preceding discussion has implicitly assumed that the amorphous system is a single amorphous phase. However, just as liquid systems may exhibit limited miscibility, a two component glassy system may also form two distinct phases. The behavior of the dextran formulation could be a result of phase separation.

Mechanisms of Stabilization

Stabilization During Freezing. During freezing, temperature is reduced, ice forms, all solutes are "freeze concentrated", and buffer salts may crystallize producing large pH shifts. These are "freezing stresses" that may induce irreversible changes in the protein, typically conformational changes, that lead to inactivation.Throughout most of the freezing process, the physical state of the protein is that of an aqueous solution, even though the solution may be far different in composition and concentration than the starting solution. Near the end of freezing, freeze concentration proceeds to a point where the environment could not reasonably be considered "aqueous" and finally forms a solute rich glass.

Since the protein environment is "aqueous" during much of the freezing process, it is perhaps understandable that solutes that stabilize the native conformation in aqueous solutions at more "normal" concentrations and temperatures frequently also stabilize during freezing, and therefore are effective "cryoprotectants". Carbohydrates and some amino acids are examples. Such solutes tend to be "excluded" from the surface of the protein[20]. The thermodynamics of this phenomenon are analogous to solute "surface excess" at the air/water interface. A negative surface excess means the solute is partly excluded from the interface, thereby increasing the surface tension of the solution. In fact, there is a good correlation between those solutes that increase the surface tension of water and those that are "excluded" from the surface of a protein [20].

The thermodynamic consequence of "solute exclusion" is to increase the chemical potential of the protein. The first basic assumption in relating this thermodynamic result to cryoprotection may be stated as: if the increase in chemical potential of the native protein caused by "solute exclusion" is denoted, $\Delta\mu_N$, the corresponding increase for the unfolded form is $k \cdot \Delta\mu_N$, where $k>1$. Thus, the free energy of unfolding would be increased by the solute, which means the native conformation is stabilized [20]. This explanation is plausible since the unfolded protein exposes more "protein surface" to the aqueous environment. Secondly, it is assumed that thermodynamic stability, i.e., free energy of unfolding, is related to degradation. Of course, they are not related unless degradation (i.e., irreversible change), by whatever mechanism, is much faster in the unfolded state than in the native state. Finally, it is assumed that the solution state concepts outlined above are valid for the freeze concentrate generated during freezing. This assumption is plausible early in the freezing stage where the freeze concentrate is still a relatively dilute aqueous solution, but is questionable as the system approaches a highly concentrated glass. While these assumptions are not trivial, a large body of experimental data generally support the "excluded solute" concept as a significant, if not dominant, factor during freezing [20].

Stabilization During Drying and Storage. Provided freeze drying is performed properly, the physical state of the protein during both drying and storage is that of a rigid glass containing some water, the only difference being the exact water content of the glassy phase. From this perspective, one could postulate that the fundamental mechanisms governing stability during drying also govern stability during storage of the dried product. However, instability during drying develops on a very short time scale. Further, while it seems likely that storage stability is generally a pure kinetic phenomena, stability during drying may involve thermodynamic control. That is, solutes that enhance storage stability may simply slow the rate of degradation by imposing a high activation barrier on a reaction that remains thermodynamically spontaneous (i.e., $\Delta G < 0$ for degradation, but the rate constant is very small). During drying, the role of the stabilizer may be to shift the equilibrium between "stable" and "reactive" forms toward the "stable" form, thereby "thermodynamically" minimizing degradation (i.e., make $\Delta G > 0$ for significant denaturation during drying, thereby preventing formation of significant amounts of the more reactive denatured form). There are two main hypotheses advanced to rationalize the role of solutes in stabilizing proteins during drying and storage. The "vitrification hypothesis" is a purely kinetic argument [2,21] that applies equally well to both drying and storage stability, and the "water substitute hypothesis", as normally presented, is essentially a thermodynamic argument [22] applied mostly to drying stability but occasionally used to rationalize storage stability as well.

The "water substitute" hypothesis [23] states that good stabilizers interact with the protein as does water, promote the native conformation, and therefore stabilize during drying by replacing the water that is removed. In support of this concept, it is noted that most stabilizers are sugars which "resemble" hydrogen bonded water aggregates, and therefore can hydrogen bond to the protein as does water. Additional support comes from spectroscopic evidence *(24)*. FTIR studies suggest an interaction between sugars and proteins which is similar to the interaction between water and the protein. Further, whereas proteins freeze dried from sugar formulations yield a solid with a FTIR spectrum characteristic of the native conformation, freeze drying that produces significant loss of activity yields a solid with a highly altered IR spectrum. However, it should also be noted that not all additives that stabilize a protein during drying and/or during storage are water "look-alikes". For example, L-asparaginase is inactivated during the drying stage, and while carbohydrates (i.e., water "look-alikes") do stabilize, PVP and TMG stabilize equally well, yet neither can be considered a "water substitute".

A strict interpretation of the "water substitute" hypothesis cannot be used to rationalize storage stability. First, if a "water substitute" is required for stability, why is not water itself a stabilizing factor? Indeed, if water is required for stability, why freeze dry at all? In addition, the generality of the assumption that "native means stable" is open to question, particularly for chemical degradation reactions. As an example of instability in the "native" state, we consider the case of hydrated crystalline zinc insulin. A protein in a highly hydrated crystalline state would certainly be regarded as closer to the native conformation than the protein in a dry amorphous solid. Yet, crystalline insulin is much <u>less</u> stable than amorphous insulin and degradation in the crystal <u>increases</u> with increasing hydration (Figure 4). This generalization is valid for both deamidation and covalent dimer formation. Figure 4 compares deamidation rates of Asn(A-21) in crystalline and amorphous insulin at various moisture levels. Here, a batch of crystalline insulin (bulk crystals) was suspended and titrated to pH 7.1. The crystals were separated from the supernatant, washed, and dried, yielding the "crystalline pH 7.1" batch. The supernatant containing dissolved insulin was then freeze dried to give the "amorphous, pH 7.1" batch. The amorphous solid is much more stable than the crystalline forms, particularly at high moisture content.

The "vitrification hypothesis" *(2, 21)* states that stabilizers are good glass formers, but are otherwise inert. The rationale for this view is based upon the fact that "glassy" means extremely high viscosity and the <u>assumption</u> that all motion relevant to stability correlates with the reciprocal of viscosity. Thus, motion that allows a degradation process essentially ceases below the glass transition temperature, and "glassy" is equivalent to stable. It is well known *(2,21)* that as a liquid approaches its glass transition temperature, the temperature dependence of the viscosity obeys the Williams-Landel-Ferry (WLF) relationship rather than the Arrhenius equation. The assumption that all relevant motion correlates with viscosity predicts a stability enhancement of about 6 orders of magnitude as a system moves from 30°C above the T_g to near the T_g *(21)*. Arrhenius behavior, with an activation energy of 25 kcal/mol, would predict a stability enhancement of only 1.5 orders of magnitude. Support for the "vitrification hypothesis" comes from the observation that excipients that stabilize during drying and/or storage are generally amorphous and good glass formers. Further, several studies show a sharp decreases in mobility *(25)* and reactivity*(26,27)* as a glass transition is approached. While many observations are at least qualitatively consistent with the vitrification hypothesis, apparent exceptions do exist. Drying stability of catalase does not correlate with glass transition temperatures of the excipients, and storage stability of hGH is not enhanced by a dextran formulation, even though this

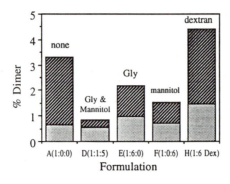

Figure 3. Storage Stability of Freeze Dried hGH at 25°C: Effect of Formulation. The formulations are: A(no excipient); D (hGH:glycine: mannitol, in a weight ratio of 1:1:5); E (hGH:glycine in a 1:6 weight ratio); F (hGH:mannitol in a 1:6 weight ratio); H (hGH:dextran40 in a 1:6 weight ratio). Mannitol is mostly crystalline, and glycine is mostly crystalline(E), or amorphous(D). Formulations A and H are 100% amorphous. All samples were freeze dried in the same experiment with product temperatures between -30°C and -35°C during primary drying and 30°C in secondary drying. Residual moisture contents ($gH_2O/ghGH$) are 0.098(A), 0.065(D), 0.079(E), 0.050(F), and 0.061(H). Key: light shading = initial assay (after freeze drying), dark shading = increase in % dimer after storage for 1 month at 25°C.

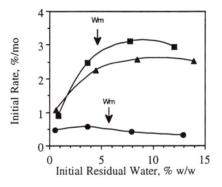

Figure 4. Asn(A-21) Deamidation of Crystalline and Amorphous Insulin. Wm is the % water content at monolayer coverage as measured by a BET analysis of water sorption isotherms.Key: circles = freeze dried amorphous at pH 7.1; triangles = crystals at pH 7.1; squares = "bulk" crystals (crystallized near the isoelectric point). The initial rate, in % Asn(A-21) deamidation per month, is calculated from the limiting slope of a quadratic fit of the time dependence of Asn(A-21) deamidated product. *D. Rigsbee and M. J. Pikal, unpublished observations*

system shows a high glass transition temperature *(15)*. Further, the basic assumption that motion relevant to stability always couples with the glass transition is probably an oversimplification.

A number of studies demonstrate that not all motion is coupled to the glass transition. As one example, Figure 5 shows "mobilities" for three different kinds of motion in liquid crystal polymers that exhibit a glass transition at 35°C. The "mobilities" are derived from D-NMR correlation times. Complete coupling of mobility with the glass transition would imply more than 9 orders of magnitude reduction in mobility between $1000/T = 2.7$ and the glass transition temperature, assuming validity of the WLF relationship with "universal constants" *(21)*. While chain fluctuation mobility appears to be strongly coupled to the glass transition, chain rotation is only moderately coupled, and ring flip motion is nearly completely decoupled. Ring flip "mobility" shows Arrhenius behavior above and below the glass transition with essentially the same activation energy. The only impact of the glass transition on ring flip is a sharp but small decrease in mobility as the temperature decreases through T_g. It appears that mobility involving motion on a larger scale (i.e., more displacement and/or more free volume generation) correlates best with viscosity. Consistent with this conclusion, diffusion of small molecules in polymers also seems to be largely decoupled from the glass transition. Diffusion of water in PVP is a relevant example *(29)*.

WLF kinetics coupled with the "vitrification hypothesis" predicts that stability correlates with the difference between storage temperature and the glass transition temperature *(21,27)*. Thus, in an experiment where both temperature and glass transition temperature are varied, stability is a function of only one variable, $T-T_g$. While this prediction is not consistent with all available data, it is consistent with studies of both enzymatic hydrolysis and ascorbic acid oxidation in frozen maltodextrin systems*(26)*. Degradation in a solid "monoclonal antibody:vinca alkaloid conjugate" system is also qualitatively consistent with the WLF model for reaction rate*(27)*. Rates of dimerization, hydrolysis of the antibody-vinca linkage, and degradation of the vinca moiety (mostly oxidation) were obtained at two temperatures and three moisture contents (i.e., three T_g's). The rate normalized to the rate at the glass transition temperature is a function only of $(T -T_g)$ (Figure 6), as predicted by the WLF theory.

Note that all three degradation reactions form a single curve (straight line), indicating that the WLF constants are a characteristic of the solid system and independent of the nature of the degradation pathway, as suggested by WLF kinetics. It should be noted, however, that the use of the "universal" WLF constants *(21)* predicts a stronger dependence on $(T-T_g)$ than observed, giving $\ln(R/R_g) \approx 15$ at $T-T_g=30$°C.

Tentative Conclusions on Mechanisms. Freezing stability involves "solute exclusion", with perhaps some role of immobilization by vitrification as the freeze concentrate passes into the glassy state. However, stability during drying and storage are best explained by features of both the "water substitute" and the "vitrification" hypotheses. The data suggest that stabilization of a labile protein during drying and storage requires an excipient that: (1) is a chemically inert glass former, (2) forms a single phase with the protein, which requires a "moderate" interaction with the protein surface to resist phase separation but yet not denature the protein, (3) couples all relevant modes of motion to the glass transition so that high viscosity <u>does</u> mean stability. The need for enhancement of thermodynamic stability (i.e., via "water substitution") during drying is uncertain.

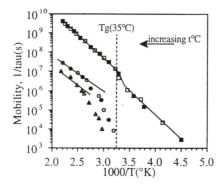

Figure 5. Coupling of Mobility with the Glass Transition in a Polymer System. Mobility is the reciprocal of the correlation time for the type of motion indicated. Correlation times were evaluated from deuteron NMR relaxation data. Key: triangles = reorientation or "fluctuation" of the chain axis; circles = rotation about the chain axis; squares = 180° ring flips of side chain aromatic rings. Open and closed symbols refer to different degrees of deuteration. *(adapted from ref. 28)*

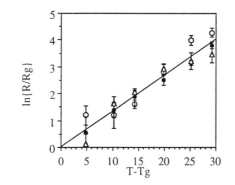

Figure 6. Stability of a Freeze Dried Monoclonal Antibody: Vinca Alkaloid Formulation. Desacetylvinblastine hydrazide is linked to the KS1/4 monoclonal antibody via aldehyde residues of the oxidized carbohydrate groups on the antibody. The formulation is conjugate: glycine:mannitol in a 1:1:1 weight ratio. Storage temperatures are 25°C and 40°C for samples with moisture contents of: 1.4%, 3.0%, and 4.7%. Key: open circles = dimer formation; filled circles = free vinca generation (hydrolysis); triangles = vinca degradation; solid line = best fit to the WLF equation. The WLF equation is: $\ln(R/R_g) = C_1 (T - T_g)/(C_2 + T - T_g)$, where R is the reaction rate, and R_g is the reaction rate at T_g. C_1 and C_2 are the "WLF constants". The solid line is for, $C_2 \gg (T - T_g)$, and $C_1/C_2 = 0.134$. *(reproduced with permission from ref. 27)*

Literature Cited

1. Pikal, M., in "Encyclopedia of Pharmaceutical Technology", Swarbrick, J.; Boylan, J., Eds, Marcel Dekker, New York, NY. 1992, Vol 6, pp 275-303.
2. Franks, F.; Hatley, R.; Mathias, S.; BioPharm **1991**,4, 38-42,55.
3. Pikal, M.; Shah, S.; Internatl. J. Pharm. **1990**, 62, 165-186.
4. Eckhardt, B.; Oeswein, J.; Bewley, T.; Pharm. Res. **1991**, 1360-1364.
5. Pikal, M.; BioPharm, **1990**, 3, 26-30.
6. Murase, N.; Franks, F.; Biophys. Chem. **1989**, 34, 293-300.
7. Pikal, M.; J. Parenter. Sci. Technol.**1985**, 39, 115-138.
8. Roy, M.; Pikal, M, J. Parenter. Sci. Technol., **1989**, 43, 60-66.
9. Pikal, M.; Shah, S.; Roy, M.; Putman, R.; Internatl. J. Pharm. **1990**, 60, 203-217.
10. Pikal, M.; Dellerman, K.; Roy, M.; Develop Biol. Standard, **1991**, 74, 21-38.
11. Greiff, D.; Cryobiology **1971**, 8, 145-152.
12. Hsu, C.; Ward, C.; Pearlman, R.; Nguyen, H.; Yeung, D.; Curley, J.; Develop Biol. Standard **1991**, 74, 255-271.
13. Pristoupil, T.; Kramlova, M.; Fortova, H.; Ulrych, S.; Haematologia **1985**, 18, 45-52.
14. Moreira, T.; Cabrera, L.; Gutierrez, A.; Cadiz, A.; Castellano, M.; Acta Pharm. Nord. **1992**, 4, 59-60.
15. Pikal, M.; Dellerman, K.; Roy, M.; Riggin, R.; Pharm. Res. **1991**, 8, 427-436.
16. Hageman, M.; Drug Dev. Ind. Pharm. **1988**,14, 2047-2070.
17. Carpenter, J.; Crowe, J.; Arakawa, T.; J. Dairy Sci. **1990**, 73, 3627-3636.
18. Tanaka, R; Takeda, T.; Miyajima, K.; Chem. Pharm. Bull. **1991**, 1091-1094.
19. Hellman, K.; Miller, D.; Cammack, K.; Biochim. Biophys. Acta **1983**, 749, 133-142.
20. Arakawa, T.; Kita, Y.; Carpenter, J.; Pharm. Res. **1991**, 8, 285-291.
21. Slade, L.; Levine, H.; Crit. Rev. Food Sci. and Nutrit. **1991**, 30, 115-359.
22. Prestrelski, S.; Arakawa, T.; Carpenter, J.; Arch. Biochem. Biophys. **1993**, 303, 465-473.
23. Crowe, J.; Crowe, L.; Carpenter, J.; BioPharm **1993**, 6, 40-43.
24. Prestrelski, S.; Tedeschi, N.; Arakawa, T.; Carpenter, J.; Biophys. J. **1993**, 65, 661-671.
25. Roozen, M.; Hemminga, M.; Walstra, P.; Carbohydrate Res. **1991**, 215, 229-237.
26. Lim, M.; Reid, D.; In "Water Relationships in Food", Levine, H.; Slade, L.,Eds, Plenum Press, New York, N.Y., 1991, 103-122.
27. Roy, M.; Pikal, M.; Rickard, E.; Maloney, A.; Develop Biol. Standard, **1991**, 74, 323-340.
28. Kohlhammer, K.; Kothe, G.; Reck, B.; Ringsdorf, H.; Ber. Bunsenges. Phys. Chem. **1989**, 1323-1325.
29. Oksanen, C.; Zografi, G.; Pharm. Res. **1992**, 9, Abstract PT6142, S-150.

RECEIVED April 19, 1994

Chapter 9

Interactions of Stabilizers with Proteins During Freezing and Drying

John F. Carpenter[1], Steven J. Prestrelski[2], Thomas J. Anchordoguy[1], and Tsutomu Arakawa[2]

[1]School of Pharmacy, University of Colorado Health Sciences Center, Denver, CO 80262
[2]Amgen, Inc., Amgen Center, Thousand Oaks, CA 91320

During the course of processing, shipping and storage, proteins are often subjected to either freezing or drying stress, or both. The preservation of proteins during freezing and drying is fundamentally different for each process. Protection of labile proteins during freezing appears to be due to the same thermodynamic mechanism that accounts for solute-induced stabilization in nonfrozen aqueous solution. Namely, preferential exclusion of the solute leads to stabilization of the native state of the protein. Protection of proteins during drying by sugars is due to the sugar hydrogen bonding to the dried protein and serving as a water replacement. Recovery of enzyme activity after freeze-drying and rehydration correlates directly with the maintenance of the native protein structure in the dried state. Sugars protect labile proteins by inhibiting lyophilization-induced unfolding. Finally, certain polymers (e.g., polyvinylpyrrolidone) protect freeze-dried multimeric enzymes during freeze-drying and rehydration. These compounds cannot serve as "water replacements" in the dried state. Rather they inhibit the freezing- and drying-induced dissociation. The fully polymerized oligomers are more resistant to these stresses.

With increasing numbers of applications for recombinant and naturally-derived proteins, the need for stable protein formulations is growing. When the inherent stability of the protein, and/or the logistics of product shipping and use, preclude storage as an aqueous solution, the protein is usually freeze-dried. Often, the focus during formulation development is on the terminal stress of dehydration and on the effects of additives on the stability of the protein in the dried solid. However, freezing stress also occurs during lyophilization, as well as at other points during processing, shipping and storage. Freezing can occur either by design or by accident. For example, a protein may be very stable in a liquid formulation at 4°C, but if this same protein accidently freezes during shipping, then the product can be

0097–6156/94/0567–0134$08.00/0

destroyed.   Planned freezing includes situations where the protein is stored frozen and thawed by the end-user, lyophilized proteins are rehydrated by the end-user and then aliquoted and frozen, and frozen storage is used as an interim measure for holding batches of proteins, prior to the lyophilization of a large lot.

Thus, for the development of successful lyophilized formulations, both freezing and drying stresses must be taken into consideration.   This review will focus on stress-specific preservation of proteins and the mechanisms by which additives protect the structure and function of labile proteins during freezing and drying.   That is, we are concerned primarily with proteins that are acutely sensitive to the freezing and/or drying stresses and that can be irreversibly damaged in the absence of a stabilizing additive.

Another related category of proteins are those that unfold during freeze-drying, but resume their native conformation upon rehydration (see chapter by Prestrelski et al., for further discussion of dehydration-induced structural changes).   Our view is that, in this case, stabilizing additives can be used to prevent the reversible denaturation and that the native protein may be more resistant to degradation during long-term storage of the dried product.   This suggestion is mostly speculative at this time, because only recently has this type of acute drying-induced structural transition been identified.   Finally, certain proteins are intrinsically resistant to the acute effects of lyophilization and retain their native conformation in the dried state. However, without the proper additives, these proteins may degrade chemically during long-term storage.   We will not address this level of stabilization since it will be reviewed rigorously by others in this volume.   However, we wish to emphasize, that for more labile proteins, the initial preservation of the native conformation in the dried formulation appears to be a fundamental requirement for subsequent long-term storage stability.

Mechanism for Protein Preservation During Freezing

A wide variety of compounds will protect labile proteins during freeze-thawing (reviewed in ref. *1-4*).   These include sugars, amino acids, polyols, methylamines, synthetic polymers (e.g., polyethylene glycol, PEG), other proteins (e.g., bovine serum albumin, BSA) and even inorganic salts (e.g., potassium phosphate and ammonium sulfate).   For most proteins, the cryoprotectant must be at a concentration of several-hundred millimolar to confer maximum protection. Exceptions are polymers such as PEG, BSA and polyvinylpyrrolidone (PVP), which at concentrations of even less than 1% (wt/vol) fully protect sensitive enzymes such as lactate dehydrogenase (LDH) or phosphofructokinase (PFK).

Based on the results of freeze-thawing experiments with LDH and PFK and a review of the literature on protein freezing, we have determined (*1*) that protein cryopreservation can be explained by the same universal mechanism that Timasheff and Arakawa have defined for solute-induced protein stabilization in nonfrozen, aqueous solution (*3-5*).   Detailed explanations of this mechanism can be found elsewhere (*3-5*).   For the purpose of the current review a brief summary will suffice.

Timasheff, Arakawa and their colleagues have observed experimentally that there is a deficiency of stabilizing solutes (e.g., sugars and polyols) in the immediate vicinity of the protein, relative to the bulk solution, and that the protein is preferentially hydrated. That is, the solutes are preferentially excluded from contact with the surface of the protein. The presence of the preferentially excluded solutes in a protein solution creates a thermodynamically unfavorable situation, because the chemical potentials of both the protein and the additive are increased. Consequently the native structure of monomers or the fully polymerized form of oligomers are stabilized because denaturation or dissociation, respectively, would lead to a greater surface area of contact between the protein and the solvent, and therefore, exacerbate this thermodynamically unfavorable effect.

An overview of the key thermodynamic aspects of this mechanism follows (reviewed in 3-5). Setting component 1 = principal solvent (here water), component 2 = protein, and component 3 = solute (e.g., sucrose or PEG), the preferential interaction of component 3 with a protein is expressed, within close approximation, by the parameter, $(\delta m_3/\delta m_2)_{\mu 1, \mu 3}$, at constant temperature and pressure, where μ_i and m_i are the chemical potential and molal concentration of component i, respectively. A positive value of this interaction parameter indicates an excess of component 3 in the vicinity of the protein over the bulk concentration (i.e., preferential binding of the solute). A negative value for this parameter indicates a deficiency of component 3 in the protein domain. Component 3 (the solute) is preferentially excluded and component 1 (water) is in excess in the protein domain.

The preferential interaction parameter is a direct expression of changes in the free energy of the system induced by component 3, and has the relation:

$$(\delta \mu_2/\delta m_3)_{m2} = - (\delta m_3/\delta m_2)_{\mu 1, \mu 3} (\delta \mu_3/\delta m_3)_{m2} \qquad (1)$$

Equation 1 indicates that those compounds that are excluded (i.e., $(\delta m_3/\delta m_2)_{\mu 1, \mu 3} < 0$) from the surface of the protein will have positive values of $(\delta \mu_2/\delta m_3)_{m2}$; they will increase the chemical potential of the protein, rendering the system more thermodynamically unfavorable. In the presence of excluded solutes, the exclusion will be greater for the denatured form of the protein than for the native form because the former has a larger surface area, as indicated by: $(\delta m_3/\delta m_2)^D < (\delta m_3/\delta m_2)^N < 0$. Consequently, the increase in chemical potential is greater for the denatured form than for the native form in the presence of a preferentially excluded solute, as indicated by: $(\delta \mu_2/\delta m_3)_{m2}^D > (\delta \mu_2/\delta m_3)_{m2}^N > 0$.

This effect of the solute leads to an increase in the free energy difference between the native and denatured forms, thus stabilizing the native state. Ultimately, it is this difference between the effects of the solute on chemical potential of the native versus the denatured state that determines if a compound will serve as a protein stabilizer. As long as the degree of exclusion is greater for the denatured state, the solute will be a protein stabilizer.

The opposite is seen for potent protein denaturants such urea and guanidine HCl.

These solutes bind preferentially to both the native and the denatured form of the protein (*3-5*) and hence decrease the chemical potential of the protein. Since the number of available binding sites is increased upon unfolding of the protein, an increase in preferential solute binding occurs as indicated by: $(\delta m_3/\delta m_2)^D > (\delta m_3/\delta m_2)^N > 0$. There is a concomitant decrease in protein chemical potential, which is greater for the denatured state: $(\delta \mu_2/\delta m_3)_{m2}^D < (\delta \mu_2/\delta m_3)_{m2}^N < 0$. This serves to lower the free energy difference between the two states, and when the native state becomes the higher energy state, protein denaturation should result.

As noted above, the same principles apply to the influence of solutes on the degree of assembly of multimeric proteins. Preferentially excluded solutes tend to induce polymerization and stabilize oligomers since the formation of contact sites between constituent monomers serves to reduce the surface area of the protein exposed to the solvent. Polymerization reduces the thermodynamically unfavorable effect of preferential solute exclusion. Conversely, preferential binding of solute induces depolymerization since there is greater solute binding to monomers than to polymers.

Since it is not possible to measure preferential interactions between solutes and proteins in frozen samples, we cannot state that cryoprotectants are actually preferentially excluded from the frozen proteins. However, preferential interactions of solutes with proteins in nonfrozen, aqueous solutions can be used to predict the cryoprotective capacity of solutes. For example, the degree of perturbation of protein chemical potential arising in the presence PEG and sucrose, respectively, can be used to explain why PEG is a much more potent cryoprotectant than sucrose.

The data for one case, which are shown in Table I, will serve to illustrate this point. The increase in chymotrypsinogen chemical potential ,$(\delta \mu_2/\delta m_3)_{m2}$, in the presence of either of two different molecular weights of PEG (e.g., Mr = 400 or 6000) is greater than that noted in the presence of the sucrose; even though the PEG is excluded to a lesser degree, on a per mole of solute basis. Comparing the two PEG molecules indicates that the larger the PEG the less it is excluded on a mole basis, but the more that it increases protein chemical potential.

The basis for these observations can be explained by examining Equation 1. The other major component in determining the effect of solute on protein chemical potential is the self-interaction parameter for the solute, $(\delta \mu_3/\delta m_3)_{m2}$. The value for this parameter is several-fold greater for PEG 400 and almost three orders of magnitude greater for PEG 6000, than that for sucrose. The self-interaction parameter is given by:

$$(\delta \mu_3/\delta m_3)_m = [(RT/m_3) + RT(\delta \ln \gamma_3/\delta m_3)m_2] \qquad (2)$$

where γ_3 is the activity coefficient of the solute and R is the universal gas constant (*5-7*). The molal concentrations needed for preferential exclusion of PEG are very small and the activity coefficient of PEG is quite large, relative to values for sucrose. Therefore, the self-interaction parameter for PEG is very large compared to that for sucrose. In addition, as the size of PEG increases there is a great increase in such nonideality (Table I).

Table I. Parameters for Solute Interactions with Chymotrypsinogen

Solute	Conc.	$(\delta m_3/\delta m_2)_{\mu 1, \mu 3}$	$(\delta \mu_3/\delta m_3)_{m2}$ [a]	$(\delta \mu_2/\delta m_3)_{m2}$ [b]
Sucrose[c]	1.27 m	-10.35	0.56	5.7
PEG 400[d]	10% w/v (0.27 m)	-6.87	2.42	16.6
PEG 6000[d]	1% w/v (0.0017 m)	-0.62	480.00	297.6

[a] kcal (mol of solute)$^{-1}$ (mol of solute in 1000 kg H20)$^{-1}$
[b] kcal (mol of protein)$^{-1}$ (mol of solute)$^{-1}$
[c] Data taken from ref. 6
[d] Data taken and calculated from ref. 7

When preferential exclusion is considered on a mass specific basis, $(\delta g_3/\delta g_2)$, the degree of exclusion increases with increasing PEG molecular weight. The exclusion is due to steric hindrance of PEG interaction with the protein (7). Also, the nonideality for PEG and its perturbation of protein chemical potential increase with molecular weight. Based on these values and the above discussion, one would predict that the larger the PEG the more effective it should be at protecting labile enzymes during freeze-thawing. The data presented in Figure 1 support this hypothesis. LDH is completely protected during freeze-thawing by PEG MW 8000 at concentrations of ≥ 0.01 % (wt/vol). In contrast, full protection in the presence of PEG MW 400 is not realized until the concentration is at least 2.5% (wt/vol). On a weight percentage basis, PEG MW 8000 is 250-fold more potent as a cryoprotectant. On a molar basis, the higher molecular weight PEG is 5000-fold more potent. Thus, we conclude that cryoprotection of proteins by PEG is due to preferential exclusion, which is in turn due to steric hindrance of the interaction of PEG with the protein.

Since degree of stabilization correlates directly with the increase in protein chemical potential in the presence of a solute, it is not surprising that PEG is much more effective than sugars at protecting labile enzymes during freezing. Interestingly, this correlation does not hold for high temperature denaturation experiments. Sugars increase the melting temperature for proteins, but PEG decreases protein stability at high temperature (6,8). This effect has been ascribed to increased hydrophobic interaction of PEG with proteins as temperature is increased, which leads to preferential binding to the denatured form (reviewed in ref. 9).

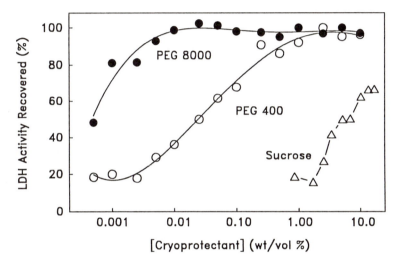

Figure 1. Effects of polyethylene glycols and sucrose on lactate dehydrogenase stability during freeze-thawing. The enzyme was prepared in 10 mM potassium phosphate buffer (pH 7.5 at 23°C) at a concentration was 25 μg/ml. Samples were frozen by plunging into liquid nitrogen and thawed in air at room temperature. Data shown for sucrose were taken from ref. *1*.

Unfortunately, much of the above discussion must be speculative because many of the data needed for a more rigorous evaluation currently are not available. Future studies should include a detailed examination of the effects of a several compounds (e.g., PEG, sucrose and glucose) on the stability of a protein during freeze-thawing and the determination of preferential interaction of these compounds with the same protein in aqueous solution. Then, it may be possible to correlate quantitatively a solute's influence on a protein's chemical potential with the capacity of the same solute to protect the protein during freeze-thawing. In addition, if preferential exclusion is the basis for recovering activity after freezing and thawing, then the protein's native structure should be preserved in the frozen sample. That is, the additive prevents denaturation, rather than fostering post-thaw refolding. The infrared spectroscopic method developed by Remmele and colleagues (see chapter in this volume) should greatly aid in these studies, since the protein's structure can be monitored in real-time during an entire freeze-thawing cycle.

Preservation of Dried Proteins: The Water Replacement Hypothesis

Sugars have been used for decades to preserve the structure and activity of proteins during drying and rehydration. For example, in 1935 Brosteaux and Eriksson-Quensel described the protection of a number of proteins, which were dried in the presence of sucrose, glucose and lactose (*10*). Therefore, some of the favored excipients for current formulations of proteins have a long history of successful use for the acute stabilization of proteins. In contrast, other agents, such as mannitol, that are often employed as "inert" bulking agents confer little protection to labile proteins and can induce additional damage during freeze-drying (e.g., *11*). In this section, we review some of the recent empirical data comparing effectiveness of additives as stabilizers and the mechanisms that have been proposed to explain preservation of labile proteins during dehydration.

We have compared a wide variety of compounds for their capacity to protect PFK during dehydration. In contrast to our results with freeze-thawing (*12*), we found that only disaccharides (e.g., sucrose, trehalose or maltose) are effective at protecting PFK during freeze-drying or air-drying (*13,14*). It appears that protection of proteins against extreme dehydration stress is a fundamentally different process than cryopreservation (cf. *15*). The empirical data support this statement. Also, this conclusion is supported theoretically because the thermodynamic arguments of the preferential exclusion mechanism are not applicable when dehydration progresses to the point that the water of hydration for the protein is removed. Finally, freezing and lyophilization expose the protein to different levels of dehydration stress. During freezing, there is some degree of dehydration arising as bulk water forms ice. However, the water of hydration for the protein remains unfrozen (*16*). This water is removed during lyophilization. Hence, in contrast to freezing, the terminal stress against which the protein needs to be protected during extreme dehydration is removal of its hydration shell.

More than two decades ago, Crowe proposed that the trehalose or sucrose accumulated in dehydration-tolerant organisms might serve as a replacement for water lost around polar and charged residues of biomolecules in the dried organisms (*17*). This idea has come to be known as the "water replacement hypothesis" and has been tested extensively with isolated biological membrane preparations and artificial membrane vesicles. The supporting evidence for membrane systems has recently been reviewed and will not be considered here (*18,19*).

There are also several studies supporting the contention that sugars protect labile proteins during drying by hydrogen bonding to polar and charged groups as water is removed and, thus, preventing drying-induced denaturation of the protein. For example, we have found, using solid-state Fourier transform infrared (FTIR) spectroscopy, that the respective spectra of trehalose or lactose dried in the presence of lysozyme or BSA are very similar to those noted for the sugars in aqueous solution and are much different from the spectra of sugars dried alone (*20*). In the dried state, the hydrogen bonding of the sugar to the protein serves to disrupt interactions between sugar molecules in the same fashion that hydrogen bonding to water disrupts these interactions in aqueous solution.

Our early FTIR spectroscopic studies also indicate that dehydration-induced transformations of the Amide I and Amide II bands of lysozyme can be partially and fully inhibited, respectively, if the protein is dried in the presence of trehalose or lactose (*20*). In addition, a band at 1583 cm^{-1}, which is due to hydrogen bonding of water to carboxylate groups, is not present in the spectrum of dried lysozyme (*20*). However, when lysozyme is dried in the presence of a disaccharide, the carboxylate band is retained in the dried sample, indicating that the sugar is hydrogen bonding in the place of water.

Two more recent studies on enzyme preservation provide further support for the water replacement mechanism. Tanaka and colleagues have found that the capacity of a saccharide to protect catalase during freeze-drying is inversely related to the size of the saccharide molecule (*21*). They suggest that as the size of the saccharide increases, steric hindrance interferes with hydrogen bonding between the saccharide and the dried protein. Lippert and Galinski have studied stabilization of freeze-dried LDH or PFK by a cyclic amino acid known as ectoine (1,4,5,6-tetrahydro-2-methyl-4-pyrimidincarboxylic acid). They found that stabilization is greatly enhanced if there is a hydroxyl group present in the molecule (*22*). They suggest that this is because hydrogen bonding between the hydroxyectoine and the dried protein is needed for preservation.

The newest and most compelling information on the mechanism of protein preservation by sugars comes from studies that used FTIR spectroscopy, combined with spectral resolution-enhancement techniques, to look at effects of dehydration on protein conformation. Prestrelski and colleagues have found, by using second derivative transformation of protein infrared spectra, that drying leads to three different types of changes in protein structure (*23* and a chapter in this volume). Very labile proteins (e.g., LDH) are denatured in the dried state and remain denatured and inactive after rehydration. Proteins such as lysozyme are unfolded in the dried state, but refold during rehydration. Finally, certain proteins (e.g., granulocyte colony stimulating factor) maintain their native conformation in the dried state and during rehydration.   When the proteins are dried in the presence of sucrose, all three classes retain their native structure in the dried solid.   The results with LDH demonstrated that the recovery of activity after rehydration correlates directly with the degree to which the protein's spectrum (i.e., its conformation) in the dried state resembles the spectrum for the aqueous, native protein.   Thus, the mechanism by which sugars preserve protein activity during freeze-drying and rehydration is by preventing unfolding during lyophilization.   Finally, these researchers also found that the carboxylate band for alpha-lactalbumin is not present in the spectrum of the dried protein, but is maintained if sucrose is added prior to freeze-drying, indicating that the sugar hydrogen bonds to the dried protein in the place of water (23).

The most relevant data from this work, in terms of the water replacement hypothesis, come from studies on the effects of freeze-drying on a model polypeptide, poly-L-lysine (23).   This peptide assumes different conformations in solution, which have been well-characterized with FTIR spectroscopy, depending on the pH and temperature.   At neutral pH, poly-L-lysine exists as an unordered peptide.   At pH 11.2, the peptide adopts an alpha-helical conformation.   Poly-L-lysine assumes a beta-sheet conformation in the dried state, regardless of its initial conformation in aqueous solution.   The preference for beta-sheet in the dried state appears to be a compensation for the loss of hydrogen bonding interactions with water.   The beta sheet allows for the highest degree of hydrogen bonding in the dried peptide.   If poly-L-lysine is freeze-dried in the presence of sucrose, the transition from random coil or alpha helix to beta sheet is inhibited, and the solution structure is retained in the dried state.   Thus, by hydrogen bonding to the dried peptides, sucrose serves as a water replacement and prevents the adoption of the favored beta-sheet conformation.

Despite the evidence supporting the water replacement mechanism, some researchers contend that hydrogen bonding by sugars to dried protein is not needed at all for protection of proteins (e.g., 24,25).   Based on a materials science approach, they suggest that formation of an amorphous solid is all that is needed for preservation of proteins in dried samples.   The proteins are proposed simply to be mechanically immobilized in the glassy, solid matrix in the dried state.   The restriction of translational and relaxation processes is thought to prevent protein unfolding, and spatial separation between protein molecules is proposed to prevent aggregation.

We have found that formation of an amorphous solid is required for preservation of labile proteins by sugars, and that the failure of compounds such as mannitol to protect these proteins is due to crystallization during lyophilization (11).   However, several recent studies with dextran document that glass formation alone is not sufficient for stabilization.   For example, Pikal et al. have found that damage to freeze-dried human growth hormone was increased in the presence of dextran, even though the dried samples were fully amorphous (26).   Tanaka and colleagues have found that the capacity of carbohydrates to protect freeze-dried catalase decreased with increased carbohydrate molecular weight (*21*). Dextrans were the largest and least effective of all the carbohydrates tested, and the larger the dextran molecule the less it stabilized catalase. Although they did not determine whether their dried samples were amorphous, it has been shown that as the molecular weight of the carbohydrate is increased, the glassy state is formed more readily during drying (*27*). However, the larger the carbohydrate molecule, the more steric hindrance could interfere with hydrogen bonding of the carbohydrate to the dried protein.

One potential alternative explanation is that dextran is not protecting the proteins during the freezing phase (see below). This hypothesis appears unlikely because it has been shown previously that dextrans are effective cryoprotectants for catalase, with the degree of protection actually increasing with increasing dextran molecular weight (*28*). In addition, we have recently found (Zhang, Prestrelski, Arakawa and Carpenter, unpublished data) that both LDH and PFK are stabilized during freezing by dextran, but are not protected during freeze-drying. There is no increase in activity of PFK recovered after rehydration and LDH activity is even less when the enzyme is freeze-dried with dextran. The infrared spectra of both proteins freeze-dried in dextran indicate that they are denatured in the dried solid.

In conclusion, although it is necessary for stabilizing carbohydrates to remain amorphous simply to allow hydrogen bonding to occur, glass formation alone appears not to be sufficient for acute protection of dried proteins. As will be discussed in a later section, the water replacement mechanism may be adequate to explain protein preservation by sugars, but certain polymers (e.g., PVP) also protect multimeric enzymes during freeze-drying. However, even in this case, it is not necessary to invoke mechanical immobilization by an amorphous solid as a mechanism for protection (see below).

Stress-specific Stabilization of Proteins during Freeze-drying

At least one apparent dilemma arises from the argument that sugars stabilize dried proteins by hydrogen bonding in place of water. If only this interaction were needed for protein preservation, then mono- and disaccharides should provide similar protection. However, this is not the case. When PFK is freeze-dried in the presence of 0.2-0.4 M trehalose, over 60% of the initial activity is recovered after rehydration (*13*). In contrast, when similar amounts of glucose are used, the recovery is less than 5% (*13*). PFK and other labile proteins are sensitive to both the freezing and the subsequent dehydration stresses encountered during lyophilization (*1-4,12,13*). Glucose does not protect during freeze-drying because it provides minimal stabilization during freezing, whereas trehalose is effective at protecting the protein during both freezing and dehydration (*12,13*).

Thus, a given additive could be highly effective at stabilizing a protein against dehydration stress. However, if this same compound does not provide cryoprotection to the protein, preservation during freeze-drying will not be realized. Since cryopreservation derives from the preferential exclusion of a solute from the protein's surface, increasing the concentration of the additive will increase protein stabilization during freezing (13,20). However, sugars (at high initial concentrations) can crystallize during freeze-drying and hence not be available to hydrogen bond to, and protect, the dried protein (20).

To resolve this apparent dilemma, the freezing and drying stresses to the protein must be treated separately. Recently, Carpenter et al. have developed a two-component system for stress-specific stabilization during lyophilization. In this stabilization scheme, PEG is used as a cryoprotectant and various carbohydrates can be used to protect during dehydration (11). PEG alone completely stabilizes either LDH or PFK during freeze-thawing. However, it provides little or no protection during dehydration, because it crystallizes during lyophilization. When small amounts (e.g, 10 - 100 mM initial concentration) of trehalose or glucose are added, which alone are ineffective at protecting these enzymes, excellent stabilization is noted during freeze-drying. Thus, under conditions where cryoprotection is provided by PEG, glucose is essentially as effective as trehalose in stabilizing dried enzymes (i.e., LDH and PFK). We propose that PEG provides cryoprotection due to preferential exclusion and the sugar protects during drying by serving as a water replacement.

In a complementary structural study of stress-specific stabilization using FTIR spectroscopy, Prestrelski et al. found that the recovery of activity after rehydration correlates directly with the ability of the additives to preserve the native structure of the enzymes in the dried state (29). Full activity recovery and maintenance of essentially aqueous structure in dried samples are only noted when a combination of PEG and sugar is employed. Based on these results Prestrelski et al. have proposed a model of the conformational events during lyophilization, which is discussed in detail in separate chapter of this volume. Briefly, this model proposes that, in order to recover structure and function after rehydration, the native structure of labile proteins must be retained, both upon freezing and during subsequent dehydration. The appropriate cryoprotectant is required for the initial structural preservation and a specific stabilizer against drying is needed for the terminal stress during lyophilization. In some instances (e.g., with disaccharides), a single additive can serve both protective functions. Rigorous testing of this model will require use of the FTIR method described by Remmele et al. (chapter in this volume), which allows one to monitor protein structure throughout the entire lyophilization cycle.

Stabilization of Multimeric Enzymes by Polymers during Lyophilization

Hellman et al. have found that PVP stabilizes the tetrameric enzyme, L-asparaginase, during freeze-drying (30). Steric hindrance should minimize the ability of PVP to hydrogen bond effectively to the charged and polar groups on the dried protein's surface. Therefore, the water replacement hypothesis cannot explain stabilization of enzymes by PVP during freeze-drying. We have recently found that PVP will also stabilize both PFK and LDH during freeze-drying (Anchordoguy and Carpenter, unpublished data). In addition, we have found that BSA protects both enzymes during lyophilization. With LDH, increased recovery can also be realized by simply increasing the initial concentration of the enzyme.

The addition of a PVP or BSA, or increasing protein concentration, should shift the equilibrium between molecular forms of the enzymes (i.e., tetramers, dimers and monomers) towards the fully polymerized form. As noted above, due to steric hindrance, bulky polymers are preferentially excluded from the surface of proteins in aqueous solution. This interaction thermodynamically stabilizes the polymerized form of multimeric enzymes. The same argument can also be used to explain the enhanced stability noted at increasing protein concentrations, if one views the remainder of the protein population as being preferentially excluded from a given individual protein molecule. In addition, by mass action increasing the enzyme concentration will favor polymerization.

Furthermore, adding bulky polymers or increasing enzyme concentration increases the resistance of the enzyme to the freezing stress that proceeds sublimation. Thus, our hypothesis is that the tetrameric forms of the enzymes are able to withstand drying stress and that damage is due to dissociation and unfolding of the constituent monomers during freezing and drying. Unfolded monomers could then form aggregates upon thawing or rehydration. This process is analogous to the dissociation-unfolding-aggregation sequence that arises during denaturation of multimeric proteins in solution (cf. *31*). To test this hypothesis, we have employed a mixture of two LDH isozymes (i.e., rabbit muscle and porcine heart LDH's), that can be separated electrophoretically on native gels and identified by activity staining. Chilson et al. have previously shown that hybrid tetramers are formed during freeze-thawing (*32*). These hybrids arise because freezing induces dissociation, and upon thawing non-denatured monomers can reassociate into the hybrid tetramers, as well as homotetramers.

In support of our hypothesis, we have found that PVP prevents hybrid formation during both freeze-thawing, and freeze-drying and rehydration (Anchordoguy and Carpenter, unpublished data). Thus, PVP inhibits dissociation during freezing and drying. The tetramers are not damaged and are active upon rehydration. These are just preliminary studies with a single enzyme, but they suggest that, as is the case with high temperature denaturation in aqueous solution (cf. *31*), constituent monomers are much less resistant to freezing or drying stress while separated than when in an intact multimer.

One of the challenges for future research efforts in this field is to try to quantitate this polymerization-dependent difference in monomer stability. In addition, a direct assessment of quaternary structure in the frozen and dried states is needed to define more clearly the role of dissociation in damage to multimeric enzymes and the role of inhibition of this process in preservation of these proteins. One promising technique is irradiation target-size analysis, which can be used to determine protein molecular weight in frozen or dried samples (e.g., *33*). The combination of data obtained with this method and the structural information available from FTIR spectroscopy should allow great advances in our understanding of the structural transitions arising during freezing and drying and how inhibition of these transitions impact on recovery of functional proteins upon rehydration.

Acknowledgments

This paper is dedicated to Professor Serge N. Timasheff. Portions of the research discussed in this review were funded by Grant RR05831 from NIH.

Literature Cited

1. Carpenter, J.F.; Crowe, J.H. *Cryobiology* **1988**, *25*, 244-255.
2. Carpenter, J.F.; Arakawa, T.; Crowe, J.H. *Devel. Biol. Standard.* **1991**, *74*, 225-239.
3. Arakawa, T.; Kita, Y.; Carpenter, J.F. *Pharm. Res.*, **1991**, *8*, 285-291.
4. Arakawa, T.; Prestrelski, S.J.; Kenney, W.; Carpenter, J.F., *Advanced Drug Delivery Reviews* **1993**, *10*, 1-28.
5. Timasheff, S.N.; Arakawa, T. In: *Protein Structure, a Practical Approach*; Creighton, T.E., Ed.; IRL Press: New York, NY, 1988; pp. 331-345.
6. Lee, J.C; Timasheff, S.N. *Biochemistry* **1981**, *256*, 7193-7201.
7. Bhat, R.; Timasheff, S.N. *Protein Science* **1992**, *1*, 1133-1143
8. Lee, L. L.-Y.; Lee, J.C. *Biochemistry* **1987**, *26*, 7813-7819.
9. Arakawa, T.; Carpenter, J.F.; Kita, Y.A.; Crowe, J.H. *Cryobiology* **1990**, *27*, 410-415.
10. Brosteaux, J.; Eriksson-Quensel, I.B. *Archives Physique Biologique* **1935**, *32*, 209-226.
11. Carpenter, J.F.; Prestrelski, S.J.; Arakawa, T. *Arch. Biochem. Biophys.* **1993**, *303*, 456-464.
12. Carpenter, J.F.; Hand, S.C.; Crowe, L.M.; Crowe, J.H. *Arch. Biochem. Biophys.* **1986**, *250*, 505-512.
13. Carpenter, J.F.; Crowe, L.M.; Crowe, J.H. *Biochim. Biophys. Acta* **1987**, *923*, 109-115.
14. Carpenter, J.F.; Martin, B.; Crowe, L.M.; Crowe, J.H. *Cryobiology* **1987**, *24*, 455-464.
15. Crowe, J.H.; Carpenter, J.F.; Anchordoguy, T.J.; Crowe, L.M. *Cryobiology* **1990**, *27*, 219-231.
16. Kuntz, I.D.; Kauzmann, W. *Adv. Protein Chem.* **1974**, *28*, 239-345.
17. Crowe, J.H. *Amer. Naturalist* **1971**, *105*, 563-573.
18. Crowe, J.H.; Crowe, L.M.; Carpenter, J.F. *BioPharm* **1993**, *6(3)*, 28-37.
19. Crowe, J.H.; Crowe, L.M.; Carpenter, J.F. *BioPharm* **1993**, *6(4)*, 40-43.
20. Carpenter, J.F.; Crowe, J.H. *Biochemistry* **1989**, *28*, 3916-3922.
21. Tanaka, R.; Takeda, T.; Miyajima, R. *Chem. Pharm. Biol.* **1991**, *39*, 1091-1094.
22. Lippert, K.; Galinski, E.A. *Appl. Microbio. Biotechnol.* **1992**, *37*, 61-65.
23. Prestrelski, S.J.; Tedeschi, N.; Arakawa, T.; Carpenter, J.F., *Biophysical J.* **1993**, *65*, 661-671.
24. Franks, F.; Hatley, R.H.M.; Mathias, S.F. *BioPharm* **1991**, *4(9)*, 38-42.
25. Levine, H; Slade, L. *BioPharm* **1992**, *5(4)*, 36-40.
26. Pikal, M.J.;Dellerman, K.M; Roy, M.L.; Riggin, R.M. *Pharm. Res.* **1991**, *8*, 427-436.
27. Levine, H.; Slade, L.; In: *Physical Chemistry of Foods*; Shartzberg, H.S.; Hartel, R.W., Eds.; Dekker: New York, NY; 1992, pp. 83-221.

28. Ashwood-Smith, M.J.; Warby, C. *Cryobiology* **1972**, *9*, 137-140.
29. Prestelski, S.J.; Arakwa, T.; Carpenter, J.F. *Arch. Biochem. Biophys.* **1993**, *303*, 465-473.
30. Hellman, K.; Miller, D.S.; Cammack, R.A. *Biochim. Biophys. Acta* **1983**, *749*, 133-142.
31. Chilson, O.P.; Costello, L.A.; Kaplan, N.O. *Fed. Proc.* **1965** *24(2)*, S55-S65.
32. Jaenicke, R. *Prog. Biophys. Molec. Biol.* **1987**, *49*, 117-237.
33. Kempner, E.S.; Miller, J.H.; McCreery, M.J. *Anal. Biochem.* **1986**, *156*, 140-146.

RECEIVED April 19, 1994

Chapter 10

Structure of Proteins in Lyophilized Formulations Using Fourier Transform Infrared Spectroscopy

Steven J. Prestrelski[1], Tsutomu Arakawa[1], and John F. Carpenter[2]

[1]Amgen, Inc., Amgen Center, Thousand Oaks, CA 91320
[2]School of Pharmacy, University of Colorado, Denver, CO 80262

The stability of different proteins to lyophilization is highly variable. Certain proteins are completely denatured upon freeze-drying and reconstition while others appear essentially unaffected. Is there a structural basis for these effects, and if so, could structure be used as basis for rational development of protein formulations with optimal stability? Recent studies of the conformation of lyophilized proteins using Fourier-transform infrared spectroscopy have demonstrated that drying of proteins induces significant and measurable conformational transitions. The extent of these transitions is variable and appears to be protein dependent. Upon rehydration, several proteins show irreversible denaturation while others readily adopt the pre-lyophilized conformation. Studies of a model homopolypeptide poly-l-lysine, also demonstrate that dehydration can serve as a driving force for conformational change. Stabilizers foster retention of the native state of the proteins during freeze-drying. Further, stabilizers prevent lyophilization-induced conformational transitions in poly-l-lysine. Upon reconstition, proteins freeze-dried in the presence of stabilizers showed no evidence of denaturation. Additional studies with a labile enzyme (lactate dehydrogenase) demonstrated a direct correlation between retention of native structure during freeze-drying and recovery of activity after rehydration. These studies point to the utility of Fourier-transform infrared spectroscopy in development of optimally stable lyophilized protein formulations.

An understanding of the effects of dehydration on protein structure is essential to the development of optimally stable freeze-dried protein formulations. It has long been established that water is essential for formation of native protein structures (1,2). However, the effects of dehydration upon protein conformation have remained a point of controversy. Kuntz and Kauzmann (3) demonstrated from theoretical considerations that dehydration should result in significant conformational changes. Several sorption isotherm studies of progressive rehydration of dried proteins employing vibrational spectroscopy, NMR and hydrogen exchange techniques have indicated hydration related changes in conformation and flexibility (4,5,6). In particular, differences in the rehydration behavior between α-lactalbumin and lysozyme were attributed to

0097–6156/94/0567–0148$08.36/0

differences in the dried state conformations of these homologous proteins. Dehydration-induced changes in conformation-sensitive vibrational modes observed by using Raman spectroscopy have also supported this claim *(7)*. However, vibrational effects due solely to removal of water could not be ruled out. Further, it has been observed that dehydration can completely and irreversibly inactivate some enzymes *(8)*. This inactivation occurs presumably through loss of the native structure. Hanafusa *(9)* has observed such dehydration-induced changes in tertiary and quaternary structure upon rehydration of dried myosin and catalase, respectively.

In contrast, other investigators have developed a hydration model for lysozyme which does not involve significant changes in conformation *(10,11,12)*. Spectral changes observed in conformation-sensitive modes were ascribed solely to changes in hydration. The strongest evidence provided for the absence of a conformational change during dehydration comes from an electron spin resonance (ESR) study. In the ESR study, carried out at -160°C, two spin probes placed on lysozyme were found not to shift more than 1.0Å as a function of hydration levels *(11)*. The controversy concerning the effect of dehydration on protein conformation has not been reconciled, in part, due to the lack of methodology for detailed examination of protein structure in both the aqueous and dehydrated states. For example, previous studies have been inconclusive because spectroscopic differences observed upon dehydration were difficult to interpret directly in terms of conformation *(6,11)*. Further, a majority of the early studies *(see Ref. 12)* have focused upon a single and unusually stable protein, lysozyme, for which all dehydration-induced spectral shifts are fully reversible.

The recent development of combining Fourier-transform infrared (FTIR) spectroscopy with mathematical band narrowing techniques has provided the capacity for detailed conformational studies of proteins in aqueous solution *(13,14)*. Further, examination of the structure of proteins in the dehydrated state using Fourier-transform infrared spectroscopy has provided a new tool for structural analysis of lyophilized proteins. Like the majority of studies of protein conformation in solution, these studies have focused upon the amide I vibrational mode (ca. 1700-1600 cm^{-1}). The amide I vibration corresponds to peptide carbonyl stretching mode and is sensitive to the backbone conformation *(13)*.

The Effect of Lyophilization on Protein Structure

In an initial study of several proteins, Prestrelski *et al. (15)* demonstrated that dehydration via lyophilization induces significant, measurable conformational changes in proteins. The observed conformational changes are highly variable and appear to be protein dependent. Several examples of the infrared spectroscopic changes observed in the amide I region upon dehydration are given in Figure 1. These proteins were lyophilized as described in Prestrelski *et al. (15)*. (Briefly, the proteins were freeze-dried from 20 mg/ml solutions in 10 mM Tris, pH 7.0, 0.1 M NaCl. Vials of the solutions were placed on the lyophilizer shelf which was precooled to -50°C. Primary drying was carried out at -45°C for 24 hr. Secondary drying was carried out at -10°C and +20°C each for 24 hr. Ramping between the different shelf temperatures was at a rate of 1°C/hr. The final level of hydration was determined to be 0.04 g H$_2$O/g protein.)

Figure 1A shows the second derivative infrared spectra of recombinant human basic fibroblast growth factor (bFGF) in aqueous solution and after lyophilization from this same solution. In the spectrum of dehydrated bFGF, a peak observed in the aqueous spectrum at 1657 cm^{-1} in the aqueous spectrum is significantly reduced, and a new band at 1614 cm^{-1} is resolved. The remaining peaks shift in frequency from 2 to 5 cm^{-1} upon dehydration. The spectral changes observed for bFGF are typical of those observed for most proteins. All proteins studied to date, to varying degrees, exhibit similar broadening and shifts in frequency. Figure 1B shows the second derivative infrared spectrum of aqueous and freeze-dried α-casein. α-casein is a protein which has no regular secondary structure in the solution state. The ir spectra of α-casein

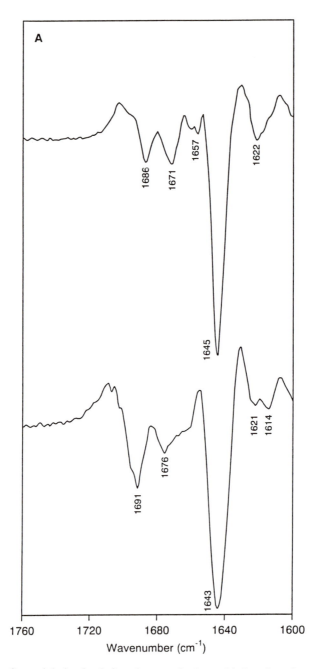

Figure 1. Second derivative infrared spectra in the amide I region of several proteins in the aqueous (upper curves) and dehydrated (lower curves) state: A) basic fibroblast growth factor (bFGF), B) α-casein and C) granulocyte colony stimulating factor (G-CSF). Reproduced with permission from Ref. 15.

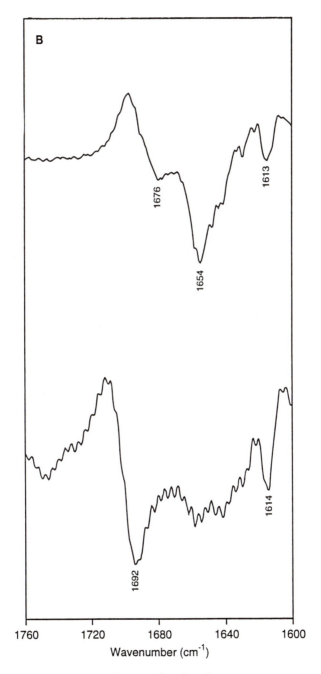

Figure 1. Continued.

Continued on next page

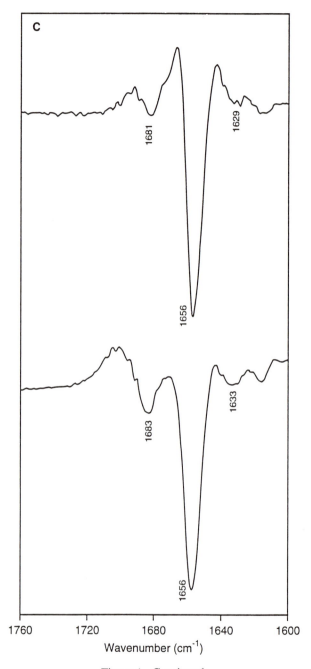

Figure 1. Continued.

indicate that this protein undergoes a large conformational transition upon dehydration. The spectrum of the aqueous protein has the characteristics of protein with little ordered structure, as expected. However, this spectrum is greatly altered after dehydration and shows a large fraction of β-sheet as evidenced by the predominant bands near 1615 and 1690 cm^{-1}. Thus, dehydration has induced a large conformational change. At the opposite extreme is the behavior of the conformation of recombinant human granulocyte colony stimulating factor (G-CSF) upon lyophilization. The spectra of this protein in the aqueous and dehydrated states (Fig. 1C) indicate that dehydration induces little change in conformation.

For all of the proteins studied the predominant spectral change observed upon dehydration is broadening of the individual amide I components, indicative of a general disordering of the protein backbone. This disordering is likely static (multiple stable conformations) rather than dynamic (increased motion) because the protein's mobility is greatly decreased upon dehydration. The interpretation of the infrared band broadening arising from static disorder is consistent with the results of two solid state nuclear magnetic resonance studies *(16,17)*. These studies examined freeze-dried lysozyme as a function of hydration and determined that the distribution of conformational states observed in the dry protein becomes much narrower with increasing hydration. Thus, the dry protein can adopt a broad distribution of conformational states which is narrowed as water is replaced. The solid state NMR results for lysozyme also indicate that the conformational changes are localized, small in magnitude and do not involve a major unfolding of the protein upon dehydration.

In addition to band broadening the infrared spectra of several show relative peak intensity differences as well as loss of spectral feature(s) upon dehydration. Shifts in peak positions are also observed. From knowledge of the relationship between amide I components and protein structure *(13,14)*, these spectral changes are indicative of conformational changes. However, the solid state NMR studies to date give no indication of large-scale conformational transitions. It is important to note that the NMR studies to date have focused on lysozyme. In FTIR studies, spectra of dried lysozyme showed evidence of band-broadening and spectral shifts but no evidence of a major transition which is apparently consistent with the solid state NMR studies. It is obvious that further studies are necessary to fully compare the results from infrared spectroscopy and NMR. NMR studies of a broad group of proteins are necessary to provide a more complete picture of the structural alterations induced upon drying of proteins.

A spectral feature often observed upon dehydration of proteins is a new band near 1615 cm^{-1}. The frequency of this band suggests that intra- or intermolecular β-sheets are forming upon dehydration, similar to β-sheets formed upon unfolding and aggregation of proteins in solution *(18,19)*. Bands at similar frequencies are also observed for the β-sheet form of poly-l-lysine which is also induced upon dehydration (see below). These bands are particularly strong in proteins (recombinant human γ-interferon (γ-IFN) (not shown) and α-casein) observed to undergo extensive aggregation upon rehydration suggesting that the dehydrated structures are precursors for the aggregated forms observed upon rehydration.

Quantitative Analysis.   To assess and compare the degrees of conformational change induced upon drying some quantitative measure is necessary. An important attribute of the spectra of the dried proteins is their degree of similarity to the aqueous spectra. The spectral correlation coefficient calculated from two amide I ir second derivative spectra provides a measure of their overall spectra similarity, and thus the similarity in conformation *(15)*. Calculation of this value between the native (aqueous) state of the protein and that when dehydrated gives a relative measure of the unfolding induced upon dehydration. Table I lists the coefficients for the proteins in Figure 1 in the dehydrated states. Data for dehydrated forms of two additional proteins rhγ-interferon and bovine α-lactalbumin are also provided. As is observed typically, dehydration results in variable changes in the amide I infrared spectra. The correlation coefficients for proteins lyophilized in the absence of stabilizers ranges from 0.696 for

Table I. Correlation coefficients for second derivative spectra of
dehydrated and rehydrated proteins [a]

Dehydrated Additive	bFGF	α-Casein	G-CSF	γ-IFN	α-Lactalbumin
Buffer Only	.892[b]	.696	.907	.743	.774
Sucrose	.972	.843	.923	.929	.898
Lactose	.954		.927	.858	.843
Mannose	.951		.933	.849	.876
Glucose	.965		.917	.914	.893
Galactose	.952		.841	.762	.783
Mannitol	.954	.833	.850	.719	.805
Myoinositol	.932		.795	.551	.801
PEG-8000	.932		.902	.858	.739
Rehydrated					
Buffer only	.909	.741	.991	.795	.984
Sucrose	.992	.952	.995	.992	.989

[a] Source: Reproduced with permission from ref. 15. Copyright 1993 Biophysical Society.

[b] The spectra of several samples of a single protein were collected in duplicate under identical conditions. The mean difference in the computed correlation coefficients for the two spectra relative to the same aqueous protein spectrum was 0.011, σ=.0085, n=4.

α-casein, which undergoes a large transition, to greater than 0.9 for G-CSF, which is little affected by dehydration.

Rehydration of Dried Proteins. The next question to be addressed concerns the behavior of the dried, unfolded proteins upon rehydration. The spectral correlation coefficients for the rehydrated proteins relative to the initial aqueous spectra are listed in Table I. The aqueous amide I ir spectra of basic FGF, γ-IFN and α-casein after rehydration are still altered relative to the pre-lyophilized aqueous spectra as evidenced by r values significantly less than 1.0. This result indicates that at least partial, irreversible conformational changes occurred upon lyophilization and rehydration. γ-IFN and casein also exhibited extensive aggregation and precipitation upon rehydration and the rehydrated α-casein was essentially insoluble. The dehydration-induced conformational changes appeared reversible, however, for α-lactalbumin and G-CSF, which give r values near 0.99 for the rehydrated proteins relative to the initial aqueous spectra. These results indicate that while the observed dehydration-induced conformational changes are sometimes irreversible, some proteins appear to be inherently stable to the stresses introduced during freeze-drying. These proteins are observed to partially unfold during dehydration, and they possess the capacity to refold to form the native state upon rehydration. The irreversible spectral changes observed for several of the proteins provide support for the contention that the spectroscopic differences observed upon dehydration are the result of conformational changes. If drying-induced spectral changes are due solely to water removal, they should be fully reversible upon rehydration.

In summary, three classes of behavior are observed for proteins upon dehydration and rehydration. First, a protein can be resistant to conformational change during drying and therefore retain the native conformation upon rehydration such as G-CSF. Second, a protein may unfold during dehydration but refold to the native state upon rehydration as is observed for α-lactalbumin and lysozyme. Third, a protein may unfold during dehydration and not regain the native conformation resulting in irreversible conformational changes and denaturation as is observed for γ-IFN and α-casein. Thus, the inherent stability of a protein to survive dehydration and subsequent rehydration must be related to its capacity to resist conformational changes during dehydration and/or its capacity to refold into the native structure upon rehydration. With the present information, it is not possible to relate the observed stabilities with a type of degree of conformational change observed upon dehydration, nor is it possible to relate the observed stability with particular categories of proteins.

Correlation of Dehydration-induced Structural and Functional Alterations. An apparent manifestation of the observed dehydration-induced loss of structure in some proteins is the observation that several labile enzymes completely lose their biological activity upon drying and rehydration *(8)*. This structural basis effect has been examined in detail using FTIR. The enzyme lactate dehydrogenase was as a model. When freeze-dried from the buffer solution only the FTIR spectrum of LDH indicates a high degree of unfolding as evidenced by a correlation coefficient essentially equal to zero (-0.003). Thus, it appears that dehydration induces a large scale unfolding of LDH, which correlates with an essentially complete loss of activity upon rehydration. Based on the failure of LDH to regain activity after rehydration, it is apparent that the dehydration induced conformational changes are irreversible.

Lyophilization of Poly-l-lysine - a Model for Conformational Change

To understand more fully the nature of the dehydration-induced spectral changes in proteins, the behavior of poly-l-lysine during lyophilization has been examined. Poly-l-lysine provides a suitable model for conformational studies because in solution it adopts an α-helical, β-sheet, or unordered conformation, depending on the pH and temperature and these solution conformations have been extensively characterized using FTIR spectroscopy *(20)*. Figure 2 shows the second derivative amide I spectra of aqueous and lyophilized poly-l-lysine under several initial conditions of pH and temperature. At neutral pH, poly-l-lysine exists as an unordered polypeptide as indicated by the strong peak near 1649 cm^{-1}. Freeze-drying of poly-l-lysine induces a transition from an unordered polypeptide to a highly ordered β-sheet (Fig. 2A). Further, lyophilization from a pH 11.2 solution, where the polypeptide adopts an a-helical conformation in solution, also induces a transition to a β-sheet, although a small amount of α-helix is apparently still observable as indicated by the peak near 1654 cm^{-1} (Fig. 2B). Heating a pH 12.0 solution of poly-l-lysine results in formation of a β-sheet. In a sample lyophilized after this treatment, the β-sheet is preserved upon dehydration (Fig. 2C) Apparently the preferred conformation in the dried state is β-sheet regardless of the initial conformation in aqueous solution. The changes in this spectrum upon dehydration are very small and limited to small frequency shifts (3-4 cm^{-1}) for each of the two

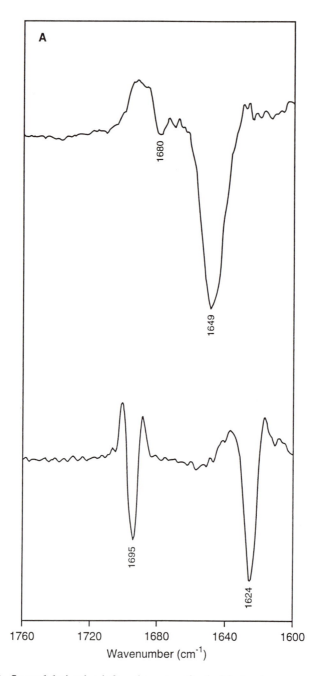

Figure 2. Second derivative infrared spectra of poly-l-lysine in aqueous solutions (upper curves) and after lyophilization (lower curves) from these solutions A) pH = 7.5 Tris buffer, B) pH = 11.2 NaOH solution and C) pH = 12.0 and heated at 75°C for 30 min. Reproduced with permission from Ref. 15.

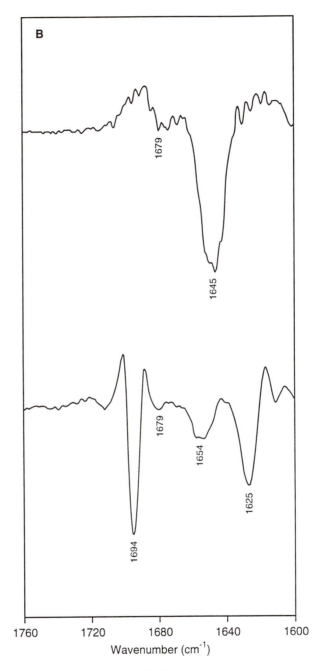

Figure 2. Continued.

Continued on next page

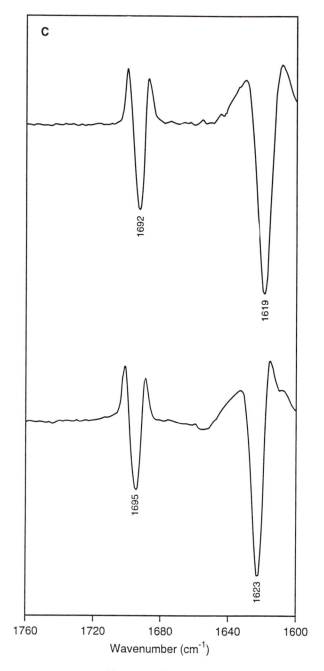

Figure 2. Continued.

peaks which are indicative of the β-sheet conformation. No changes in band shape, width or relative peak intensities are apparent upon dehydration. Thus, this result demonstrates that the spectroscopic changes observed upon dehydration are due solely to conformational changes. Further, spectral changes due to removal of water from the polypeptide, independent of a conformational change, are very small. This observation agrees with results indicating that local association effects have the greatest influence and that solvent factors have little effect on carbonyl stretching vibrations such as the amide I mode *(21,22)*. The results of the poly-l-lysine experiments provide strong evidence that the spectral changes in the amide I region observed upon dehydration of proteins are predominantly related to conformational changes, and that the effect of water removal independent of a conformational change is minimal.

The dehydration-induced conformational transitions observed for poly-l-lysine appear to be due to compensation for the loss of hydrogen bonding interactions with water. In solution, a random coil has its peptide hydrogen bonds satisfied by water molecules. Upon dehydration, these hydrogen bonds are lost, and to compensate for this loss, the polypeptides form intermolecular hydrogen bonds resulting in the observed β-sheet conformation for the dried polypeptide. Further, in the absence of water, the partial charges of hydrogen bonding groups are less screened due to the lower dielectric environment, increasing the electrostatic attraction between dehydrated peptides. Thus, in the dried state, the hydrogen bonding interaction energy between amide groups should be stronger than in aqueous solution. The conformational transition observed for α-helical poly-l-lysine is also consistent with this mechanism. At low hydration levels, the β-sheet conformation is energetically more favorable than the a-helix because the β-sheet, which has a higher degree of intermolecular hydrogen-bonding, requires a lower degree of solvation *(23)*. As the hydration shell is removed from the helical polypeptide, a transition to β-sheet is induced.

These proposed mechanisms for dehydration-induced conformational transitions can be extrapolated to proteins. During dehydration, the protein rearranges its conformation to maximize intra- and inter-chain hydrogen bonding to replace lost hydrogen bonds to water. In addition to hydrogen bonding several different types of interactions are present in proteins that are not in the poly-l-lysine model, including hydrophobic interactions. It is apparent that the effects of dehydration on different proteins are significantly more complex than the poly-lysine model. It is clear that significant conformational rigidity exists in some proteins because only a small conformational change is observed after dehydration (e.g., G-CSF, Fig. 1C). (It is interesting to note the similarity in the behavior of the proteins and poly-l-lysine in the formation of β-sheet structures upon dehydration.) In contrast, many proteins are unstable during lyophilization and after reconstitution lose all or part of their structure. Thus, while it is reasonable to focus upon the hydrogen bonding interactions in attempting to explain the dehydration induced conformational changes, it is clear that these changes are more complicated and more studies are necessary to completely account for all interactions in the aqueous and dehydrated states.

Effects of Additives on the Structure of Dried Proteins

Studies of labile enzymes have shown that certain additives can stabilize proteins during lyophilization and rehydration *(8)*. Carbohydrates, disaccharides in particular, appear to be the most effective stabilizers. Based on the observation that lyophilization induces (sometimes irreversible) conformational transitions in proteins, it is reasonable to speculate that the stabilizing effect of these solutes involves the protein's conformation. Figure 3 shows the infrared spectra of the

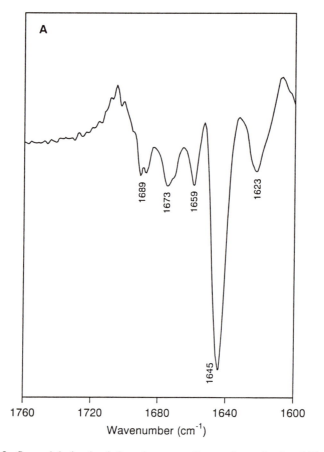

Figure 3. Second derivative infrared spectra of several proteins lyophilized in the presence of 200 mg/ml sucrose. A) bFGF, B) α-casein and C) G-CSF. Reproduced with permission from Ref. 15.

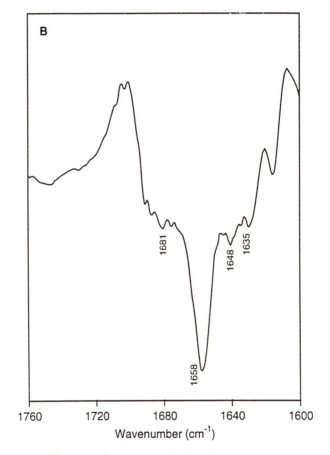

Figure 3. Continued. *Continued on next page.*

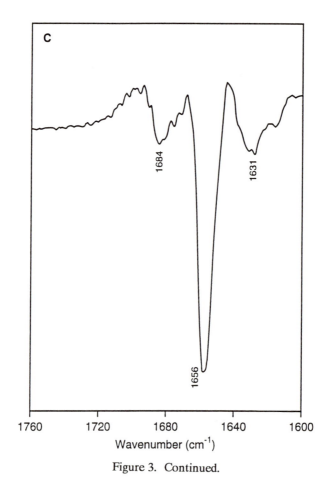

Figure 3. Continued.

proteins shown in Figure 1 lyophilized from sucrose (200 mg/ml) solutions. Sucrose has been shown to be particularly effective at protecting labile enzymes during dehydration *(8)*. In all cases, the dried state spectra of proteins lyophilized in the presence of sucrose resemble closely the respective spectra of the same proteins in aqueous solution (as shown in Figure 1). This effect is quantitatively demonstrated by the spectral correlation values listed in Table I. Lyophilization in the presence of sucrose results in an r value much closer to 1.0 than the value resulting from lyophilization in the absence of any additives. Thus, the stabilizer preserves the native or aqueous conformation of the protein during dehydration.

In addition to sucrose, the effects of numerous disaccharides, monosaccharides and polyhydric alcohols on protein spectra in the dried state have been examined. The data listed in Table I indicate that the effects of the additives examined falls into three classes. First, several additives, disaccharides in particular, result in spectra that are very similar to the aqueous spectra. The r values for the spectra of proteins lyophilized with disaccharides range from ~0.85 to 0.97, significantly higher than r values for the respective unprotected proteins. Second, certain additives have an intermediate or no significant effects upon the spectra of the dried proteins. For example, under conditions employed here, PEG has minimal effect on the spectra of dried G-CSF. Finally, certain additives further denature the proteins upon dehydration as evidenced by a decreased correlation coefficient relative to the protein dried from buffer alone. For example, γ-IFN lyophilized from a buffer solution has a correlation coefficient of 0.743 relative to the aqueous protein spectra, but when lyophilized from a buffer solution containing myo-inositol, γ-IFN has correlation coefficient of 0.551. Finally, it is evident from these results that the effect of a given additive varies with the protein examined indicating that the protein itself plays an important role in determining the protein-additive interaction.

Numerous studies have emphasized the importance of the physical state of buffer and stabilizer components in stabilization of proteins during freeze-drying *(24,25,26)*. While these reports clearly indicate that the amorphous or crystalline nature of additives are important in achieving optimal protein stability, the results presented here indicate that the effects of the physical characteristics of a given additive cannot be extrapolated to all proteins. That is, one cannot assume that the observed effects of a given additive in a given protein system are universally applicable. The results observed for γ-IFN provide an illuminating example. Mannitol is often used as a bulking agent in preparing lyophilized proteins with the assumption that because it crystallizes upon freezing it is therefore inert with respect to the protein. However, the correlation coefficients for the spectra of γ-IFN (Table I) indicate that under conditions employed here, mannitol, and other crystallizing components such as myo-inositol, are destabilizing and induce further unfolding during dehydration. Thus, focusing upon the non-protein components, while clearly necessary, cannot alone be used to develop optimally stable lyophilized formulations. It is clear that examination of the protein's conformation provides essential information.

Effects of Stabilizers on Poly-l-lysine. The effects of the various additives on the observed dehydration-induced conformational transitions of poly-l-lysine have also been examined. Several additives (sucrose, lactose and maltose) inhibited the random coil to β-sheet transition (spectra not shown). In contrast, several other additives that do not protect protein structure upon dehydration (e.g. mannitol, myo-inositol, PEG) also do not prevent conformational transitions of poly-l-lysine. Galactose had an intermediate effect where a mixture of random coil and β-sheet is observed after dehydration. Additionally, lyophilization with sucrose also resulted in retention of the α-helical form of poly-l-lysine during dehydration. These results further support the conclusion that the mechanism of protein stabilization by additives during lyophilization is through maintenance of the native conformation in the dehydrated state.

Effects of Stabilizers upon Rehydration. In addition to the effects observed for additives on the structure of dried proteins, additives also have effects during rehydration of dried proteins. Rehydration of dehydrated sucrose/protein mixtures resulted in spectra essentially identical to spectra of the mixtures before lyophilization. This is in contrast to results for proteins dried without stabilizers. Further, no aggregation or precipitation formation was apparent for proteins which did so when rehydrated in the absence of stabilizers. The r values for the spectra of proteins lyophilized in the presence of sucrose and rehydrated are all close to 0.99 (Table I) indicating that they are essentially identical to the pre-lyophilized conformation. Thus, by preserving the native conformation effectively upon dehydration, the stabilizer fosters retention of the native structure after rehydration.

Effects of Stabilizers on LDH Structure and Activity Recovery. The effect of various stabilizers for their capacity to preserve the structure and activity of LDH has also been examined. Figure 4 is a plot of the percentage activity recovered after lyophilization and reconstitution of 1 mg/ml LDH in the presence of 25 mg/ml of various additives versus the spectral correlation coefficients for the dried state spectra of LDH lyophilized under identical conditions. Also plotted is a best fit to a linear function. The coefficient of determination, r^2, for these points is 0.97. Thus, a strong correlation is apparent between the preservation of the native structure during dehydration and retention of enzymatic activity upon reconstitution. As discussed previously, lyophilization of LDH in the absence of stabilizers results in a loss of native structure and enzymatic activity upon rehydration. In contrast, lyophilization in the presence of stabilizers results in preservation of the native structure, as indicated by increasing r values, and enhanced recovery of enzymatic activity after reconstitution. Spectral correlation coefficients near 0.8 result in essentially complete recovery of activity and lower values result in intermediate levels of activity recovery. The results for polyethylene glycol are the only exception. However, cloudiness was observed upon rehydration of the PEG containing sample indicating precipitation. It is possible that the high concentration of PEG, a strong protein precipitant *(27)*, induced precipitation of the protein during rehydration, resulting in the lower activity value. It is concluded from these results that 1) the mechanism by which dehydration induces loss of biological activity in proteins is through irreversible unfolding of the native protein structure and 2) the protective effect observed for stabilizers is through preservation of the native structure during dehydration. More recent studies (Carpenter, J.F. and Anchordoguy, T., University of Colorado, unpublished data) have indicated that freeze-drying also results in the dissociation of the LDH tetramer. Stabilizers were observed to prevent this dissociation.

Stress-specific stabilization of freeze-dried proteins

More recent experiments *(28,29)* have examined the structure of two labile enzymes, lactate dehydrogenase and phosphofructokinase, upon lyophilization and in conjunction with a stress-specific stabilization scheme which uses both an effective cryoprotectant (polyethylene glycol) and a small amount of carbohydrate to protect during dehydration. (These studies are reviewed more extensively in this volume by Carpenter *et al.*) It was observed that only when both the freezing and the drying stresses were protected against by using both types of stabilizers was the native structure observed for the dehydrated proteins. In contrast, when only one (or none) of the stresses is protected against, a non-native structure is also observed. Further, a correlation between the structure of the dehydrated proteins and the recovery of activity after

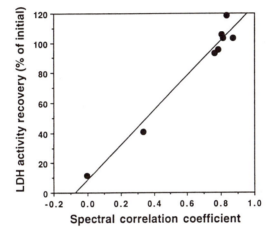

Figure 4. The percentage of initial activity recovered after rehydration plotted against the spectral correlation coefficients for LDH dried with different additives. The line plotted is the best fit from linear regression analysis, $r^2=0.97$. (The error in the activity values is ~1-3%. The experiments before and after lyophilization are performed several days apart, owing to the length of the lyophilization cycle. Activity values greater than 100% most likely arise from small changes in substrate concentration and/or temperature.)

rehydration is observed with this scheme. When the native structure of the enzyme is preserved in the dried state, essentially complete activity is recovered upon rehydration. These results further demonstrate that for labile proteins, preservation of the native structure during lyophilization is requisite for recovery of biological activity after reconstitution.

Protein conformational states during freezing, drying and rehydration

Studies of protein conformation in the aqueous, dehydrated and rehydrated states have begun to develop a clearer picture of the conformational events which occur during lyophilization. Figure 5 is a schematic of the effects of freezing, drying and rehydration on protein conformation and the effects of stabilizers during each of these steps. Starting with a native protein in aqueous state, N_{H_2O}, freezing in the presence of suitable cryoprotectant, which prevents freezing-induced unfolding, results in preservation of the native structure in the frozen state, N_{frozen}. In contrast, freezing in the absence of cryoprotectant results in a non-native conformation in the dried state, U_{frozen}. Freezing can induce denaturation in proteins through several mechanisms (8). Because some of the solution components including the bulk water crystallize and the others, including the protein, are concentrated many fold, high-salt concentrations, pH changes, and protein aggregation can all lead to protein unfolding during freezing. The cryoprotectant will shift the equilibrium, denoted K_1 in Figure 5, toward the native state during the freezing process and preserve the structure after freezing. However, a particular cryoprotectant will not necessarily have a stabilizing effect during dehydration.

During subsequent dehydration, a compound which can serve as a water substitute and hydrogen bond to the protein will maintain the structure observed after the freezing process. Starting with N_{frozen}, this should lead to formation of N_{dry}. Conversely, use of low concentrations of carbohydrates alone will be ineffective during freezing, but will prevent further damage during dehydration, resulting in conversion of U_{frozen} to U_{dry}. Use of an additive that stabilizes only during freezing or drying also results in a non-native structure, U_{dry}, after freeze-drying. This is supported by the observation that the proteins are generally less unfolded when a single stabilizer is used than when no stabilizer is used (29). In some cases, such as high sugar concentrations, a single compound can serve as both cryo- and lyoprotectant. Nonetheless, the conclusion is the same. To maintain the native structure during lyophilization, proteins must be protected against both the freezing and drying stresses.

Next, the effect of the dried state conformation must be considered during rehydration. Upon rehydration, a protein which is in a native conformation in the dried state simply exchanges the water substitute (typically a carbohydrate or polyol) for water and remains in the native state. In contrast, proteins unfolded in the dried state must refold to regain biological activity upon rehydration. As illustrated in Figure 5, refolding may compete with other pathways which lead to the formation of irreversibly denatured forms, resulting in incomplete recovery of the native structure and biological activity. For all proteins studied to date, there is at least some degree of unfolding upon lyophilization if no stabilizer is added. However, certain proteins appear to refold completely to the native structure upon rehydration, and consequently do not lose biological activity. The rates of refolding and formation of irreversibly denatured forms are denoted by k_1 and k_2, respectively, in Figure 5. Ultimately, the relative rates of k_1 and k_2 determine the recovery of biological activity after rehydration. These rates may also be affected by the presence of other components of the solution including hydrogen ion concentration, salts, and/or specific cofactors.

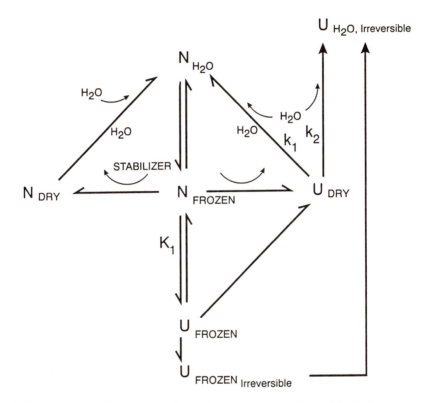

Figure 5. Schematic representation of the conformation of a protein during freezing, drying, and rehydration. N, native, U. unfolded, K1, conformational equilibrium formed upon freezing, k_1, rate constant for refolding, k_2, rate constant for formation of irreversibly denatured forms. Reproduced with permission from Ref. 29.

Based on the varied results observed for different proteins upon lyophilization and rehydration, it is apparent that proteins vary in their degree of sensitivity to freeze-drying, as they do against stresses encountered in solution. This variability brings up the concept of the *intrinsic stability* of a protein to withstand a given stress, in this case, freezing and dehydration. In general, small monomeric proteins are more stable than multimeric proteins, particularly multimeric intracellular enzymes. Hence, a stable monomeric protein may have a sufficient level of intrinsic stability to withstand freeze-drying. Similarly, if the protein is inherently stable to freezing, then only a relatively low concentration of carbohydrate, which is insufficient to stabilize labile proteins completely, should be sufficient to protect such a protein. In contrast, multimeric proteins are usually much more labile. Freezing alone is sufficient to dissociate the protein into constituent monomers *(ref.)*. Further, these proteins appear to be much more sensitive to denaturation when dissociated, most likely because they must both refold and reassociate to regain activity as in the case of LDH. Also, we should note that without water, the polar groups in the protein must satisfy their hydrogen bonding requirement by interacting intra- or intermolecularly with other groups. The unfolded molecule is then more susceptible to aggregation upon rehydration. As illustrated in Figure 5, aggregation is a potential pathway which competes with refolding, leading to an irreversibly denatured protein.

One question which arises concerns the nature of the dehydration-induced conformational changes that are indicated by the altered infrared spectra. That is, does a certain population of protein molecules completely (irreversibly) denature during lyophilization or do all molecules unfold partially and only a certain fraction properly refold? The observed spectra correspond to the average of different conformational states which exist. Thus, it is not possible to answer this question with the present information. The lack of ability to characterize proteins in the dried state, other than the spectroscopic methods presented, prohibits studies which may provide more details. Clearly, more studies are necessary to develop a complete picture of the conformational events during lyophilization and reconstitution. Recently developed techniques for *in situ* examination of protein conformation during lyophilization (see Remmele *et al.*, this volume) provide further details.

The Effect of the Conformation of Dried Proteins on Long-term Stability

The results presented thus far have focused upon stabilization of proteins during the lyophilization cycle itself. This process is the first step in developing stable lyophilized formulations. However, the ultimate goal is to develop formulations which have extended shelf stability. While the effect of conformation on the stability of proteins during lyophilization has been demonstrated, the effect of conformation on long-term shelf stability has yet to be fully examined. Several studies have examined the effects of stabilizers and other lyophilization parameters upon stability *(30-33)*, however, these studies have not examined the structure of the dried proteins. On first inspection, it is expected that (all other factors being equivalent) a protein which retains the native conformation during lyophilization will have superior stability characteristics relative to a protein which is partially unfolded for several reasons. A partially unfolded protein is likely to have a smaller barrier to further unfolding. Although the mobility of proteins is greatly diminished upon dehydration, some mobility is necessary to explain conformational and chemical degradation upon storage. Furthermore, an unfolded protein will have a greater exposure of its side chains leading to greater reactivity for groups which are normally buried in the native protein. Additionally, unfolded proteins are more likely to interact with other proteins through H-bonding and hydrophobic interaction, leading to aggregate formation. It is difficult, however, to speculate on the nature of hydrophobic interactions in the dehydrated state. Clearly, extensive studies are necessary to determine the effects of protein conformation on long-term stability. Infrared spectroscopy should prove to be a valuable tool in these examinations.

Literature Cited

1. Kauzmann, W. *Adv. Protein Chem.* 1951, 14, 1-63.
2. Edsall, J.T. and H.A. McKenzie. *Adv. Biophys.* 1983, 16, 53-183.
3. Kuntz, I.D. and W. Kauzmann. *Adv. Protein Chem.* 1974, 28, 239-345.
4. Poole, P.L. and J.L. Finney. *Int. J. Biol. Macromol.* 1983, 5, 308-310.
5. Poole, P. L. and J. L. Finney. *Biopolymers* 1983, 22, 255-260.
6. Poole, P.L. and J.L. Finney. *Biopolymers* 1984, 23, 1647-1666.
7. Yu, N.-T. and B.H. Jo. *Arch. Biochem. Biophys.* 1973, 156, 469-474.
8. Carpenter, J.F., J.H. Crowe and T. Arakawa. *J. Dairy Sci.* 1990, 73, 3627-3636.
9. Hanafusa, N. In "Freezing and Drying of Microorganisms" (T. Nei, Ed.) 1969, pp. 117-129. Univ. of Tokyo Press, Tokyo.
10. Careri, G., E. Gratton, P.-H. Yang and J.A. Rupley. *Nature* 1980, 284, 572-573.
11. Rupley, J. A., P.-H. Yang and G. Tollin. In "Water in Polymers", Rowland, S. P. ed., ACS Symp. Ser. 1980, 127, 111-132.
12. Rupley, J.A. and G. Careri. *Adv. Protein Chem.* 1991, 41.
13. Byler, D.M. and.H. Susi. *Biopolymers* 1986, 25, 469-487.
14. Surewicz, W.K. and H.H. Mantsch. *Biochim. Biophys. Acta.* 1988, 952, 115-130.
15. Prestrelski, S.J., Tedeschi, N., Arakawa, T. and Carpenter, J.F. *Biophys. J.* 1993, 65, 661-671.
16. Kennedy, S.D. and Bryant, R.G. *Biopolymers* 1990, 29, 1801-1806.
17. Gregory, R.B., Gangoda, M., Gilpin, R.K. and Su, W. *Biopolymers* 1993, 33, 513-519.
18. Purcell, J.M. and Susi, H. *J. Biochem. Biophys. Meth.* 1984, 9, 193-199.
19. Byler, D.M. and Purcell, J.M. *Proc. SPIE, Int. Opt. Engl. Soc. (Fourier-Transform Spectroscopy)* 1989, 1145, 415-417.
20. Jackson, M., P.I. Haris, and D. Chapman. *Biochim. Biophys. Acta* 1989, 98, 75-79.
21. Bellamy, L.J. and R.L. Williams. *Trans. Faraday Soc.* 1959, 55, 14-18.
22. Cutmore, E.A. and H.E. Hallam. *Spectrochim. Acta.* 1969, 25A, 1767-1784.
23. Barlow, D.J. and P.L. Poole. *FEBS Letters* 1987, 213, 423-427.
24. Franks, F. *Process Biochem.* 1989, February, iii-vii.
25. Hatley, R.H.M. and F. Franks. *J. Thermal. Anal.* 1991, 37, 1905-1914.
26. Chang, B.S. and C.S. Randall. *Cryobiology* 1992, 29, 632-656.
27. Arakawa, T. and S. N. Timasheff. *Biochemistry* 1985, 24, 6756-6762.
28. Carpenter, J.F., Prestrelski, S.J. and Arakawa, T. *Arch. Biochem. Biophys.* 1993, 303, 456-464.
29. Prestrelski, S.J., Arakawa, T. and Carpenter, J.F. *Arch. Biochem. Biophys.* 1993, 303, 465-473.
30. Izutsu, K., Yoshioka, S. and Rakeda, Y. *Int. J. Pharm.* 1991, 71, 137-146.
31. Townsend, M.W., Byron, P.R. and DeLuca, P.P. *Pharm. Res.* 1990, 7, 1086-1091.
32. Pikal, M.J., Dellerman, K.M., Roy, M.L. and Riggin, R.M. *Pharm. Res.* 1991, 8, 427-436.
33. Hora, M.S., Rana, R.K. and Smith, F.W. *Pharm. Res.* 1992, 9, 33-36.

RECEIVED June 10, 1994

Chapter 11

Real-Time Infrared Spectroscopic Analysis of Lysozyme During Lyophilization

Structure–Hydration Behavior and Influence of Sucrose

Richard L. Remmele[1], Cecil Stushnoff[1], and John F. Carpenter[2]

[1]Department of Horticulture and Biochemistry, Colorado State University, Fort Collins, CO 80523
[2]School of Pharmacy, University of Colorado Health Sciences Center, Denver, CO 80262

Spectral changes for a 49.4 mg/ml solution of lysozyme with 10% sucrose in D_2O were followed in real time during the course of lyophilization by using infrared spectroscopy. A low-temperature, single reflection, horizontal, attenuated total reflection (SRHATR) accessory, connected to a lyophilizer provided the capability to probe the effects of dehydration on sugar, protein, and water structure directly during lyophilization. At -45°C, changes were observed in the O-D stretch band which were consistent with a resumption of crystal growth, presumably initiated during non-equilibrium freezing. Sucrose bands appeared insensitive, but the amide bands showed a transition in frequency following the conclusion of this event. The final freeze-dried product exhibited the spectral traits (in the amide region) of a hydrated protein structure and upon rehydration had complete recovery of activity. The data support the water substitution hypothesis and provide physicochemical information about non-equilibrium freezing and the consequences of warming on protein structure and hydration.

It is now possible to monitor changes in biomedia at the molecular level during lyophilization in real time by using internal reflection infrared spectroscopy. Low temperature internal reflection spectroscopy offers advantages over classical transmission experiments. Obscurities that can arise as a consequence of increased absorption as water turns to ice and distortions due to light scattering within the sample matrix are minimized. Components in biomedia which are highly absorbing (i.e., water, ice) can be analyzed without problems associated with signal saturation. A reproducible path length can be achieved with this method, unlike transmission methods where a decrease in density as water becomes ice can result in an expansion of the path length sufficient to contribute to absorption measurement errors. Additionally, the spacial properties of the electric field in the

0097–6156/94/0567–0170$08.54/0

analyte medium permit absorption in all directions which is not possible with classical transmission experiments. We have combined the technology of internal reflection infrared spectroscopy with lyophilization and present information concerning the process of dehydration and its effects on the infrared spectra of lysozyme in the presence of sucrose.

Influence of Hydration on Protein Structure and Stability

Infrared spectroscopy offers a means to probe the protein structure (1-4) and the influence of hydration on the protein (5-7). Prestrelski, et al. (chapter 10, this series) have provided evidence for changes in protein structure as determined from infrared spectra of freeze-dried products. Changes in protein structure that constitute an irreversible non-native state in the lyophilized product are assumed to contribute to the loss of activity. Hence, it is plausible to characterize the effectiveness of a given lyoprotectant by its ability to preserve the native structure in the dried state. To expand this further, a freeze-dried state may exist depicting a conformation different from the native hydrated structure, but which is fully reversible giving rise to complete recovery of activity upon rehydration.

During lyophilization, several barriers for potential instability to a given protein must be overcome to establish optimum conditions for freeze-drying. First, there is the impact of freezing where changes in pH, increased solute concentration, and ice formation may lead to irreversible denaturation (8-10). Second, the composition of the product changes during drying where, in addition to the removal water, other components may be removed by precipitation (crystallization) (11). Third, the prospect of devitrification during the drying process may also contribute to irreversible conformational changes during lyophilization (12). Finally, collapse of the lyophilized cake can prove deleterious to protein stability and change the properties of the final product (12).

Carbohydrates have been shown to be effective at protecting labile proteins from inactivation by lyophilization (13-15). Their ability to provide stability in the dried state has been ascribed to their properties of substituting for lost water and maintaining the native protein structure in the dried solid (13). Even though considerable progress has been made toward optimizing conditions and improving our conception of the role solute stabilizers play in preserving functionality of labile proteins during freeze-drying, there is a lack of information regarding the complex nature of physicochemical changes during the lyophilization process. This study monitors the behavior of protein, stabilizing solute (sucrose), and water throughout the process of lyophilization by Fourier Transform Infrared Spectroscopy (FTIR).

Materials and Methods. Lysozyme, 3X crystallized, was purchased from Sigma (product L-2879) and used without any further purification. Deuterium oxide obtained from Sigma (product D-450l) was 99.9% deuterium enriched and used in the preparation of the 49.4 mg/ml protein solution. J. T. Baker analyzed reagent sucrose (product 4072-01) was used to prepare the 10% (w/v) solutions (with and without lysozyme). Enzyme assay and concentration of the solution were determined spectrophotometrically as described in an earlier publication (16). Residual moisture was determined by Karl Fischer titration (17) and measurements were reliable to within $\pm$ 0.28%.

A custom-built single reflection, horizontal, attenuated total reflectance accessory adapted for lyophilization experiments was used in the investigation. As shown in Figure 1, the sample compartment was connected to a Virtis lyophilizer. A dewar positioned above the sample cooled the compartment to the desired temperature. Two accessible openings to the sample compartment (one connected to the lyophilizer, the other capped) provided a closed environment for lyophilization. Within the sample compartment, during lyophilization, the sublimation front was not detected by the infrared until it reached the depth of penetration of the evanescent wave. Therefore, the infrared radiant energy was sampling a thin slice of the total material that was close to the Ge crystal surface. Temperature was controlled by adjusting the amount of liquid nitrogen in the dewar (for cooling) and compensating with heating elements embedded within the crystal holder. The temperature of the sample within the domain from the crystal surface to nearly a micron above the surface was assumed to equal the temperature of the germanium crystal. The temperature was measured by an iron-constantan (Type J) thermocouple in contact with the crystal. Initially, the sample compartment was cooled to -100°C under atmospheric pressure. After the desired temperature was attained, a stopcock connecting the sample compartment and the lyophilizer was opened to evacuate the sample chamber and provide access to the condenser in the lyophilizer. The time-temperature profile for the experiment was achieved as shown in Figure 2. The plateaus in the profile denoted periods where temperature was maintained to within $\pm$ 1°C.

The collection of the spectral data was carried out by using a Bio-Rad FTS-7 infrared spectrometer equipped with a highly sensitive Hg/Cd/Te detector. All spectra were measured at a resolution of 4 cm^{-1} and a total of 1024 interferograms were co-added for each spectrum (constituting a time interval of 15 minutes per spectrum). Nine-point Savitsky-Golay smoothing was used in the region of the protein bands. Band positions were repeatable to within $\pm$ 2 cm^{-1}.

Theory. In principle, a standing wave normal to the reflecting surface exists at the interface where the incident and totally reflected electromagnetic fields meet (18,19). This standing wave is the evanescent wave (see Figure 1) and exhibits the frequency of the incident wave, but its electric field amplitude exponentially decays into the rarer (sample) medium as described by the equation:

$$E = E_0 \, e^{-z/dp} \qquad (1)$$

where E_0 is the electric field amplitude for the evanescent wave at the reflecting interface (between the internal reflection element (IRE) and the sample), z is the distance the wave travels into the rarer medium (normal to the surface) and dp is the depth of penetration defined as the distance required for the amplitude to fall to e^{-1} of its value at the interface (19, 20). The depth of penetration is related to the wavelength by:

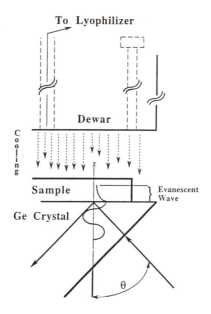

Figure 1. Schematic of the sample chamber showing the "z" axis of the evanescent wave (corresponding to Equation 1).

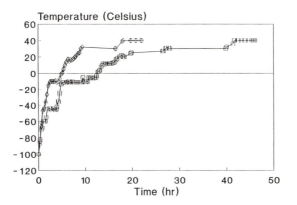

Figure 2. Time-temperature profile for the experiments of lysozyme + 10% sucrose (□) and 10% sucrose (◊) alone in D_2O.

$$dp = \lambda/(2\pi n_1 [\sin^2\theta - (n_2/n_1)^2]^{1/2}) \tag{2}$$

where λ is the wavelength of the radiant energy, n_1 is the refractive index of the IRE (which in this study is Ge), θ is the angle of incidence relative to the perpendicular drawn at the internal surface (which is fixed at 45°) and n_2 is the refractive index of the rarer medium. The electric fields exist in all spacial directions at the reflecting surface (19). For this reason there exists a difference between internal reflection and transmission such that dipoles will absorb energy in internal reflection regardless of their orientation, while in transmission dipoles oriented parallel to the direction of propagation will not absorb energy. The relationship between absorbance, A, of a band in an ATR (attenuated total reflection) spectrum and the electric field amplitude is presented in the equation:

$$A = (n_2/n_1)\cdot(\alpha/\cos\theta)\cdot\int_0^t E^2 dz \tag{3}$$

where α is the absorption coefficient per unit thickness, t. By substituting Equation 1 into Equation 3 and integrating, the expression becomes:

$$A = (n_2/n_1) \cdot (\alpha E_0^2/\cos\theta) \cdot (dp/2) \cdot [1 - \exp(-2t/dp)] \tag{4}$$

For a sample which is infinitely thick, the last term reduces to 1 and the expression becomes:

$$A = (n_2/n_1) \cdot (\alpha E_0^2/\cos\theta) \cdot (dp/2) \tag{5}$$

In our experiments, the sample solution occupies a volume of 0.5 ml and a thickness in excess of 1 mm (1000 μm). At this thickness, equation 5 may be used to evaluate the absorbance of the sample medium. Based on Equation 5, at relatively long wavelengths an increase in the depth of penetration (Equation 2) will be favored and will contribute to an increase in absorption within the sample medium. The sampling area varies as $1/\cos\theta$ and will enhance absorption where the light beam interacts with a larger amount of the sample (i.e., oblique incidence) rather than at normal incidence. For Ge at a fixed angle of incidence of 45°, this parameter becomes a constant of $\sqrt{2}$. Index matching, where the factor n_2/n_1 approaches unity, will improve the interaction with the sample (21) and augment absorbance. The electric field amplitude (squared) in the rarer medium consists of polarized components, parallel and perpendicular to the plane of incidence (19, 21). Both electric field components decrease with increasing angle of incidence (21) and therefore vary opposite to sampling area.

To better understand how low temperature might influence absorbance, the refractive indices (n_1 and n_2) in the 27°C to -107°C regime were evaluated. Germanium at 25°C was reported to have a refractive index of 4.005 at 10.6 μm (22). The wavelength dependence for the refractive index of Ge at 27°C, reported by Adams (23), is illustrated in Figure 3. At wavelengths that are shorter than 8

μm (1250 cm^{-1}) there was a dramatic increase in the refractive index. The data in Figure 3 for -107°C were estimated based on the known value at 10.6 μm (22) and Sellmeyer's dispersion formula (24). The refractive index of water was used to assess changes in n_2 at low temperature and to approximate changes in aqueous media. Dispersion in the refractive index (oscillating positively and negatively) was characteristic in the region of water absorption bands. Refractive index data were reported for liquid water (25°C) in the mid-infrared region which exhibits small oscillations in the region of the O-H stretch and H-O-H bend regions (25, 26). The oscillation in the refractive index was previously determined as shown in Table I (26, 27). In comparison to liquid water, the dispersion of the refractive index for ice substantially increases in magnitude in the region of the O-H stretch, whereas in the H-O-H bend region there is a dramatic decrease (27, 28). An average refractive index in the region from 900 to 2800 cm^{-1} of 1.219 was used for water at -107°C and this value was slightly lower than the value at -93°C reported by Tsujimoto, et al., (27). With these data it was possible to evaluate the effect on n_2/n_1 at low temperature. The values $n_2 = 1.307$, $n_1 = 4.010$ at 25°C and $n_2 = 1.219$, $n_1 = 3.967$ at -107°C, provided an $n_2/n_1 = .326$ and .307, respectively. From these data, it can be concluded that lowering temperature slightly reduces the n_2/n_1 ratio. This factor would have a slightly negative influence on absorbance concomitant with lower temperatures according to Equation 5.

For bulk materials (infinitely thick) which consist of weak absorptions (characteristic of the protein solutions in this study), simple expressions have been derived for the effective thickness of penetration by the parallel ($d_{e\parallel}$) and perpendicular ($d_{e\perp}$) components of the electric field within the sample (21). The effective thickness from an ATR measurement is defined to represent the equivalent thickness obtained by a transmission measurement (19). Harrick's equations (18, 21) for $d_{e\parallel}/\lambda$ and $d_{e\perp}/\lambda$ are given by,

$$\frac{d_{e\parallel}}{\lambda} = \frac{(n_2/n_1)\cos\theta(2\sin^2\theta - (n_2/n_1)^2)}{\pi(1 - (n_2/n_1)^2)[(1 + (n_2/n_1)^2\sin^2\theta - (n_2/n_1)^2](\sin^2\theta - (n_2/n_1)^2)^{1/2}} \tag{6}$$

and

$$\frac{d_{e\perp}}{\lambda} = \frac{(n_2/n_1)\cos\theta}{\pi(1 - (n_2/n_1)^2)(\sin^2\theta - (n_2/n_1)^2)^{1/2}} \quad (3) \tag{7}$$

A comparison of $d_{e\parallel}$, $d_{e\perp}$ and dp as a function of wavenumber (from 600 to 4000 cm^{-1}) for 25°C and -107°C is displayed in Figure 4. It is important to note that $d_{e\parallel}/\lambda$ and $d_{e\perp}/\lambda$ are measures of the strength of interaction of the evanescent field with the rarer medium (21). There is little difference in the depth of penetration for the two temperatures (Figure 4). Therefore, dp can be considered constant within this range of temperatures. Moreover, theory suggests that internal reflection studies conducted within this temperature range will not exhibit artifacts in the spectra as a result of dp (Equation 5). However, for $d_{e\parallel}$ there is a noticeable temperatures permitting the acquisition of useful quantitative information.

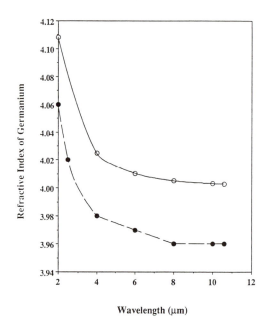

Figure 3. Variation of germanium refractive index with wavelength at 27°C (○ solid line) and -107°C (● broken line).

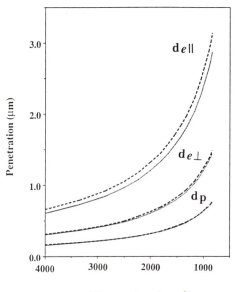

Figure 4. Variation of penetration depth with frequency (cm⁻¹) for effective thickness components, $d_{e\parallel}$, $d_{e\perp}$ and dp were performed at either 25°C (broken line) or -107°C (dotted line).

Table I. Optical Constants of Liquid and Frozen Water

Region	ν(cm⁻¹)	λ(μm)	n_2(25°C)	κ(25°C)	n_2(-93°C)	κ(-93°C)	n_2(-123°C)	κ(-123°C)	α(25°C)	α(-93°C)	α(-123°C)
O-H stretch	3846	2.60	1.242	.0025	1.16	.0023	1.16	.0021	.015	.013	.012
	3704	2.7	1.188	.0160					.088		
	3636	2.75	1.157	.05099	1.09	.0022	1.06	.011	.2696	.108	.052
	3571	2.80	1.142	.1007					.5161		
	3448	2.90	1.201	.2231					1.1611		
	3390	2.95	1.292	.2306	.62	.400	.65	.41	1.2691	1.056	1.135
	3226	3.10	1.467	.1309					.7784		
	3125	3.20	1.478	.0625	1.86	.600	2.07	.600	.3628	4.382	4.877
	2941	3.40	1.42	.0137	1.39	.030	1.47	.030	.0719	.154	.163
	2778	3.60	1.385	.0037	1.33	.083	1.33	.009	.01798	.038	.042
	2703	3.70	1.374	.0026					.0121		
	2632	3.80	1.364	.0025					.0113		
	2564	3.90	1.357	.0028					.0122		
	2500	4.00	1.351	.0034					.0144		
H-O-H bend	1818	5.50	1.298	.0089	1.22	.022	1.24	.022	.0264	.061	.062
	1786	5.60	1.289	.011					.0318		
	1754	5.70	1.277	.0159					.0448		
	1724	5.80	1.262	.0261					.0714		
	1695	5.90	1.248	.0498					.1324		
	1667	6.00	1.265	.0846					.2241		
	1639	6.10	1.319	.0993	1.19	.051	1.25	.070	.2698	.125	.180
	1539	6.50	1.339	.0293	1.21	.061	1.25	.070	.0758	.143	.169
	1429	7.00	1.317	.0243	1.23	.051	1.25	.050	.0574	.113	.112

Note: n and κ values at 25°C were obtained from ref. 26 and those at -93°C and -123°C were from ref. 27

difference between the 25°C and -107°C curves. These findings suggest that $d_{e\parallel}$ is more sensitive to a small decrease in n_2/n_1 with low temperature than $d_{e\perp}$ or dp. A decrease in the n_2/n_1 parameter results in a decrease in the effective thickness. Additionally, the relationship, $d_{e\parallel} = 2d_{e\perp}$ is valid throughout the range of frequencies shown in Figure 4. Interestingly, the depth of penetration is less than a μm and decreases nonlinearly from about 0.7 μm (600 cm^{-1}) to 0.17 μm (4000 cm^{-1}).

The absorption coefficient, α, as it relates to Lambert's Law is given by:

$$\alpha = 4\pi n\kappa/\lambda \qquad (8)$$

where n is the real part of the spectral complex refractive index and κ is the imaginary part known as the absorption index (29). Values of κ and α (for the O-H stretch and H-O-H bend regions) are displayed in Table I (26, 27). The impact of temperature on the absorption coefficient of water in the vicinity of the O-H stretch and H-O-H bend is illustrated in Figures 5a and b. The changes in the O-H stretch are characteristic of the phase change from liquid water to ice. A distinct change in the O-H stretch maximum from 3390 to 3125 cm^{-1} is observed. In addition, a dramatic increase in the relative intensity of the O-H stretch is exhibited as a result of the transition toward ice. The intensity of the O-H stretch for ice is nearly five times that of liquid water based on the absorption coefficient (Figure 5a). Thus, ice absorbs more strongly than liquid water in the O-H stretch region. It is also noteworthy that the absorption coefficient near 3125 cm^{-1} increases further on going from -93°C to -123°C.

In contrast to the O-H stretch region, the H-O-H bending region displays a significant reduction in the band intensity at -93°C which slightly increases at -123°C, but in both cases the reduction is significantly lower than the absorption associated with liquid water at 25°C. Hence, the transition from liquid water to ice serves to reduce the absorption coefficient near 1640 cm^{-1}. The bands associated with ice in the bending region show signs of broadening. The use of D_2O in place of water permits a window of transparency in the region of the amide I band of proteins, thereby eliminating the complexity associated with this region during lyophilization. It becomes apparent that the nature of the absorption coefficient of water with low temperature varies with phase behavior (changing in frequency and intensity). Near 3390 cm^{-1} the absorption coefficient is nearly constant and exhibits little change as a consequence of phase behavior.

The theoretical perspective presented offers insight to aid in experimental design and the evaluation of data during lyophilization. The penetration depth dictates that the sample be in close contact with the Ge crystal surface. For meaningful quantitative measurements, it is important that the sample remain in contact with the surface during the experiment. The inherent differences between internal reflectance and transmission measurements involving path length (i.e., wavelength and refractive index dependence) and absorbance behavior (i.e., electric field interactions with the sample), suggest caution when directly comparing data from these two techniques. With internal reflection infrared spectroscopy a reproducible and consistent path length can be achieved at low

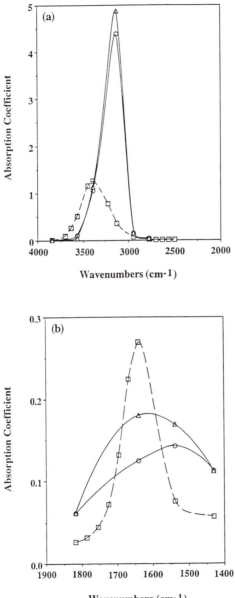

Figure 5. Absorption coefficient as a function of frequency for (a) O-H stretch and (b) H-O-H bend regions. The drawn contours estimate band shapes. The data represent 25°C (□), -93°C (○), and -123°C (Δ).

Results and Discussion. The important spectral changes for lysozyme with 10% sucrose, from onset of lyophilization to the final freeze-dried product at 40°C after 46 hr, are displayed in Figure 6. All spectra presented were normalized to the amide I band (near 1649 cm⁻¹). An interesting event occurred during the process at -45°C (2.37-2.62 hr). Dramatic changes in the O-D stretch region (2600 cm⁻¹ to 2200 cm⁻¹) produced an increase in band intensity and differences in overall band contour. Distinct peaks near 2357 and 2434 cm⁻¹ and a shoulder component near 2507 cm⁻¹ became predominant features in the band contour. The temperature was maintained at -45°C over the course of 1.5 hr until no further changes were observed in the O-D stretch band (intensity or contour). At this point, the event was assumed completed and subsequently the temperature was permitted to increase. Consecutive scans from -24°C (4.60-4.85 hr) to -13°C (4.95-5.20 hr) exhibited the removal of the predominant peaks and a decrease in relative intensity of the O-D stretch band. The resulting features in the overall contour appeared broad and representative of bulk water.

O-D Stretch Region. We interpret the changes in the spectra at -45°C to result from a resumption of crystal growth upon warming, as a consequence of initial non-equilibrium freezing (30). We assume that under the conditions of this experiment protein and sucrose were in a matrix absent of any eutectic behavior. The crystalline phase in this case was D_2O ice. This hypothesis is supported by comparing the spectra of D_2O ice at -50°C (shown in Figure 6 as a dashed contour) with the -45°C spectra (2.37 -3.87). The predominant peaks and shoulder features were present although differences in relative peak intensities also existed. The differences in peak intensity between these spectra may have arisen from exchange between the hydroxyl moiety of the sugar and deuterium which could have contributed to some underlying fraction of the O-D stretch band (31). Additionally, there was a contribution from the protein amide A' (5) which also existed in the region and which collectively could have resulted in the differences in the peak intensities near 2434 cm⁻¹ and 2357 cm⁻¹ compared to D_2O ice. The transition occurred at low temperature where ice could form most readily.

Lastly, as water (H_2O) became ice there was a substantial increase in the absorption coefficient. This effect may explain the increase in the overall absorbance of the O-D stretch band. Similar spectral behavior has been reported by Bertie and Whalley (32) during annealing experiments with D_2O at -93°C where an increase in intensity and two peaks were observed in the O-D stretch region. They ascribed these effects to devitrification of initially deposited vitreous or partly vitreous ice. We suggest that the changes in the O-D stretch at -45°C were a result of resumed crystal growth approaching completion of non-equilibrium freezing.

Ice was sublimed away from -24°C to -13°C after 0.35 hr. The remaining composition at -13°C appeared to be primarily amorphous (based on the broadness of the overall O-D stretch band contour and the similarity of the band features to that of liquid D_2O). The changes in the O-D stretch from -13°C to the completion of the experiment are shown in Figure 7. These data have all been normalized to

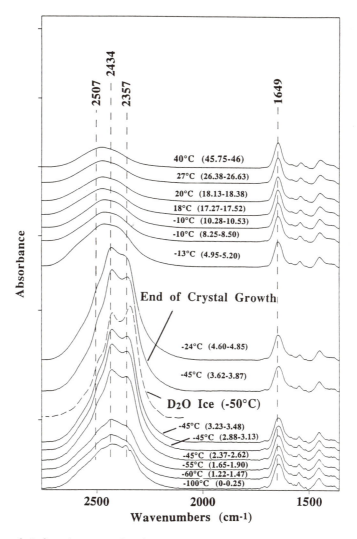

Figure 6. Infrared spectra showing the O-D stretch and amide regions of the lysozyme-sucrose sample at key points during the lyophilization process. The time intervals (hr) for collecting data are in parentheses. Positions marked indicate spectral features of D_2O ice in the sample matrix and the amide I band positions in a 49.4 mg/ml solution.

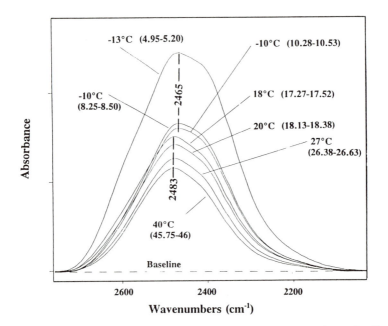

Figure 7. Expanded view of changes in the O-D stretch band from the final stages of primary drying to the end of the experiment.

the amide I and show a decrease in overall band area as lyophilization proceeded. A substantial loss in water occurred from -13°C to -10°C (8.25-8.50 hr) and then at -10°C (10.28-10.53 hr) further loss in moisture was not observed (characterized by lack of additional reduction in the O-D stretch band area at subzero temperatures). It is tempting to attribute the spectral appearance of the O-D stretch at -13°C to the completion of primary drying and to the onset of secondary drying. This would require water loss from the amorphous phase below -10°C. Although this may be possible, the position of the band maximum in D_2O liquid occurs near 2483 cm^{-1} and the band position from -13 to -10°C is near 2465 cm^{-1}. The 18 cm^{-1} difference in the band maxima as drying proceeded above -10°C could also be explained as the removal of residual ice within the amorphous matrix. Thus when the ice was completely removed the band maximum shifted to the position of liquid D_2O (similar to spectra at 18°C and higher temperatures). This would require completion of primary drying after -10°C (10.28 - 10.53 hours).

The next period of noticeable loss was at 18°C which continued through 20°C (18.13-18.38 hr). It becomes clear at this point in the sequence of the experiment that loss of moisture was considerably slower above 0°C (on the order of hours for small changes in band area). The next major loss occurred at 27°C after 8.25 hr. followed by the final product some 19.37 hr later. We suggest that changes in band area, were due to primary drying, until constant composition was attained at -10°C. Beyond this point in the progress of the experiment, additional loss of moisture was a result of secondary drying. However, we also acknowledge the possibility of secondary drying of the amorphous phase occurring between -13 and -10°C. Perhaps a combination of both primary and secondary drying was responsible for the change in the O-D stretch band area.

As a consequence of non-equilibrium freezing, the sample matrix had not maximally freeze-concentrated the solute (34). Differential scanning calorimetry (DSC) studies of 20% sucrose have shown a glass transition at about -50°C followed by a small exotherm near -40°C when rapidly cooled and annealed (37). By not maximally freeze-concentrating the lysozyme-sucrose solutes, unfrozen water in the amorphous fraction may have contributed to a lower glass transition temperature as well as to the resumption of crystal growth during rewarming. Sucrose-containing solutions are known to maximally freeze-concentrate to 80% (37). Hence, in order to optimize the current lyophilization scheme, slow cooling rates favoring equilibrium freezing and maximum freeze-concentration of the solutes would be expected to yield a higher glass transition temperature and eliminate an increase in ice mass during warming. Maintaining the temperature at or just below the glass transition temperature during primary drying should prevent collapse of the cake structure (38). Higher subzero glass transition temperatures contribute to increased rates of sublimation (due to a higher vapor pressure of ice) and should improve efficiency in overall freeze-drying conditions.

The final lyophilized product (containing lysozyme-sucrose) visually appeared as a brittle, homogeneous, transparent glass. After 3.5 months, stored in a sealed glass vial in a desiccator (25°C), the sample was analyzed for moisture and activity. Upon rehydration complete activity was recovered. Karl Fischer water titration measurements revealed 7% moisture remaining within the sample. In contrast, the protein sample lyophilized without sucrose which had a final moisture content of 3.7% and also exhibited no loss in activity upon rehydration (Remmele, R. L.; Stushnoff, C. *Biopolymers*, in press.). The appearance of the final product suggested that collapse of the cake structure occurred and perhaps accounts for the difficulty in removing more moisture from the sample. The collapse probably occurred after resumed crystallization when the temperature was permitted to increase above Tg'. Collapse is known to result from surface tension-induced viscous flow of the amorphous phase after the sublimation front has moved past that region of the sample (38). The glass transition temperature (Tg') reported in a phase diagram for sucrose-water occurs at approximately -40°C (11). This is higher than the glass transition temperature from the infrared data we observed, approximately -55°C based on the temperature of resumed crystallization. The lower glass transition temperature was consistent with failure to maximally freeze-concentrate the solute. At -24°C, sublimation was still taking place, but the ice-like features in the O-D stretch band disappeared at -13°C exhibiting the appearance of an amorphous phase. Hence, nearing the completion of sublimation, viscous flow probably occurred, collapsing the pore structure within the sample matrix and creating an impervious condition within the sample during further drying.

Sucrose Bands. Figure 8 shows the changes in the sucrose bands which correspond to the spectra of Figure 6. It is important to realize that the composition is changing as a result of water loss throughout the process of the experiment. It is difficult to assign all the observed infrared bands associated with sucrose because the complex vibrations arising from combination bands overlap with fundamental modes and interact with one another causing distortions in band shapes (13, 33). However, general features throughout the process can be evaluated. The major sucrose bands at 971, 1040, and 1139 cm^{-1} are sharper at -100°C compared to the same bands at 40°C (Figure 8). It is well known that measurements of infrared spectra on samples of oligosaccharides at low temperature often result in increased peak heights and narrower bands permitting greatly improved resolution (31). On drying, the bands shifted to lower frequency and no significant changes were observed in band position at the end of resumed crystal growth.

Protein Bands. An expanded view of the protein infrared bands (corresponding to Figure 6) is shown in Figure 9. The observed changes in the amide I showed a propensity to shift toward higher frequency in contrast to the amide II which shifted toward lower frequency as lyophilization progressed. This shifting pattern has been reported before from infrared investigations of lyophilized lysozyme in the presence and absence of trehalose (13). It appears to arise as a consequence of dehydration. However, there is a possibility that changes in conformation may also be involved. Recent solid-state ^{13}C NMR studies of lyophilized lysozyme indicate that the protein can adopt a broad distribution of conformational states which are small in magnitude (34, 35). The shifting pattern of the amide I and II bands for lysozyme-sucrose relative to the lysozyme without sucrose (no evidence of resumed crystallization) are shown in Figure 10a and 10b. Interestingly, a discontinuity exists in the shifting pattern for both the amide I and

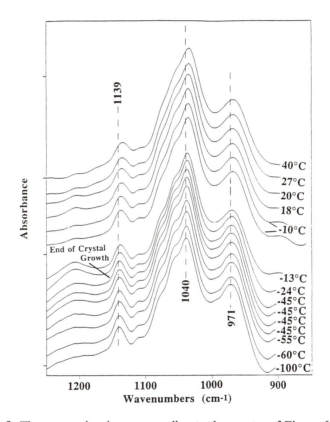

Figure 8. The sucrose bands corresponding to the spectra of Figure 6 during the lyophilization experiment.

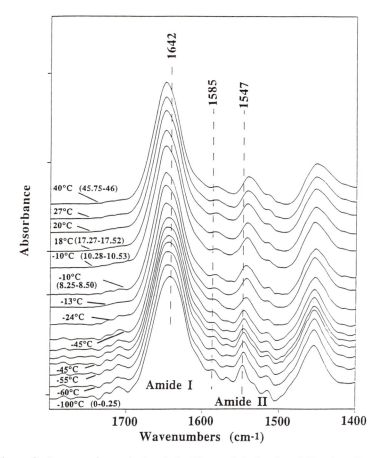

Figure 9. Spectra of protein bands in Figure 6 during lyophilization. Protein recovers complete bioactivity upon reconstitution of the lyophilized powder.

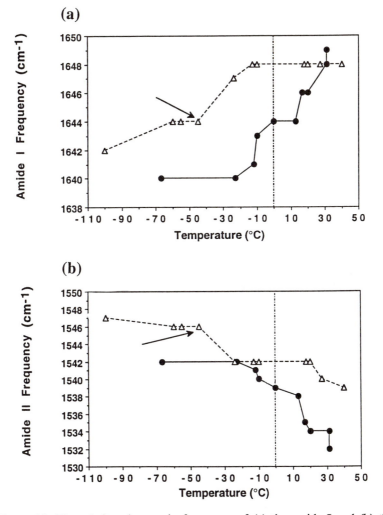

Figure 10. The relative changes in frequency of (a) the amide I and (b) the amide II bands with temperature for lysozyme in the presence (Δ, dashed line) and absence (●, solid line) of sucrose. The vertical dashed line marks the 0°C temperature and the arrows indicate the end of resumed crystal growth (noted in Figure 6).

II at the completion of resumed crystallization. Perhaps an explanation for the shifting event at the conclusion of crystallization, where the temperature was allowed to increase, is movement within the sample matrix as a result of viscous flow fostered by an increase in the interaction of sucrose with protein. In addition, the amide II in the presence of sucrose did not shift to the low frequency observed for the sample without sucrose. Studies carried out on lyophilized products of lysozyme with 10% trehalose agree with this behavior in the amide II band (13). The amide I after primary drying did not exhibit a change in position during secondary drying in the presence of sucrose. However, the solution without sucrose did exhibit a shift in frequency throughout the secondary drying process.

A comparison of the amide bands for a fully hydrated 49.4 mg/ml solution at 25°C to those of final lyophilized products with and without sucrose present is shown in Figure 11. The spectrum of the hydrated protein in solution is characterized by an amide I band maximum at 1649 cm^{-1}, a carboxylate band near 1584 cm^{-1}, and the amide II near 1541 cm^{-1}. The lyophilized product with sucrose (lower solid contour) exhibited an amide I band maximum at the same frequency as the aqueous solution and it also possessed the carboxylate band and an amide II maximum close to that of the hydrated protein in solution. The presence of the carboxylate band in the sucrose-containing dried product, suggests that sucrose substituted for the water by hydrogen bonding to the dried protein (13).

Conversely, the lyophilized product without sucrose (lower dashed contour) had a broader amide I band with a maximum at the same frequency as the hydrated solution, a carboxylate band which nearly disappeared and a broader amide II band shifted to lower frequency near 1532 cm^{-1}. In an earlier lyophilization study of lysozyme in D$_2$O (Remmele, R. L.; Stushnoff C. *Biopolymers*, in press.), the amide II band exhibited the broadening and low frequency shift at a hydration level near 12% (before the carboxylate band began to decrease). In the lyophilized lysozyme-sucrose product, with a moisture content of 7%, the amide II of lyophilized lysozyme did not exhibit these alterations suggesting that H-bonding of the protein with hydroxyls of sucrose was responsible for the observed differences.

The experiment with sucrose was repeated with attention to preserving the cake structure by completing sublimation below Tg'. A 4% hydration level was achieved. In the region of the amide bands, spectral features were similar to those of the initial study with 7% hydration (Figure 11). The carboxylate band was distinctly present at 1582 cm^{-1} and the amide I was at 1647cm^{-1}. The amide II was narrower and at a higher frequency than the lyophilized lysozyme without sucrose, at a hydration level of 3.7%. The spectral differences between these two lyophilized products provided clear evidence that sucrose in the matrix was involved in maintaining the structure of the protein. These observations argue in favor of the hypothesis that sucrose substitutes for water in the desiccated state (13). If all the water at 4% hydration was associated with the protein and excluded from the sugar, this would amount to 29 molecules of water per molecule of lysozyme. This amount of water is insufficient to completely hydrate the charged groups on the protein (5).

Comparing the spectral features of the amide I and II in the aqueous solution to those of the lyophilized product with sucrose (Figure 11), the broadness

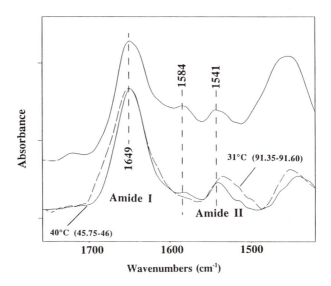

Figure 11. Comparison of spectra in the amide region for 49.4 mg/ml hydrated lysozyme (top trace) to the final lyophilized products with (lower solid trace) and without (lower dashed trace) sucrose.

observed in liquid solution (at 25°C) may be explained as an average protein structure consisting of a high degree of translational mobility and less contact between protein molecules. Upon freezing, as the solute and protein were freeze concentrated, the bands sharpened and the broadness was reduced. We attribute this change to a higher degree of association among protein molecules, characterized by a loss in mobility and resulting in part from lower temperature and increased viscosity in the amorphous phase. Therefore, it seems plausible that the band features in the lyophilized state arise from a higher degree of protein-protein association induced by sucrose.

Differences observed in the spectral features of lysozyme in solution (with and without sucrose) at 25°C suggest that sucrose affected the protein prior to freezing. The amide I and II bands were found to be narrower with 10% sucrose. This may be corroborating evidence in support of a more compact structure due to co-solute exclusion (Carpenter et al., Chapter 9, this series). Additionally, from the data presented in Figure 10a and b, a significant difference in band position for both the amide I and II was observed when sucrose was in the frozen matrix as shown at -70°C. It can be inferred from these findings that the presence of sucrose seems to have inhibited the amide I shift to lower frequency and to have shifted the amide II to higher frequency on freezing, relative to the initial positions before freezing (Figure 11). We hypothesize that sucrose induces an increase in the amount of association between protein molecules in solution, which continues into the frozen state.

Based upon the preceding discussion, during secondary drying as water is removed from the amorphous phase, hydrogen bonding of surface groups needed to preserve the protein structure is provided (at least partially) by the hydroxyl moiety of the sucrose molecules. In order for the sucrose to substitute for lost water, it would be necessary for it to be in close proximity to the surface of the protein in the freeze-concentrated amorphous phase. It follows that the role of sucrose before freezing may be an important link to its role as a stabilizer during the drying process.

Future Perspectives

Efforts are being made to perform experiments under more accurate measurement conditions within the sample chamber. Vacuum pressure inside the sample compartment and temperature on the top surface of the sample (beyond the depth of penetration of the evanescent wave) will be monitored throughout future lyophilization experiments. It is hoped that optimizing the sample environment will permit better control over the lyophilization process when using this technique. Experiments are currently underway to directly evaluate compositional changes during the process by determining hydration levels via quantitative changes in the O-D and O-H stretch bands. Investigations addressing protein-surface interactions which contribute to stabilizing/destabilizing effects during lyophilization are also planned. This technology provides a means to design and optimize lyophilization protocols from a protein structure viewpoint. It is anticipated that this capability will facilitate enhanced product viability and augment our ability to improve formulations for lyophilization.

Acknowledgments

Support for this research was provided in part by the Colorado Agricultural Experiment Station, project 6901; the USDA NSSL, and USDA CGRO 90-02419.

Table of Symbols

A	- absorbance
α	- absorption coefficient
dp	- depth of penetration for an evanescent wave
$d_{e\parallel}$	- parallel component of effective thickness
$d_{e\perp}$	- perpendicular component of effective thickness
E	- electric field amplitude of an electromagnetic wave
E_0	- electric field amplitude for the evanescent wave at the reflecting surface
K	- imaginary portion of the complex refractive index (absorption index)
λ	- wavelength of radiant energy
n	- real portion of the complex refractive index
n_1	- refractive index of the internal reflection element
n_2	- refractive index of the rarer medium
θ	- angle of incidence at the reflection interface
t	- thickness of rarer medium layer
z	- distance the electromagnetic wave travels into the rarer medium

Literature Cited

1. Susi, H.; Byler, D. M. *Biochem. Biophys. Res. Commun.* **1983**, *115*, 391-397.
2. Casal, H. L.; Kohler, U.; Mantsch, H. H. *Z. Naturforsch.* **1987**, *42C*, 1339-1342.
3. Surewicz, W. K.; Moscarello, M. A.; Mantsch, H. H. *Biochemistry* **1987**, *26*, 3881-3886.
4. Yang, P. W.; Mantsch, H. H. *Biochemistry* **1987**, *26*, 2706-2711.
5. Careri, G.; Giansanti, A.; Gratton, E. *Biopolymers* **1979**, *18*, 1187-1203.
6. Poole, P. L.; Finney, J. L. *Biopolymers* **1983**, *22*, 255-260.
7. Poole, P. L.; Finney, J. L. *Biopolymers* **1984**, *23*, 1647-1666.
8. Pikal, M. J. *Biopharm.* **1990**, *3*, 26-30.
9. Tamiya, T.; Okahashi, N.; Sakuma, R.; Aoyama, T.; Akahane, T.; Matsumoto, J. J. *Cryobiology* **1985**, *22*, 446-456.
10. Koseki, N.; Kitabatake, N.; Doi, E. *J. Biochem.* **1990**, *107*, 389-394.
11. Franks, F.; Hatley, R. H. M.; Mathias, S. F. *Pharm. Technol. Int.* **1991**, *3* 24-34.
12. Chang, B. S.; Randall, C. S. *Cryobiology* **1992**, *29*, 632-656.
13. Carpenter, J. F.; Crowe, J. H. *Biochemistry* **1989**, *28*, 3916-3922.

14. Crowe, J. H.; Carpenter, J. F.; Crowe, L. M.; Anchordoguy, T. J. *Cryobiology* **1990**, *27*, 219-231.

15. Arakawa, T.; Kita, Y.; Carpenter, J. F. *Pharmaceutical Res.* **1991**, *8*, 285-291.

16. Remmele, R. L. Jr.; McMillan, P.; Bieber, A. *J. Protein Chemistry* **1990**, *9*, 475-485.

17. May, J. C.; Grim, E.; Wheeler, R. M.; West, J. *J. Biological Stand.* **1982**, *10*, 249-259.

18. Harrick, N. J. *J. Opt. Soc. Am.* **1965**, *55*, 851-857.

19. Harrick, N. J. *Internal Reflection Spectroscopy;* John Wiley & Sons, Inc.: New York, NY, 1967; pp 13-65.

20. Ohta, K.; Iwamoto, R. *Applied Spectroscopy* **1985**, *39*, 418-425.

21. Harrick, N. J.; duPre, F. K. *Applied Optics* **1966**, *5*, 1739-1743.

22. Hoffman, J. M.; Wolfe, W. L. *Applied Optics* **1991**, *30*, 4014-4016.

23. Adams, J. H. In *Encyclopedia of Chemical Technology;* Mark, H. F., Othmer, D. F., Overberger, C. G., Seaborg, G. T., Grasson, M., Eckroth, D., Eds.; John Wiley & Sons, Inc.: New York, NY, 1980, Vol.11; pp 791-802.

24. Bottcher, C.J.F. *Theory of Electrical Polarization*; Elsevier publishing Co.: New York, NY, 1952; pp 253-260.

25. Bertie, J. E.; Eysel, H. H. *Applied Spectroscopy* **1985**, *39*, 392-401.

26. Hale, G. M.; Querry, M. R. *Applied Optics* **1973**, *12*, 555-563.

27. Tsujimoto, S.; Konishi, A.; Kunitomo, T. *Cryogenics* **1982**, *22*, 603-607.

28. Wood, B. E.; Roux, J. A. *J. Opt. Soc. Am.* **1982**, *72*, 720-728.

29. Simon, I. *J. Opt. Soc. Am.* **1951**, *41*, 336-345.

30. MacKenzie, A.P. *Phil. Trans. R. Soc. Lond. B.* **1977**, 278, 167-189.

31. Michell, A. J. *Aust. J. Chem.* **1967**, *27*, 1257-1266.

32. Bertie, J. E.; Whalley, E. *J. Chem. Phys.* **1964**, *40*, 1637-1645.

33. Mathlouthi, M.; Koenig, J. L. In *Advances in Carbohydrate Chemistry and Biochemistry;* Tipson, R. S., Horton, D., Eds.; Academic Press, Inc.: New York, NY, 1986, Vol. 44; pp 7-89.

34. Kennedy, S. D.; Bryant, R. G. *Biopolymers* **1990**, *29*, 1801-1806.

35. Gregory, R. B.; Gangoda, M.; Gilpin, R. K.; Su, W. *Biopolymers* **1993**, *33*, 513-519.

36. Roos, Y.; Karel, M. *J. Food Sci.* **1991**, *56*, 1676-1681.

37. Roos, Y.; Karel, M. Int. *J. Food Sci. Technol.* **1991**, *26*, 553-566.

38. Pikal, M. J.; Shah, S. *International Journal of Pharmaceutics* **1990**, *62*, 165-186.

39. Careri, G.; Gratton, E.; Yang, P. H.; Rupley, J. A. *Nature* **1980**, *284*, 572-573.

RECEIVED April 19, 1994

Chapter 12

Recovery of Type A Botulinal Toxin Following Lyophilization

Michael C. Goodnough[1] and Eric A. Johnson[1-3]

Departments of [1]Food Microbiology and Toxicology and [2]Bacteriology, Food Research Institute, University of Wisconsin, 1925 Willow Drive, Madison, WI 53706

Type A botulinum toxin is diluted to very low concentrations (ng/ml) for medical use and preserved by lyophilization in a mixture of human serum albumin and sodium chloride at a slightly alkaline pH. This commercial process results in considerable loss of activity. In this study, conditions were found that gave >90% recovery of the toxicity following lyophilization of solutions containing 20-1,000 mouse 50% lethal doses (1-50 ng of toxin complex). Full recovery of starting toxicity was obtained upon drying 0.1 ml when the pH was maintained below 7.0 and serum albumins or other protein excipients were used as stabilizers without sodium chloride. Possible mechanisms of toxin inactivation were examined and may include aggregation, deamidation, and peptide bond hydrolysis.

Type A Clostridium botulinum neurotoxin is currently used in the treatment of hyperactive muscle disorders and dystonias (1,2,3). Disorders approved by the Food and Drug Administration for treatment in the United States include blepharospasm, strabismus, and hemifacial spasm. Other dystonias being treated with the toxin on an investigational basis in North America and some countries in Europe and South America include adult- onset spasmotic torticollis, aberant regeneration of the seventh facial nerve, myofacial pain syndromes, and others (1). Treatment of patients involves injecting very small quantities (nanograms) of the toxin directly into the affected neuromuscular region causing a decrease in muscle hyperactivity.

C. botulinum consists of a diverse group of organisms which are distinguished by their ability to produce a potent proteinaceous neurotoxin. There are seven antigenically different serotypes recognized, termed A, B, C_1, D, E, F, and G. All of the botulinal organisms are characterized as gram-positive, obligately anaerobic, endospore forming rod-like bacteria. The endospores of C. botulinum as well as those of a close relative, Clostridium tetani, are found worldwide in soils, marine, and freshwater environments (4).

Neurotoxins produced by Clostridium botulinum are large molecular weight proteins (approximately 150 kDa) (4) which cause a flaccid paralysis of muscles by binding to motor neuron endplates and inhibiting release of the neurotransmitter acetylcholine. The neurotoxins produced are a part of a complex with non-neurotoxic

[3]Corresponding author

0097–6156/94/0567–0193$08.00/0

proteins (5). High quality type A toxin complex has a specific toxicity of 3×10^7 mouse intraperitoneal (IP) 50% lethal doses per mg (1 IP LD_{50} = 1U) (5). One U of toxin is defined as the amount necessary to kill 50% of a population of mice when injected intraperitoneally. Type A toxin produced by C. botulinum is part of a complex of at least seven different noncovalently bound proteins one of which is neurotoxic (Fig. 1, lane 1; arrow indicates neurotoxin). In culture medium this toxin complex associates into mainly trimers with a molecular weight of approximately 900 kDa (6).

Type A toxin complex may be dissociated chromatographically yielding pure neurotoxin with a molecular weight of approximately 145 kDa (7) (Figure 1, lane 3). The pharmacologically important neurotoxin has been well characterized (5, 8). The purified neurotoxin molecule itself consists of two separate peptide chains of 93 and 52 kDa (Figure 1, lane 4) that are connected by a disulfide link between cysteine residues 430 and 454 (9). The two peptide chains are designated the heavy and light chains and have biologically distinct activities. The heavy chain is responsible for binding to the motor neuron target and facilitates internalization of the light chain through the cell membrane, while the light chain itself has neurotoxic activity (4).

After some 23 years of development as a pharmaceutical, a single batch of crystalline toxin complex was licensed by the Food and Drug Administration for medical use in the United States (5). This batch (#79-11) was produced at the University of Wisconsin-Madison, Food Research Institute in 1979 by Dr. E. J. Schantz. There are some anomalies with this particular batch of toxin kept in our laboratory which are considered elsewhere (Johnson and Goodnough, In *A Handbook of Dystonias* ; J. Tsui, Ed; Marcel Dekker, Inc.: New York, NY, 1993, in press). The current commercial product is made by combining type A neurotoxin complex with 5.0 mg/ml human serum albumin (HSA) and 9.0 mg/ml sodium chloride at a pH of 7.3 and drying 0.1 ml per vial in a flash-evaporation cycle to obtain 100 $\pm$ 30 active U of toxin, 0.5 mg of HSA, and 0.9 mg of sodium chloride per vial. This product has a saline concentration of 0.9% when reconstituted in 1.0 ml of dH_2O. The current formulation results in a considerable loss of activity (up to 90 %) after drying and reconstitution (10). This process causes formation of inactive toxin that could serve as a toxoid and incite antibody formation. Antibodies have been detected in patients resistant to treatment (1, 2, 11). Our research has been aimed at improving recovery of active toxin following lyophilization, thereby reducing the amount of material required to obtain the requisite 100 U per vial and reducing the chances of neutralizing antibody formation.

Materials and Methods.

Bacterial strains and culture production. The Hall A strain of type A C. botulinum is currently used to produce crystalline type A complex for medical use. This strain was originally obtained from Dr. J. H. Mueller at Harvard University and was further screened for high toxin titers by Dr. E. J. Schantz and coworkers at Fort Detrick, MD. Stock cultures of C. botulinum Hall A were grown statically in 15 ml Hungate tubes containing 10 ml of cooked meat medium + 0.3 % dextrose (CMM; Difco Laboratories, Detroit, MI) under an anaerobic atmosphere (80% N_2, 10%CO_2, 10%H_2) at 37°C for 24 h and frozen at -20°C until use. CMM cultures of the Hall A strain give toxin titers in excess of 10^6 U/ml in 48-72 hr.

For toxin production, cultures of Hall A were grown statically in 12-15 liter volumes of toxin production medium (TPM) consisting of 2.0% casein hydrolysate (Sheffield Laboratories, NY), 1.0% yeast extract (Difco), and 0.5% dextrose, pH 7.3, for 5-7 days at 37°C. Cultures of the Hall A strain show heavy growth in this medium during the first 24-48 hours followed by autolysis of the culture which is evident as a clearing and settling over the next 48-72 hr.

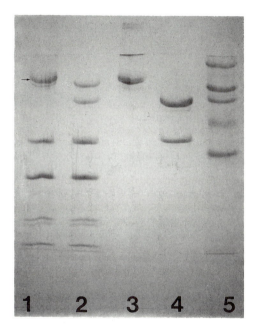

Figure 1. SDS-PAGE of type A toxin complex and purified type A neurotoxin. Lanes 1 (unreduced) and 2 (reduced with 0.5% w/v dithiothreitol) contained 4-5μg of type A complex. Lanes 3 (unreduced) and 4 (reduced) contained 4-5 μg of purified type A toxin. Lane 5 contains molecular weight markers (rabbit myosin, 205 kDa, E. coli ß-galactosidase, 116 kDa, rabbit phosphorylase b, 94 kDa, bovine serum albumin, 66 kDa, chicken egg albumin, 45 kDa); 5μg of protein total.

Electrophoresis. Protein samples were examined electrophoretically using the
Pharmacia Phast system according to the manufacturers instructions using precast gels
obtained from Pharmacia and run according to the manufacturers instructions.

Toxin assays. Toxin titers were estimated in mice by using the intravenous method of
Boroff and Fleck (12) and the intraperitoneal method of Schantz and Kautter (13) in 18-
22 g, female ICR strain mice. Botulinum toxin for titration was dissolved in 50 mM
sodium phosphate, pH 6.8, and then further diluted as required in 30 mM sodium
phosphate, 0.2 % gelatin, pH 6.4.

Botulinal toxin preparations. Type A toxin complex for use in drying studies was
purified from culture broth by using both the FDA approved method involving
precipitation and crystallization of the toxin complex and by a method using preparative
column chromatography. Briefly, to prepare crystalline type A toxin, the 5-7 day
culture was acidified with sulfuric acid to pH 3.4 and the resulting precipitate that
contained the toxin complex was extracted and purified through additional precipitations
and crystallizations according to the method of Duff et al. (14). Type A crystalline toxin
used for these studies was crystallized twice in the presence of 0.9 M $(NH_4)_2SO_4$ and
had a 260/278 nm absorbance ratio of 0.53. The extinction coefficient of type A toxin
complex at 278 nm is 1.65 for a 1 mg/ml solution of the toxin complex in a 1 cm light
path. The ratio of the absorbances at 260 and 278 nm is used as a measure of the
relative quantities of protein and nucleic acid which are present. A 260/278 nm ration of
0.5 indicates that the preparation contains approximately 0.1% nucleic acid or less (5).
Type A toxin complex was also purified chromatographically by the method of Tse et al.
(6). Extracts of the acid precipitated material were dialyzed against 50 mM sodium
citrate, pH 5.5, and chromatographed on a 1 liter DEAE-Sephadex A-50 column (Sigma
Chemical Co., St. Louis, MO) equilibrated with the same buffer. One-tenth the column
volume or less was chromatographed in a single passage with the toxin complex eluting
in the first column volume without a salt gradient being applied. Detection of protein
was done by absorbance at 278 nm. Fractions from this protein peak which had a
260/278 nm absorbance ratio of less than 0.6 were pooled and precipitated by the
addition of $(NH_4)_2SO_4$ at approximately 60% saturation (39 g/100 ml). This material
was electrophoretically equivalent to the crystalline material on sodium dodecylsulfate-
polyacrylamide gels (SDS-PAGE) and showed a higher specific toxicity than the
crystalline toxin (>30 U/ng for the chromatographed material compared to
approximately 20 U/ng for the crystalline toxin). Type A neurotoxin was purified from
the associated non-toxic proteins of the complex by chromatography on 25 ml of SP-
Sephadex C50 in 25 mM sodium phosphate, pH 7.0. Contaminating material did not
bind to the column under these conditions. The toxin was eluted with a linear 0-0.25 M
sodium chloride gradient. The resulting material showed a single band on unreduced
SDS-PAGE gels of approximately 145 kDa indicating that it was free of non-toxic
complex proteins (Figure 1, lane 3). All gels were stained with 0.1 % Coomassie
Brilliant Blue R250, destained in 5:5:2 dH_2O:methanol:acetic acid and then silver
stained according to the procedure of Hames and Rickwood (15). Upon reduction of
the disulfide bonds in the samples with dithiothreitol (0.5% w/v), the two chains of the
toxin migrated separately as the 93 kDa heavy chain and the 52 kDa light chain (Figure
1, lane 4). The purified neurotoxin had a specific toxicity of 90 U/ng.

Lyophilization and excipients. For lyophilization, toxin samples were diluted in
solutions containing the excipients to be tested. All excipients were from Sigma
Chemical Co., St. Louis, MO. The solutions were aliquoted at 0.1 ml or 0.5 ml into 2
ml glass vials (Fisher Scientific Co., Pittsburgh, PA) and the teflon lined screw cap
closures fastened loosely. The samples were then quickly frozen in liquid nitrogen.
The frozen samples were placed into a lyophilization flask which was then immersed in

liquid nitrogen. The flask was then connected to a laboratory freeze drier (Virtis Freezmobile 12, Virtis Co., Inc., Gardiner, NY). When the pressure had dropped to approximately 30 mTorr, the liquid nitrogen jacket was removed. Pressure was maintained at or below 30 mTorr and condenser temperature was maintained constant at -60°C. Samples were allowed to come to room temperature and drying continued at ambient temperature over the next 18-24 h. After drying, the flask was removed and the vials tightly capped. Vials were kept at ambient temperature and assayed for toxicity within 1-3 days.

Size exclusion high pressure liquid chromatography. HPLC was performed with a Rainin HPXL system (Rainin Instrument Co., Woburn, MA). The size exclusion columns used were a Rainin SEC column (4.6mm x 250 mm) at a flow rate of 0.25 ml/min or a Dupont Zorbax GF-450 column (9.4 mm x 250 mm) at a flow rate of 1.0 ml/min. The isocratic solvent system in both cases was 100 mM sodium phosphate, pH 7.0, and the protein was detected by absorbance at 278 nm. The molecular weight cutoff for both columns was reported by the manufacturer as $1\text{-}2 \times 10^6$ daltons.

Results.

Recovery of toxin following freezing at -20°C, -70°C, and at -200°C was examined. Only slight reductions in recovery of active toxin following freezing and thawing was noted at -20°C and -70°C as compared to -200°C. Freezing at the highest temperature tested of -20°C or at an intermediate temperature of -70°C in 50 mM sodium phosphate buffer, pH 6.8, resulted in slight loss of activity (75-90% recovery) compared to >90% recovery on freezing in -200°C liquid nitrogen (10). All samples were tested for toxicity after approximately 24 h at the freezing temperatures. Previous studies in our laboratory also showed that no detectable inactivation (≤20 %) of type A crystalline toxin (10^4 U/ml) occurred during repeated freezing and thawing at -20°C in several buffers (16). These buffers included 50 mM sodium phosphate (pH 6.2-6.8), 50 mM sodium succinate (pH 6.0), or 50 mM sodium citrate (pH 5.5). Freezing in 30 mM acetate buffer, pH 4.2, at -20°C resulted in irreversible loss of toxicity (16).

We next examined the effect of salt concentration during freezing on the recovery of toxicity. During its preparation, botulinum toxin is precipitated using ammonium sulfate at concentrations in excess of 60% saturation at room temperature with no loss of activity. This salt concentration is approximately equivalent to a 33% (w/v) salt solution. Sodium chloride at 0.9% in the commercial formulation may reach concentrations in excess of 6 M during lyophilization (17) equivalent to a 35% (w/v) solution. Freezing samples of type A toxin complex at -20°C and -70°C in 5.0 M solutions of sodium chloride at pH 6.2 did not affect toxin activity and full recovery was obtained 24 h after freezing irrespective of the freezing temperature.

Lyophilization of toxin complex at 100-1,000 U/vial (i.e. 3.3-33 ng/vial) in the absence of protein excipients gave almost complete loss of activity (Table I).

One possible mechanism of loss was aggregation but it was not possible to assess aggregation of the toxin complex with such small quantities of toxin . However, when larger amounts of purified toxin (100-250 μg) were lyophilized, there was a substantial degree of aggregation (Table II).

At higher concentrations of toxin (>100 μg/ml), a certain amount of self-protection occurred during drying as only 32% of the toxin remaining in solution was inactivated.

Table I. Effect of excipients on recovery of type A toxin complex toxicity after lyophilization

Excipients	Toxin concentration[a]	pH	%recovery[b]
sodium phosphate[c] bovine serum albumin/	50, 100, 1,000	5.0, 6.0, 6.8	<10
sodium chloride[d]	100	6.4	10
bovine serum albumin[e]	100, 1,000	6.4	88, 75
bovine serum albumin/citrate[f]	100, 1,000	5.0	>90, >90
bovine serum albumin/phosphate[g]	100, 1,000	5.5	>90, >90
bovine serum albumin/phosphate[g]	1,000	7.3	60
bovine serum albumin/phosphate[h]	1,000	5.5	>90
human serum albumin[i]	100, 1,000	6.4-6.8	>90, >90
alpha-lactalbumin[i]	1,800	6.1	>78
lysozyme[i]	1,800	5.3	>78
gelatin[i]	1,800	6.3	>78

[a]Type A mouse IP lethal doses/vial before lyophilization; [b]%recovery = (number mouse lethal doses after lyophilization/number mouse lethal doses prior to lyophilization) x 100; [c]50mM sodium phosphate; [d]bovine serum albumin (5.0mg/ml), sodium chloride (9.0mg/ml); [e]bovine serum albumin (9.0mg/ml); [f]bovine serum albumin (9.0mg/ml), 50mM sodium citrate; [g]bovine serum albumin (9.0mg/ml), 50mM sodium phosphate; [h]bovine serum albumin (9.0mg/ml), 50mM potassium phosphate ; [i]concentration = 9.0mg/ml (Adapted from ref. 10).

Table II. Aggregation of purified type A toxin upon lyophilization *

Sample	specific activity[a]	concentration (µg/ml) A$_{278}$[b]	BCA[c]
pre-lyoph	88	100	131
post-lyoph[d]	60	65	80

*in 50 mM sodium phosphate, pH 6.8, 0.1 ml fill volume. [a]U/ng determined by i.v. method of Boroff and Fleck (1966); [b]absorbance at 278 nm using an extinction coefficient for type A toxin of $E^{1\%}$=16.3 in 1 cm light path; [c]BCA assay (bicinchoninic acid assay with an average of two determinations) using bovine serum albumin as standard (Peirce Biochemicals, Rockford, IL, U.S.A.); [d]assayed after being dissolved in 1.0 ml dH$_2$O and centrifuged to remove aggregated protein.

The use of either sodium phosphate or potassium phosphate in our experiments gave equivalent results and did not affect recovery of active toxin (Table I).

In most experiments in this study, 0.1 ml of botulinal toxin was lyophilized. This amount is equal to or less than the amount needed for most automated filling equipment in commercial lyophilization laboratories. When the fill volume was increased to 0.5 ml with a subsequent reduction in serum albumin concentration to 1.8 mg/ml to maintain a post-lyophilization concentration of 0.9 mg/ml following reconstitution, a slightly lower recovery of active toxin (60-80% of initial toxicity) was obtained after lyophilization and reconstitution. The loss of activity may have been caused by the thicker frozen cake which resulted from the 0.5 ml fill in a 2 ml vial and increased resistance to water vapor escape leaving more residual moisture in the freeze-dried cake. Alternatively, the larger surface area and lower serum albumin concentration could have caused more surface denaturation. Further work is needed to clarify the process parameters which influence toxin lability including fill-volume and lyophilization cycle.

We next investigated possible mechanisms of inactivation of the toxin during lyophilization. Examination of the toxin complex by SDS-PAGE before and after lyophilization did not show evidence of peptide bond hydrolysis (Figure 2). Lanes 1 (unreduced) and 2 (reduced with 0.5% w/v dithiothreitol) contained the starting toxin complex. A typical pattern for type A toxin complex was observed with the toxin (145 kDa unreduced, 93 and 52 kDa reduced) and the associated non-toxic binding proteins (118, 50, 35, 21,8, and 20.8, and 17.5 kDa). Lanes 4 (unreduced) and 5 (reduced) contained the complex following lyophilization. The formulation used for freeze-drying contained 50 mM sodium phosphate, pH 7.3, 9.0 mg/ml sodium chloride, and 250 µg/ml of toxin complex (28 U/ng) and 0.1 ml was lyophilized in 2 ml glass vials (10). The lyophilized material had a specific activity of 7 U/ng corresponding to a loss of approximately 70% of the starting activity. Therefore, in the case of toxin complex, the loss in activity did not appear to correlated with the hydrolysis of peptide bonds.

Native gel electrophoresis of type A complex and purified toxin indicated that aggregation occurred during lyophilization. Type A toxin complex migrated as a single polypeptide on 4-15% polyacrylamide native gels. When 8M urea was added to the toxin complex, some dissociation occurred (approximately 30%). Purified type A neurotoxin (145 kDa) treated with 8 M urea migrated as a single polypeptide. Purified type A neurotoxin that had been lyophilized contained aggregates that were unable to enter the gel without first being treated with 8M urea (not shown).

Size exclusion chromatography of type A toxin complex by HPLC in 100 mM sodium phosphate, pH 7, at room temperature showed that the toxin complex before and after lyophilization eluted with the same retention time in this solvent system. Following lyophilization, peaks indicative of breakdown products were not detected and peak broadening which may indicate aggregation also was not observed. However, since the toxin complex is very nearly at the size exclusion limit of both commercially available columns (approximately $1\text{-}2 \times 10^6$ daltons in each case) used it may not be possible to detect large molecular weight aggregates by using this system.

Isoelectric focusing was used in an attempt to determine the difference in overall charge of the lyophilized, purified toxin particularly to investigate possible deamidation during lyophilization. Purified type A toxin with a molecular weight of 145 kDa, a specific toxicity of ca. 90 U/ng, and a pI of 6.1. Lyophilization caused no detectable shift in pI, although the specific toxicity dropped to 20 U/ng. These results suggest losses in specific activity need not be correlated with shifts in pI.

Discussion.

The conditions used for lyophilization in this study had a considerable effect on recovery of active botulinal toxin. The most critical factors that contributed to recovery

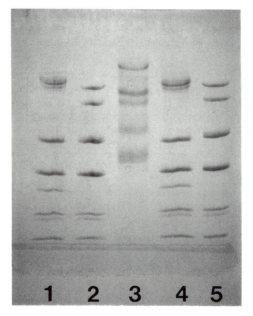

Figure 2. SDS-PAGE of type A toxin complex before and after lyophilization.
Lanes 1(unreduced) and 2 (reduced with 0.5% w/v dithiothreitiol) contained 4-5 µg
of type A toxin complex with a specific toxicity of 28 U/ng. Lane 3 contained
molecular weight markers (rabbit myosin, 205 kDa, E. coli ß-galactosidase, 116
kDa, rabbit phosphorylase b, 94 kDa, bovine serum albumin, 66 kDa, chicken egg
albumin, 45 kDa, bovine erythrocyte carbonic anhydrase, 29 kDa); 5 µg of protein
total. Lanes 4 (unreduced) and 5 (reduced) contained 4-5 µg of type A toxin
complex with a specific activity of 7 U/ng following lyophilization.

of active toxin after lyophilization were the removal of sodium chloride, addition of a protein excipient, and an acidic pH. The removal of sodium chloride and inclusion of bovine or human serum albumin gave a recovery of >90 % of the starting toxicity when the pH was maintained below 7.0. The freeze concentration and subsequent differential crystallization rates of buffer components and salts present in solution has been shown to alter the pH during lyophilization (18, 19), especially with sodium phosphate buffered systems. Protein excipients were required for recovery of toxicity: the solutions of BSA or HSA at 9.0 mg/ml resulted in high percentage recovery of active toxin at pH 6.4-6.8. Full recovery of toxin was also obtained in BSA or HSA when the pH was adjusted to 5.0 or 5.5, whereas a slight loss of 20-30% was found at pH 7.0 -7.3 (Table I). It is possible that denaturation and aggregation is occurring in systems such as sodium phosphate and sodium phosphate/sodium chloride with pH values in excess of approximately 6.2 due to isoelectric precipitation during the freeze-concentration period. An excess of serum albumin in such freeze-drying formulations may help to keep the pH above the isoelectric point of the toxin.

The differences in recovery rates dependent on pH could also be due to the tendency for increased deamidation at higher pHs. This phenomenon has been demonstrated with other proteins such as lysozyme (20), triose-phosphate isomerase (21), calmodulin (22) and others (23). Asparagine deamidates more readily than glutamine in model peptides with some of the contributing factors being the primary amino acid sequence and the tertiary structure of the protein (24, 25). Also, deamidation is more prevalent at asparagine-glycine and asparagine-serine sequences (24, 25). There are at least six asparagine-glycine sequences in type A neurotoxin with three in the light chain and three being in the heavy chain (Table III).

Table III. Positions of asparagine-glycine and asparagine-serine amino acid sequences in type A neurotoxin (adapted from ref. 9)

	Light chain	Heavy chain
Asn-Gly	-15,16-	
	-178,179-	
	-402, 403-	
		-539,540-
		-1032,1033-
		-1243, 1244-
Asn-Ser		-570,571-
		-798,799-
		-935,936-
		-954,955-
		-971,972-
		-1026,1027-
		-1093,1094-
		-1151,1152-

Other potentially labile sequences include eight asparagine-serine sequences in the heavy chain region of the toxin (9) (Table III). These sequences were deamidated in other protein systems under certain drying and storage conditions (25). Isoelectric focusing

(IEF) is one method of detecting deamidation as the pI shifts accordingly with the loss of amino groups from asparagine residues with the concomitant gain of a carboxyl group. However, IEF of purified type A toxin may be insensitive to charge modifications since the molecule is very large and a charge shift may not be observed from losses of a minimal number of amino groups (23, 25). Work is also currently under way in our laboratory using more sensitive methods to determine if deamidation is occurring during lyophilization of botulinal toxin.

Shelf stability is an important property of protein pharmaceuticals such as botulinum toxin. The present commercial formula must be stored below -10°C and ideally below -20°C following drying to retain potency. One possible reason for the instability observed at ambient temperatures is that the glass transition temperature of the excipient mixture is around -10°C and storage above this transition temperature allows the residual moisture in the amorphous phase to interact with the toxin molecule promoting degradative chemical reactions (26). One aspect of the shelf life of botulinal toxin complex that has not been investigated is whether the presence of the non-toxic binding proteins impart stability to the toxin molecule. Work in our laboratory has shown that purified type A toxin (145 kDa) in buffered solution begins to hydrolyze in a few days at 4°C resulting in fragmentation and reduction in toxicity, whereas type A complex (900 kDa) in solution at the same temperature is more stable and does not lose potency for weeks (Goodnough and Johnson, unpublished data). Recovery of activity immediately following lyophilization of purified type A neurotoxin does not seem to be dependent on the presence of these non-toxic binding proteins as full recovery is obtained by using the same formulation that is used for the toxin complex (Goodnough and Johnson, manuscript in preparation). Shelf life of the toxin in various formulations is currently under study.

Aggregation of the toxin complex is difficult to detect by using normal size-exclusion chromatography as the native complex of ~900 kDa elutes near the exclusion limit of commercially available columns. However, since peak broadening and earlier eluting peaks were not observed for the lyophilized toxins it is possible that the conditions used (i.e. 100 mM sodium phosphate, pH 7.0) promoted dissociation of aggregates to the native MW of 800-900 kDa. Further work is underway to elucidate the cause(s) of denaturation events occurring during lyophilization of botulinum toxins.

Acknowledgments.

We thank Stacy Kramer for valuable assistance with these experiments. We thank E. J. Schantz and M. E. Whitmer for valuable advice and for cooperation in certain experiments on toxin stability.

References.

1. Borodic, G. E.; Pearce, L. B; Johnson, E. A.; Schantz, E. J. *Opthamol.Clinics North Amer.* **1991**, 4, 491-503.
2. Jankovic, J; Brin, M. F. *N. Engl. J. Med.* **1991**, 324, 1186-1194.
3. Scott, A. B. In *Botulinum neurotoxin and tetanus toxin*; L. L. Simpson, Ed. Academic Press, Inc., San Diego, CA, **1989**, p. 399-412.
4. *Botulinum neurotoxin and tetanus toxin*; Simpson, L.; Ed.; Academic Press, San Diego, CA; **1989**.
5. Schantz, E. J.; Johnson, E. A. *Microbiol. Rev.* **1992**. 56, 80-99.
6. Tse, C; Dolly, J.; Hambleton, P.; Wray, D.; Melling, J. *Eur. J. Biochem.* **1982**, 122, 493-500.
7. Gimenez, J.; DasGupta, B. *J. Prot. Chem.* **1993**, 12, 349-361.

8. Tsui, J. K. C.; Eisen, A.; Stoessl, A. J.; Calne, S.; Calne, D. B. **1986**, *Lancet ii,* 245-247.
9. Binz, T.; Kurazono, H.; Wille, M.; Frever, J.; Wernars, K.; Niemann, H. *J. Biol. Chem.* **1990**, 265, 9153-9158.
10. Goodnough, M.; Johnson, E. *Appl. Environ. Microbiol.* **1992**, 58, 3426-3428.
11. Greene, P.; Shale, H.; Fahn, S.; Brin, M.; Friedman, A. *Neurology.* **1987**, 37(suppl.), 12.
12. Boroff, D. A.; Fleck, U. *J. Bacteriol.* **1966**, 92, 1580-1581.
13. Schantz, E. J.; Kautter, D. A. *J. Assoc. Off. Anal. Chem.* **1978**, 61, 96-99.
14. Duff, J., Wright, G., Klerer, J., Moore, D., Bibler, R. *J. Bacteriol.* **1957**, 73, 42.
15. Hames, B. In *Gel Electrophoresis*; Editors, D. Rickwood, B. Hames, Ed.; Oxford University Press, Essex, UK, **1990**, pp. 60-66.
16. Schantz, E. J.; Scott, A. B. In *Biomedical aspects of botulism*. Editor, G. E. Lewis Ed.; Academic Press, Inc., New York, NY, **1981**, pp. 143-150.
17. Franks, F. Freeze-drying: In *Developments in Biological Standardization;* Editors, J. May and F. Brown, Eds. Karger Publishing, Basel, **1990**, Vol. 74, pp. 9-19.
18. Van den Berg, L. *Cryobiology.* **1966**, 3, 236-242.
19. Pikal, M. *BioPharm..* **1990**, 3(9), 26-30.
20. Ahern, T.; Klibanov, A. *Science.* **1985**, 228, 1280-1284.
21. Ahern, T.; Casal, J.; Petsko, G.; Klibanov, A. *Proc. Natl. Acad. Sci.* U. S. A. **1987**, 84, 675-679.
22. Johnson, B.; Shirokawa, J.; Aswad, D. *Arch. Biochem. Biophys.* **1989**, 268, 276-286.
23. Johnson, B.; Shirokawa, J.; Hancock, W.; Spellman, M.; Basal, L.; Aswad, D. **1989**, 264, 14,262-14,271.
24. Wright, H.T. *Critical Rev. Biochem. Mol. Biol.* **1991**, 26, 1-52.
25. Liu, D. *Trends Biotechnol.* **1992**, 10, 364-369.
26. Franks, F. *Cryo-Letters.* **1990**, 11, 93-110.

RECEIVED April 19, 1994

Chapter 13

Cryopreservation of Surimi, a Fish Protein Food Base

Grant A. MacDonald[1] and Tyre C. Lanier[2]

[1]Seafood Research Laboratory, CRI Crop and Food Research Ltd., P.O. Box 5114, Nelson, New Zealand
[2]Department of Food Science, North Carolina State University, Raleigh, NC 27695–7624

Surimi, a wet, frozen protein food base refined from fish muscle, requires the addition of cryopreservative compounds to stablize its functional properties. A number of compounds have been evaluated for this role, with sucrose and sorbitol being primarily used commercially at present. A study to determine the mechanism whereby sodium lactate (SL) stabilizes and cryoprotects actomyosin (AM) is used as an example of an additive stabilizing by the preferential exclusion mechanism. Whereas low concentrations of SL stabilized AM, higher concentrations destabilized AM and the transition from stabilization to destabilization was intimately associated with either an increase or decrease in solution surface tension with SL concentration. Mechanistic studies on the cryopreservative properties of these compounds in fish muscle proteins parallels that in pharmaceutical and other non-food applications, and could lead to development of more efficacious mixtures of compounds for the cryopreservation of food proteins.

Surimi is a semi-refined concentrate of the myofibrillar proteins of fish muscle (Figure 1). Myosin, the primary myofibrillar constituent, forms a protein gel network when solubilized in saline and heated. This gel imparts texture by binding fat, water, and muscle particles and is the basis of many processed muscle foods such as frankfurters, bologna, and surimi-based shellfish analog products and Asian kamaboko (1).

Myosin molecules consist of two regions: two globular heads and a fibrous rod. The heads are the site of the ATPase activity that is important in muscle contraction, while the rod, which has no enzyme activity, is involved in the packing of the myosin molecules into a filament (2). While both sections of the myosin molecule can be involved in gel formation (3), conformational changes usually occur in the head region at a temperature somewhat lower than that at which unfolding takes place in the rod (4, 5, 6). Thus myosin ATPase activity can be used as a sensitive indicator of myosin denaturation related to unfolding of the tertiary structure in the head region. It follows that ATPase activity also is related directly to the functional property of gelation.

The myofibrillar proteins of fishes, these being poikilotherms, are more labile to denaturation than the contractile proteins of homeotherms commonly converted to

0097–6156/94/0567–0204$08.00/0

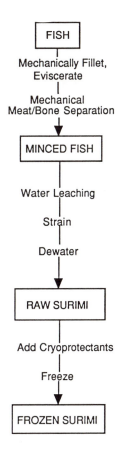

Figure 1. Schematic of manufacture of frozen surimi from whole fish.

meat for food, including beef, pork, and poultry (7). During freezing or frozen storage unfolding of myofibrillar proteins (mainly myosin) exposes nonpolar amino acids which become available for formation of hydrophobic interactions with like groups in the vicinity. This process leads to protein aggregation and loss in gelling functionality (8, 9). Thus, surimi requires the inclusion of cryoprotective components (typically sucrose and sorbitol) prior to freezing to ensure stability of the proteins. The use of cryoprotectants has allowed Japanese food processors to smooth raw material flow, increase investment in production facilities, and offer high quality products to the market on a continuous basis. As a result of the introduction of this technology in the early 1960's, the Japanese market was able to expand at a tremendous rate and production of kamaboko, a surimi-based food, increased from about 500,000 m.t. (metric tonnes) in 1960 to a peak of 1,187,000 m.t. in 1975 (10). Currently the U.S. consumes about 150 million pounds of surimi-based shellfish analog products while producing over 150,000 m.t. of surimi for domestic use and export.

Cryoprotectants of Surimi

Noguchi (11) surveyed a wide variety of chemical compounds for their ability to maintain the solubility of carp actomyosin in dilute solutions over brief periods of frozen storage. This model system was reliable in predicting the ability of compounds to cryoprotect the functionality of surimi during extended frozen storage. Besides a variety of carbohydrate compounds, including many mono- and di-saccharides and several low molecular weight polyols, many amino acids and carboxylic acids were also found to be cryoprotective. Other workers have also reported the cryoprotective action of a number of amino acids, quaternary amines, and other compounds with regard to the stability of various proteins and enzymes (12, 13, 14, 15).

The nucleotides ATP, ADP and IMP have been shown to exert a protective effect on fish actomyosin stored at -20°C while the nucleotide catabolites inosine and hypoxanthine destabilized these proteins (16). This finding may help explain why fresh fish, with consequent higher concentrations of ATP, ADP, and IMP, are more stable during frozen storage than less fresh fish (17, 18).

Watanabe et al. (19) demonstrated the cryoprotective ability of certain surfactants, particularly certain polyoxyethylene sorbitan esters and sucrose esters, in preventing loss of gel-forming ability in surimi. A cryoprotective effect has even been attributed to triglycerides. Free fatty acids, which may be released through hydrolysis of phospholipids and then denature proteins, instead interact preferentially with triglyceride, thus indirectly protecting the proteins (20).

Sucrose and/or sorbitol, typically alone or mixed 1:1 and added at 8% w/w to leached fish muscle, serves as the primary cryoprotectant in manufacture of surimi from Alaska pollock. Polyphosphate at 0.2-0.3% is also commonly added as a synergist (mechanism unknown) to the cryoprotective effect of the carbohydrate additives (21). These carbohydrates were chosen because of their relatively low cost, good availability, and low tendency to cause Maillard browning in the bright white kamaboko products typically enjoyed by the Japanese. For certain shellfish analog products, less costly dextrose and crystalline fructose are now being tested with success, since the heat process used does not lead to objectionable browning.

All these additives impart a considerable sweet taste to the surimi, which can be objectionable in certain product applications (22). Thus, there has been some effort to select non-sweet additives with a cryoprotective effect equal to that of sucrose or sorbitol. Lanier and Akahane (23) patented the use of Polydextrose (®Pfizer Inc.), a non-sweet, low calorie, bulking agent, for the cryoprotection of muscle proteins. They compared its effectiveness with that of sucrose/sorbitol and a 10 dextrose equivalent (DE) maltodextrin (also imparts no sweetness) in maintaining the salt-solubility and gel-forming properties of Alaska pollock surimi. While these three additives maintained similar high levels of solubility in the myofibrillar proteins at -28°C over several months compared to a control, the surimi containing the 10 DE maltodextrin failed to form strong and cohesive gels. These results indicated that the 10 DE maltodextrin interfered with the gelation of the surimi myofibrillar protein, in much the same way as do pre-gelatinized starch and certain gums (24, 25, 26). A more recent study (27) indicates the potential of using higher DE starch hydrolysis products as effective cryoprotectants with less interference in the gelation process of the proteins.

Many other low MW sugars and polyols which could be used as muscle cryoprotectants are presently or soon to be available. Lactitol and lactulose reportedly have low sweetness, the former having also been demonstrated to effectively cryoprotect surimi protein (28, 29, 30). Maltitol, isomalt, and hydrogenated glucose syrups could also be considered for special applications (28). High molecular weight food polymers, such as edible gums, have been proposed to function as effective cryoprotectants, but tests have failed to demonstrate their effectiveness (31, 32, 33). The reduced functionality of muscle proteins in the presence of gums may result from their competition with protein for water, or from interaction with proteins, which results in poor protein gelation. Addition of gums in the fully hydrated form without adding excess water, and the attainment of concentrations sufficient for cryoprotection, are additional problems in the practical application of gums as cryoprotectants.

Mechanisms of Cryoprotection and Cryostabilization

Sucrose and sorbitol are not only cryoprotectants, but are also known to stabilize proteins to the denaturing effects of heat (34, 35, 36, 37, 38). Similarly, sodium chloride addition, which was found to promote freeze denaturation of beef (37), also destabilizes myofibrillar proteins to heat denaturation (39, 36). Thus, the mechanism of heat stabilization by low MW carbohydrates may also explain their cryoprotective properties.

Timasheff and coworkers (40, 41) showed that solutes such as sugars and low molecular weight polyols are excluded from the surface of protein molecules in solution. The preferential exclusion of solutes from the protein surface has been largely attributed to the effects of the solute in increasing the surface tension of water (41, 42, 43, 44). However, several cryoprotective compounds are thought to be excluded by other mechanisms, such as by stearic hindrance.

This explanation of the mechanism of cryoprotection by low M.W. sugars and polyols is in direct contradiction to that forwarded by Matsumoto (45), who hypothesized a protective "coating" of the protein by cryoprotectant molecules. (44) noted that certain compounds which are known to be preferentially excluded from the surface of proteins at room temperature, such as DMSO, proline, PEG, and ethylene glycol, are effective cryoprotectants. At temperatures above ambient, however, these compounds preferentially interact with the protein surface and as a result destabilize proteins. Arakawa et al. (46) noted also that the protein denaturants urea and guanidine hydrochloride act by binding to the the protein surface. They concluded that "it does not seem likely ... that stabilization of proteins during freezing involves direct interaction with the solute."

That many of the cryoprotectant sugars and polyols do increase the surface tension of water may be important in other ways to protein stabilization. Based on experimental data, Back et al. (34) noted that "hydrophobic interactions between pairs of hydrophobic groups are stronger in sucrose or glycerol solutions than in pure water" and concluded that "this is the mechanism by which sugars and polyols in general may stabilize proteins to heat denaturation." Similarly, Melander and Horvath (47) addressed the issue of why certain salts of the Hofmeister or lyotropic series have a stabilizing effect on proteins. They were able to demonstrate that such stabilization results from a strengthing of the protein intramolecular hydrophobic interactions in the presence of these salts, and concluded that "the property of a salt that affects hydrophobic interactions is quantified by its molal surface tension increment."

Carpenter and Crowe (42) theorize that certain high MW polymers, such as polyvinylpyrrolidone, polyethylene glycol, and dextran, are good cryoprotectants because they are stearically excluded from the protein surface by their size. However, other workers explain the cryoprotective effects of many high MW polyols and glucose polymers (starch hydrolysis products) based upon their ability to raise the glass transition temperature (T_g) of a protein solution (48, 49), thereby vitrifying the system at conventional cold storage temperatures. This result is due to their propensity to entangle as well as to form hydrogen and other bonds, imparting a greater viscosity at any given concentration. Levine and Slade's data (49, 50) indicate a generally direct relationship between molecular weight (or, inversely, DE for starch hydrolysis products) and T_g.

Lim and co-workers (24, 51) recently attempted to demonstrate the vitrification mechanism in the cryoprotection of surimi. In model studies which used salt-soluble fish muscle protein to represent the surimi, they found that a particular maltodextrin type and concentration $(T_g = -10°C)$ protected the solubility of the protein in a much more temperature-dependent fashion than did sucrose. However, there was not a dramatic change in stability of the maltodextrin-containing system when comparing the response to storage temperatures just above and below the T_g of maltodextrin at that concentration, as might be expected if a sharp glass transition took place in the system. Surimi behaved similarly to the model system with respect to the stability of the proteins in the presence of these two carbohydrates. Carboxymethylcellulose was also tried in these test systems (24, 51), but it failed to protect the proteins and seemed to interfere with their heat-induced gelation.

Thus, there are fundamental differences between the mechanisms of "cryoprotection" by low MW sugars and polyols and "cryostabilization" by high MW polymers. Cryoprotectants function by altering the thermodynamics of the system to favor the native state of the protein, while cryostabilizers act to trap the protein in a glass wherein all deteriorative processes are greatly slowed.

Sodium Lactate: Experimental Evidence for the Cryoprotection Mechanism

Our recent studies using sodium lactate (SL) to stabilize and cryoprotect fish actomyosin in model systems illustrated well the mechanism whereby low molecular weight additives stabilize proteins (52, 53). In these studies, we compared the effectiveness of SL to sucrose at various concentrations. Interestingly, SL had an optimum concentration for both freeze-thaw and heat stabilization and the explanation for this anomalous behavior reinforces the key role surface tension plays in stabilizing proteins in solution.

Freeze-Thaw Studies. Denaturation of tilapia (*Tilapia nilotica X Tilapia aurea*) actomyosin was measured after freeze-thaw experiments (Figure 2). A loss of $Ca^{2+}ATPase$ activity revealed the extent of denaturation during freeze-thawing. The cryoprotective effect increased to a maximum at about 6% (w/v) SL concentration with 80% of the activity recovered. In contrast, for sucrose the level of cryoprotection increased monotonically with sugar concentration. On a percent basis, sodium lactate appeared to be about four times more effective than sucrose since 25% sucrose would be needed to give an equivalent degree of cryoprotection to the optimum for SL.

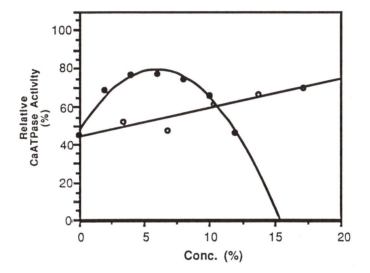

Figure 2. Cryoprotection of tilapia actomyosin by sodium lactate and sucrose. Actomyosin (2.5 mg/mL, pH 7.0 with 25 mM Tris-maleate in 0.6 M KCl) was frozen in liquid nitrogen for 3 min then held in a bath at -5°C for 60 min and thawed for 5 min at 25°C. Unfrozen control was taken as 100% Ca^{2+}ATPase activity. Results are the average of replicate experiments. Models for frozen-thawed: sodium lactate, $y = 47.2 + 10.8x - 0.9x^2$, $R^2 = 0.976$ (p<0.001); sucrose, $y = 44.1 + 1.5x$, $R^2 = 0.896$ (p<0.001). ● sodium lactate, ○ sucrose. (Reproduced with permission from ref.52. Copyright 1993 Institute of Food Technologists).

Heat Denaturation Studies.

Enzyme deactivation. In this series of experiments, actomyosin was heated at 40°C in the presence of various concentrations of solute and Ca^{2+}ATPase activity was used as a measure of the extent of protein denaturation. The E value, slope of a log K_d (first order inactivation constant) versus solute concentration, is a measure of stabilizer effectiveness (54). On a percent basis, sodium lactate (E = 4.4±0.4 X 10^{-2}%$^{-1}$) was nearly twice as effective as sucrose (E = 2.4±0.6 X 10^{-2}%$^{-1}$) (Figure 3). For SL concentrations between 20 - 30%, the results were variable, while above 30% SL actomyosin was destabilized such that no useful data were collected.

Aggregation. A further method, that of turbidity changes with thermal denaturation, was used to verify the enzyme results. The change in optical density as a function of SL concentration was determined (Figure 4a). A first derivative plot of this data with respect to solution temperature showed peaks (apparent transition temperatures, T_r) indicating temperatures at which actomyosin denatured and formed aggregates (Figure 4b). A plot of apparent T_r versus sodium lactate concentration showed stabilization of the actomyosin with increasing concentration up to 15% SL (Figure 5). Above 15% SL actomyosin was destabilized such that no useable data was collected.

There is a lower optimum concentration of SL for stabilization in the aggregation data (15%) compared to Ca^{2+}ATPase deactivation because fish actomyosins are less stable at the lower pH and protein concentration used in the aggregation study (11, 55).

The destabilizing forces that arise during freezing and thawing, such as the osmotic shock as ice is formed and alterations in pH, can be viewed simply as other types of solution-induced perturbations that also occur in unfrozen solutions. Hence, it has been suggested that solutes which stabilize proteins to heat also serve to prevent these forces from denaturing the protein in the frozen state (42). But consideration of the data from this study, in particular the lower SL concentration optimum for freeze-thaw (6%) compared to heat stabilization (25% and 15%), suggests that even if the stabilization mechanisms are similar, there are still major differences between specific stresses that occur in these two types of denaturing situations.

Overall, both freeze-thaw and heat denaturation studies showed sodium lactate was a more effective stabilizer than sucrose. However, in each case, above a certain concentration lactate actually destabilized the actomyosin. This property of lactate presented the opportunity of testing the preferential exclusion mechanism of protein stabilization.

Surface Tension of Sodium Lactate and Sucrose Solutions.

We noted above that polyols and sugars which increase the surface tension of water may act to stabilize proteins. Surface tension measurements showed that indeed the concentration of SL which corresponds to stabilization and destabilization of heated actomyosin closely correlates with concentrations where surface tension either increased or decreased (Figure 6). In comparison, surface tension of sucrose solutions increased monotonically with increasing concentration which also correlates with our previous freezing-thawing and heat denaturation studies showing a similar relationship between stability and concentration (Figure 6). Surface tensions of sucrose solutions were also lower than the corresponding SL solutions up to 20% SL.

It is possible to estimate the concentration of solute that is present at the interface compared to the bulk concentration by approximating the increasing and decreasing portions of the surface tension curve with linear models and using Gibbs adsorption isotherm (55).

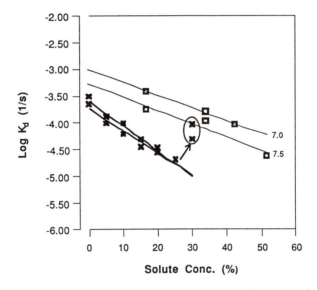

Figure 3. Apparent rate constant (K_d) for inactivation of actomyosin $Ca^{2+}ATPase$ as a function of solute concentration. Actomyosin (1.75 mg/mL, pH 7.0 and 7.5 with 20 mM Tris-maleate in 0.6 M KCl) was heated at 40°C. Arrow indicates destabilization of actomyosin as sodium lactate concentration increased from 25% to 30%.(Reproduced with permission from ref.52 Copyright 1993 Institute of Food Technologists).

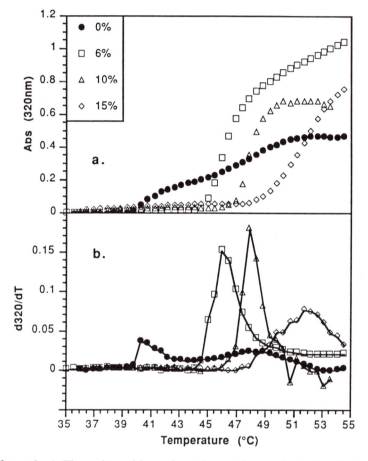

Figure 4. a). Thermal transitions of protein-protein association for tilapia actomyosin (0.5 mg/mL, pH 6.5) heated at 0.5 °C/min with various concentrations of sodium lactate. b). First derivative plot of the data. $dAbs_{320}/dT$ = differential change in absorbance. Representative data for each concentration is shown. ● 0%, □ 6%, △ 10%, ◇ 15% (Reproduced with permission from ref.53. Copyright 1993 American Chemical Society).

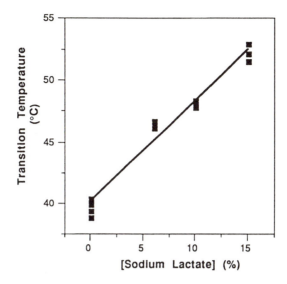

Figure 5. Thermal transition of tilapia actomyosin (0.5 mg/mL, pH 6.5) as a function of sodium lactate concentration.(Reproduced with permission from ref.53. Copyright 1993 American Chemical Society).

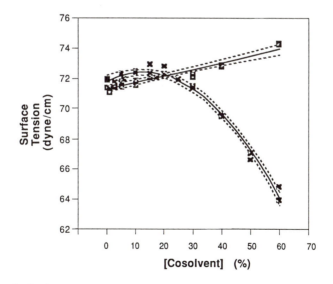

Figure 6. Surface tension of sodium lactate and sucrose solutions at 25°C as a function of concentration. Dashed lines are the 95% confidence intervals for the following models; Sodium lactate:
y = 71.8 + 0.096 [SL] (%) - 0.00375 [SL]^2 , R^2 = 0.9864 (p < 0.001).
Sucrose: y = 71.3 + 0.043 [Sucrose] (%), R^2 = 0.9031 (p < 0.001).
x sodium lactate, □ sucrose. (Reproduced from ref.62.).

$$\Gamma = - \frac{1}{RT} \frac{d\gamma}{d\ln c}$$

where Γ = adsorption (excess concentration) of solute at surface, mol m^{-2}

γ = surface tension, N m^{-1}

R = gas constant = 8.314 J deg^{-1} mol^{-1}

c = concentration of solute in bulk solution

We estimated that from 0 to 17.5% the concentration of SL at the interface was -0.47 X 10^{-10} moles/cm^2. The negative value shows that solute was excluded from the interface compared to the bulk concentration. Using the same procedure, adsorption of solute at the interface for the decreasing surface tension portion of the graph (25% - 60% SL), was estimated to be 0.026 X 10^{-10} moles/cm^2 which is low compared to values of 0.2 - 4.0 X 10^{-10} moles/cm^2 for typical detergents (57).

Preferential Exclusion Mechanism Tested. To further test the hypothesis that SL stabilizes mainly by increasing surface tension of the solvent, we measured the surface tension of solutions of different SL concentrations at the corresponding T_r calculated using the regression model developed from data in Figure 5. The resultant plot of surface tension versus concentration and temperature indicated that up to about 15% the surface tension is, within experimental error, close to constant (Figure 7). At higher SL concentrations and temperature in the region where previously actomyosin proteins were shown to be destabilized, surface tension was lower (p < 0.05). Our results were similar to findings by Lee and Timasheff (40) who calculated a similar relationship between the surface tension at the transition temperature and sucrose concentration for denaturation of three proteins. The increase in actomyosin T_r in the presence of SL therefore appears to be closely related to the need for lowering the surface tension at the protein-solvent interface to a level at which the free energy change provided by the protein expansion is sufficient to overcome the pressure of the solvent that counteracts the process. It appears, therefore, that increasing surface tension does play an important part in the mechanism by which SL stabilizes actomyosin. We then sought to better understand the basis for an increasing and then a decreasing surface tension with increasing SL concentration.

Water Activity Coefficients. In dilute solutions the water activity coefficient, a_w, is a colligative property and is only dependent on number of molecules of solvent (58). The a_w of SL solutions was characterized by a linear relationship from 0 to 20%, a transition at about 25%, and thereafter a_w continued to decrease but at a greater rate and in a non-linear manner (Figure 8). Our results compare closely to water activity values calculated from freezing point depression data of Dietz et al. (59) by using the equation;

$$a_w = e^{\frac{f.p.}{103}}$$

where f.p. = freezing point depression, °C

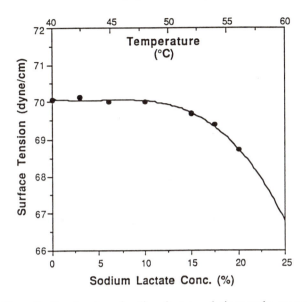

Figure 7. Surface tension of sodium lactate solutions at the corresponding T_r from tilapia actomyosin aggregation data in Figure 5. The regression model $T_r = 40.3°C + 0.81$ [SL] (%) was used to derive corresponding SL concentrations and temperatures at which the surface tension was measured (Reproduced with permission from ref.53. Copyright 1993 American Chemical Society).

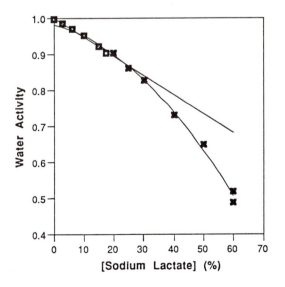

Figure 8. Water activity of sodium lactate solutions as a function of concentration. Squares represent data calculated from freezing point depression data of Dietz et al. (1941) by using equation (2) (Reproduced with permission from ref. 53. Copyright 1993 American Chemical Society).

These results suggested that some interaction between SL molecules was occurring at concentrations above 20% and that possibly a new molecular species was being formed at these levels of SL with a greater ability to affect a_w of the solution. Interestingly, the transition from ideal to non-ideal solution coincided with the transition from a positive surface tension increment to negative surface tension increment.

Lactic acid can form polymers at higher concentrations by inter-esterification, with the dimer lactoyllactate being the main polymer formed between 25 to 60% lactic acid (60). We plotted lactoyllactate concentration from published data (60) against SL concentration. The resulting model was used to predict lactoyllactate concentration and plotted against a_w (Figure 9). There was a linear relationship between a_w and lactoyllactate (predicted) for concentrations from 30 to 60% SL ($p < 0.001$). However, the correlation was nonlinear below 30% SL.

Inspection of the chemical structure of lactoyllactate (Figure 10) shows that it is amphiphilic, exhibiting one side that is more hydrophobic and the other more hydrophilic. Hence, lactoyllactate would be expected to preferentially absorb at an air-water interface and lower surface tension. This behavior would then explain the relationship between surface tension and SL concentration, since as the concentration of lactoyllactate increased with SL concentration, surface tension would be expected to decrease.

Conclusions

Lactate serves as an example of compounds which "cryoprotect" and otherwise stabilize proteins by inducing conditions which thermodynamically favor the maintenance of the native protein state. Sodium lactate is a particularly useful example of an additive stabilizing by the preferential exclusion mechanism, because higher concentrations of SL actually destabilized actomyosin and the transition from stabilization to destabilization was associated with either an increase or decrease in solution surface tension with SL concentration.

Alternatively, certain high molecular weight polymers may "cryostabilize" proteins by raising the glass transition temperature, thus ensuring a less reactive glass state in the system at conventional freezer temperatures. Perhaps in the future we might combine various compounds to create "cryoprotectant cocktails" which may be more effective than single compounds alone. We should especially explore interactions between high and low molecular weight carbohydrates and certain ions, organic acids, and/or amino acids to uncover possible synergies.

Cryoprotection of foods is a developing field with many potential new applications. The commercial success of cryoprotected fish mince, surimi, in the fishing industry has stimulated research efforts in the red meat industries to apply similar technology to gain quality or other functional benefits from frozen red meat raw materials. The USA beef industry imports about 3% of its meat from New Zealand and about 6% from Australia. This meat is frozen and recent research from this laboratory has shown that this meat could potentially be cryoprotected to improve functionality (37, 38, 61).

The study of cryoprotective mechanisms of food-grade compounds contributes to the broader inquiry into methods of stabilizing food proteins to all conceivable forms of stress, including drying and heat or extended storage. In addition to muscle-derived foods and food ingredients, functional food protein fractions from non-meat sources, such as egg, soy, and beef blood plasma, as well as the delivery of enzymes to the food processing industry, should benefit in quality from better stabilization technology.

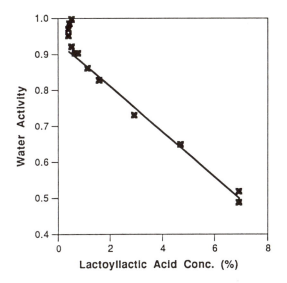

Figure 9.   Water activity of sodium lactate solutions compared to predicted lactoyllactate concentration (Reproduced with permission from ref. 53. Copyright 1993 American Chemical Society).

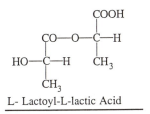

Figure 10.   Structure of lactoyllacte ester.

Literature Cited

1. Acton, J. C., Ziegler, G.R., and Gurge, D. L. 1983. Functionality of muscle constitutents in the processing of comminuted meat products. CRC Crit. Rev. Food Sci. Nutri. 18(2):99.
2. Rawn, J.D. 1989. Biochemistry. Neil Patterson Publ. Burlington, NC.
3. Samejima, K., Y. Oka, K. Yamamoto, A. Asghar and T. Yasui. 1988. Effects of SH groups, e-NH2 groups, ATP and myosin subfragments on heat-induced gelling of cardiac myosin and comparison with skeletal myosin and actomyosin gelling capacity. Agric. Biol. Chem. 52: 63.
4. Acton, J. C. and R. L. Dick. 1989. Functional roles of heat induced protein gelation in processed meat. In "Food Proteins". Eds. Kinsella, J. E. and Soucie, W. G. AOCS. Protein and Co-products Div.
5. Yamamoto, K. 1990. Electron microscopy of thermal aggregation of myosin. J. Biochem. 108(6): 896.
6. Sharp, A. and G. Offer. 1992. The mechanism of formation of gels from myosin molecules. J. Sci. Food Agric. 58: 63.
7. Connell, J. J. 1961. The relative stabilities of the skeletal muscle myosins of some animals. Biochem. J. 80: 503.
8. Haard, N.F. 1992. Biochemical reactions in fish muscle during frozen storage. In "Seafood Science and Technology." Ed. E.G. Bligh, Fishing News Books, Oxford, UK.
9. Niwa, E. 1992. The chemistry of surimi gelation. In "Surimi Technology". Eds. Lanier, T.C. and Lee, C.M. Marcel Dekker, Inc. New York, NY.
10. Okada, M. 1992. History of surimi technology in Japan. In "Surimi Technology." Eds. T.C.Lanier and C.M. Lee, Marcel Dekker Inc., New York, NY.
11. Noguchi, S. 1974. The control of denaturation of fish muscle proteins during frozen storage. D.Sc. Thesis, Sophia Univ., Tokyo.
12. Jiang, S.T., B.O. Hwang and C.T. Tsao. 1987a. Protein denaturation and changes in nucleotides of fish muscle during frozen storage. J. Agric. Food Chem. 35: 22.
13. Jiang, S.T., C.Y. Tsao and T.C. Lee. 1987b. Effect of free amino acids on the denaturation of mackerel myofibrillar proteins in vitro during frozen storage at   -20°C. J. Agric. Food Chem. 35: 28.
14. Loomis, S.H., T.J. Carpenter and J.H. Crowe. 1988. Identification of strombine and taurine as cryoprotectants in the intertidal bivalve *Mytilus edulis*. Biochem. et Biophys. Acta. 943: 113.
15. Loomis, S.H., T.J. Carpenter, T.J. Anchordoguy, J.H. Crowe and B.R. Branchini. 1989. Cryoprotective capacity of end products of anaerobic metabolism. J. Exp. Zoo. 252: 9.
16. Jiang, S.T., B.O. Hwang and C.T. Tsao. 1987c. Effect of adenosine-nucleotides and their derivatives on denaturation of myofibrillar proteins in vitro during frozen storage at -20°C. J. Food Sci. 52: 117.
17. Dyer, W.J. and J. Peters. 1969. Factors influencing quality changes during frozen storage and distribution of frozen products, including glazing, coating and packaging. In "Freezing and Irradiation of Fish", Ed. R. Kreuzer, Fishing News (Books) Ltd., London, UK.
18. Fukuda, Y., Z. Tarakita and K. Arai. 1984. Effect of freshness of chub mackerel on the freeze denaturation of myofibrillar protein. Bull. Japan. Soc. Sci. Fish. 50(5): 845.

19. Watanabe, T., N. Kitabatake and E. Doi. 1988. Protective effects of non-ionic surfactants against denaturation of rabbit skeletal myosin by freezing and thawing. Agric. Biol. Chem. 25(10): 2517.
20. Wessels, J.P.H., C.K. Simmonds, P.D. Seaman and L.W.J. Avery. 1981. The effect of storage temperature and certain chemical and physical pretreatments on the storage life of frozen hake mince blocks. Ann. Report No. 35, p. 32, Fishing Industry Research Institute, Capetown, South Africa.
21. Park, J.W., T.C. Lanier and D.P. Green. 1988. Cryoprotective effects of sugar, polyols, and/or phosphates on Alaska pollock surimi. 1988. J. Food Sci. 53: 1.
22. Sych, J., C. Lacroix, L.T. Adambounou and F. Castigne. 1991a. The effect of low- or non-sweet additives on the stability of protein functional properties of frozen cod. Intnl. J. Food Sci. Technol. 26: 185.
23. Lanier, T.C. and T. Akahane. 1986. Method of retarding denaturation of meat products. US Patent no. 4,572,838.
24. Lim, M.H., Reid, D.S. and Haard, N.F. 1990. Cryostabilizers for surimi - which to choose. Presented at Inst. Food Technologists Annual Meeting, Anaheim, Calif. June 16-20.
25. Foegeding, E.A. and S.R. Ramsey. 1986. Effect of gums on low-fat meat batters. J. Food. Sci. 51: 33.
26. Foegeding, E.A. and S.R. Ramsey. 1987. Rheological and water holding properties of gelled meat batters containing iota carrageenan, kappa carrageenan or xanthan gum. J. Food Sci. 52: 549.
27. Anderson, R.J. 1990. Personal communication. Grain Processing Corp., Muscatine, IO.
28. Sych, J., C. Lacroix, L.T. Adambounou and F. Castaigne. 1990a. Cryoprotective effects of lactitol, palatinit and Polydextrose on cod-surimi proteins during frozen storage. J. Food Sci. 55: 356.
29. Sych, J., C. Lacroix, L.T. Adambounou and F. Castaigne. 1990b. Cryoprotective effects of some materials on cod-surimi proteins during frozen storage. J. Food Sci. 55: 1222.
30. Sych, J., C. Lacroix and M. Carrier. 1991b. Determination of optimal level of lactitol for surimi. J. Food Sci. 56(2): 285.
31. da Ponte, D.J.B., J.P. Roozen and W. Pilnik. 1985a. Effects of additions on the stability of frozen stored minced fillets of whiting. I Various anionic hydrocolloids. J. Food Qual. 8: 51.
32. da Ponte, D.J.B., J.P. Roozen and W. Pilnik. 1985b. Effects of additions on the stability of frozen stored minced fillets of whiting. II Various anionic and neutral hydrocolloids. J. Food Qual. 8: 175.
33. da Ponte, D.J.B., J.M. Herfst, J.P. Roozen and W. Pilnik. 1985c. Effects of different types of carrageenans and carboxymethyl cellulose on the stability of frozen stored minced fillets of cod. J. Food Technol. 20: 587.
34. Back, J.F., D. Oakenfull and M.B. and Smith. 1979. Increased thermal stability of proteins in the presence of sugars and polyols. Biochem. 18: 5191.
35. Park, J.W. and T.C. Lanier. 1987. Combined effects of phosphates and sugar or polyol on protein stabilization of fish myofibrils. J. Food Sci. 52: 1509.
36. Park, J.W. and T.C. Lanier. 1990. Effects of salt and sucrose addition on thermal denaturation and aggregation of water-leached fish muscle. J. Food Biochem. 14: 395.
37. Park, J.W., T.C. Lanier, J.T. Keeton and D.D. Hamann. 1987a. Use of cryoprotectants to stabilize functional properties of prerigor salted beef during frozen storage. J. Food Sci. 52: 537.

38. Park, J.W., T.C. Lanier, H.E. Swaisgood, D.D. Hamann and J.T. Keeton. 1987b. Effects of cryoprotectants in minimizing physicochemical changes of bovine natural actomyosin during frozen storage. J. Food Biochem. 11: 143.
39. Wu, M. C., T. Akahane, T.C. Lanier and D.D. Hamann. 1985. Thermal transitions of actomyosin and surimi prepared from Atlantic croaker as studied by differential scanning calorimetry. J. Food Sci. 50: 10.
40. Lee, J.C. and S.N. Timasheff. 1981. The stabilization of proteins by sucrose. J. Biol. Chem. 256(14): 7193.
41. Arakawa, T. and S.N. Timasheff. 1982. Stabilization of protein structure by sugars. Biochem. 21(25): 6536.
42. Carpenter, J. F. and J.H. Crowe. 1988. The mechanism of cryoprotection of proteins by solutes. Cryobiology 25: 244.
43. Arakawa, T., J.F. Carpenter, Y.A. Kita and J.H. Crowe. 1990a. Basis for toxicity of certain cryoprotectants: A hypothesis. Cryobiology 27: 401.
44. Arakawa, T., R. Bhat and S.N. Timasheff. 1990b. Why preferential hydration does not always stabilize the native structure of globular proteins. Biochem. 29(7): 1924.
45. Matsumoto, J.J. 1979. Denaturation of fish muscle proteins during frozen storage. In "Proteins at Low Temperatures". Ed. O. Fennema. ACS Adv. in Chem. Series. Washington, DC.
46. Crowe, J.H., J.F. Carpenter, L.M. Crowe and T.J. Anchordoguy. 1990. Are freezing and dehydration similar stress vectors? A comparison of modes of interaction of stabilizing solutes with biomolecules. Cryobiology 27: 219.
47. Melander, W. and C. Horvath. 1977. Salt effects on hydrophobic interactions in precipitation and chromatography of proteins: An interpretation of the lyotropic series. Arch. Biochem. Biophys. 183: 200.
48. Levine, H. and L. Slade. 1988a. A food polymer science approach to the practice of cryostabilization technology. Comments Agric. Food Chem. 1: 315.
49. Levine, H. and L. Slade. 1988b. Principles of "cryostabilization" technology from structure/property relationships of carbohydrate/water systems. A review. Cryo-Lett. 9: 21.
50. Levine, H. and L. Slade. 1986. A polymer physico-chemical approach to the study of commercial starch hydrolysis products (SHPs). Carbohydr. Polym. 6: 213.
51. Lim, M. H. 1989. Studies of reaction kinetics in relation to thermal behavior of solutes in frozen systems. Ph. D. Thesis. Univ. California, Davis, CA.
52. MacDonald, G.A., and Lanier, T.C. 1993. Stabilization of fish actomyosin to freeze-thaw and heat denaturation by lactate salts compared to sucrose. J. Food Sci. (in press).
53. MacDonald, G.A., Lanier, T.C., Swaisgood, H.E., and Hamann, D.D. 1993. Investigation of the mechanism for stabilization of fish actomyosin by sodium lactate compared to sucrose. J. Agr. Food Chem. (in press)
54. Ooizumi, T., K. Hashimoto, J. Ogura and K. Arai. 1981. Quantitative aspect for protective effect of sugar and sugar alcohol against denaturation of fish myofibrils. Bull. Japan. Soc. Sci. Fish. 47:901.
55. Hashimoto, A. and Arai, K. 1985. The effect of pH on the thermostability of fish myofibrils. Nippon Suisan Gaikaishi 51:99
56. Tinoco, I., K. Sauer and J.C. Wang. 1985. Physical Chemistry. Principles and Applications in Biological Sciences. Prentice-Hall Inc., New Jersey, NJ.
57. Rosen, M.J. 1978. Surfactants and Interfacial Phenomena. John Wiley & Sons, New York, NY.
58. Noggle, J.H. 1985. Physical Chemistry. Little, Brown and Co., Boston, MA.

59. Dietz, A.A., E.F. Degering and H.H. Schopmeyer. 1941. Physical properties of sodium, potassium, and ammonium lactate solutions. Ind. Eng. Chem. 33(11): 1444.
60. Holten, C.H., A. Muller and D. Rehbinder. 1971. Lactic Acid. Properties and Chemistry of Lactic Acid and its Derivatives. Verlag Chemie GmbH, Weinheim.
61. Park, J.W., T.C. Lanier and D.H. Pilkington. 1993. Cryostabilization of functional properties of pre-rigor and post-rigor beef by dextrose polymer and/or phosphates. J. Food Sci. 58:467-472.
62. MacDonald, G.A. 1992. Mechanisms for the cryoprotection and stabilization of myofibrillar proteins. Ph.D. thesis, Dept. Food Science, N.C. State Univ., Raleigh NC 144 pp.

RECEIVED June 6, 1994

Chapter 14

Chemistry of Protein Stabilization by Trehalose

C. A. L. S. Colaco[1], C. J. S. Smith[2], S. Sen[1,2], D. H. Roser[1], Y. Newman[3], S. Ring[3], and B. J. Roser[1]

[1]Quadrant Research Foundation, Maris Lane, Cambridge CB2 2SY, United Kingdom
[2]Quadrant Holdings Cambridge Ltd., Maris Lane, Cambridge CB2 2SY, United Kingdom
[3]AFRC Institute of Food Research, Norwich Laboratory, Norwich Research Park, Norwich NR4 7UA, United Kingdom

Carbohydrates are widely used to stabilise dried protein and peptide drugs. Several sugars can prevent damage to proteins during drying, but the products often have a short shelf life at ambient temperatures. There is increasing evidence to suggest that this is due to a progressive chemical reaction between the reducing carbonyl groups of the sugars and the amino groups of proteins, the so-called Maillard reaction. The rational choice of an excipient for pharmaceutical use must thus consider not only the avoidance of desiccation damage but also subsequent degradative chemistries on storage. We present data from comparative studies on the long term stability of a variety of proteins, including enzymes and therapeutic peptides, dried using various carbohydrate excipients, which establish the superiority of the simple disaccharide trehalose as a stabilising excipient. These observations illustrate the chemical stability of trehalose and document the protein degradation observed in samples dried using other carbohydrates.

Stability at room temperature is the ultimate goal for pharmaceutical formulations. The dual role of water, as a nucleophile in hydrolysis reactions and as a plasticiser which increases the molecular mobility of reactive chemical species, makes aqueous protein formulations inherently less stable than their dry counterparts. This increased stability of dry protein formulations has focused attention on techniques of drying and led to the development of freeze-drying as a popular method of water removal (1-4). However, despite its widespread use, many freeze-dried products are still unstable at ambient temperatures (4-6). Detailed theoretical analyses of the physico-chemical events during freeze-drying have led to a substantial literature on the use of cryoprotectants as stabilising excipients (2, 5-7). Various carbohydrates have been advocated as stabilising excipients in freeze-drying, and these are proposed to act via the generation of an amorphous, glassy solid state in the freezing step (2-4, 7). Nevertheless, the freezing step remains a major variable, as evidenced by the equivocal values for the experimentally measured glass transition temperature of the maximally freeze-concentrated unfrozen matrix (T'g) for various carbohydrate excipients (3, 7-9), and is suggested to be the major cause of protein damage during freeze-drying (4, 6, 10). Recent attention has thus focused on the techniques of ambient temperature drying, as these not only eliminate the freezing step but are more rapid and energy-efficient in the removal of water during drying (6, 11-17).

In nature, there exist organisms that have evolved the remarkable ability to dry out at ambient temperatures, remain metabolically dormant for long periods under harsh environmental conditions and yet regain full metabolic activity when rehydrated (17, 18). These cryptobiotic or anhydrobiotic organisms are found in many phyla in both the plant and animal kingdoms and the better known examples are the brine shrimp *Artemia salina,* the soil nematode *Ditylenchus dipsaci,* the tardigrade *Adoribiotus coronifer,* the resurrection plant *Selaginella lepidophylla* and baker's yeast *Saccharomyces cerevisiae.* Studies on cryptobiosis in the bakers yeast, *S.Cerevisiae,* have shown that the synthesis of a simple disaccharide, trehalose (α-D glucopyranosyl α-D glucopyranoside), is both necessary and sufficient to account for the protection of all the biomolecules in the organism during drying (20, 21). Recent work has shown that this protection can be replicated *in vitro* by drying biological molecules in the presence of trehalose at ambient temperatures (6, 11-15, 24). Concordant with the fact that the dehydration of cryptobionts occurs at ambient or higher temperatures, this reproducible stabilisation of biomolecules dried at ambient temperatures is in sharp contrast with previously reported work using trehalose as an excipient during freeze-drying (3-7, 11-15, 22-24).

A feature of biomolecules stabilised using trehalose technology is their ability to be stored for extended periods at high temperatures without any apparent damage, as assessed by the full recovery of biological activity on rehydration (11-14). Studies of other sugars, polyhydric alcohols and oligosaccharides under identical conditions showed that this degree of stabilisation is unique to trehalose, even though some of these excipients do protect the biomolecules from damage during the drying process itself and confer more limited tolerance to high temperatures (6, 11, 14, 22-24). Some of these data are presented below, together with the results of studies on the mechanism of action of the disaccharide which suggest that the chemical inertness of trehalose may be important. These data highlight a previously ignored feature of the use of carbohydrate excipients in protein formulations, namely the spontaneous chemical reaction between stabilising agent and protein product. This feature is well recognised in the food industry, under the umbrella of the non-enzymatic browning or Maillard reaction, which is one of several causes of spoilage of many food products during storage (25-27). In these foodstuffs containing high amounts of sugar and protein the Maillard reaction is even observed during refrigerated storage at low temperatures (25-27) and is a particular problem during the processing of these food products by drying (28).

Trehalose stabilisation of biomolecules

Our early work has shown that antibodies, air-dried in the presence of trehalose, are undamaged, and full biological activity is recovered on rehydration, even after several years storage at room temperature or 37°C (11, 13, 14). Similar results were obtained with a variety of enzymes, hormones and blood coagulation factors, suggesting that this process may be generally applicable to biological molecules (11-15). As a stringent test of this technology to preserve labile biological molecules, the enzymes used in molecular biology, which are notoriously fragile and thus usually transported and stored at or below -20°C, were studied in detail (11, 14, 15). We have shown that both restriction endonucleases and DNA modifying enzymes can be dried from trehalose solutions at ambient temperatures without loss of activity (11, 14, 15). Furthermore, these dried enzymes show stability on storage for extended periods even at elevated temperatures (11, 15, Fig.1, Table 1).

Illustrated below (Fig.1) are the results obtained in an accelerated ageing study with the restriction enzyme *Pst*I vacuum-dried with supplemental heating to

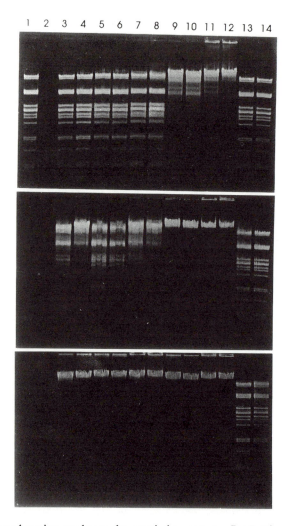

Fig.1. Accelerated ageing study on the restriction enzyme *Pst* I. Five units of fresh enzyme control (track 1) was compared, for the ability to cut bacteriophage λ DNA, with 2.5unit (even numbered tracks) or 5unit (odd numbered tracks) of enzyme dried using various carbohydrate excipients and stored for 35 days at 37°C (top panel), 55°C (middle panel) and 70°C (bottom panel). The carbohydrate excipients used were Glucopyranosyl-mannitol (tracks 3 & 4) or sorbitol (tracks 5 & 6), reduced isomaltose (tracks 7 & 8), sucrose (tracks 9 & 10), maltose (tracks 11 & 12) and trehalose (tracks 13 & 14). As can be seen only trehalose shows any stabilising effects at the two higher temperatures studied.

Table 1. Stability of PstI dried in various carbohydrate excipients

CARBOHYDRATE	CHEMICAL NAME	RED. SUGAR	TEMP °C	TIME days	ACTIVITY
MONOSACCHARIDES AND ALCOHOLS					
Glucose	α-D-glucopyranose	+	37°	1	+
			"	14	-
Sorbitol	sugar alcohol of glucose	-	"	14	+
			"	35	+
			"	70	-
Galactose	α-D-galactopyranose	+	"	1	-
Galactitol	sugar alcohol of galactose	-	"	1	-
Mannose	α-D-mannopyranose	+	"	1	-
Mannitol	sugar alcohol of mannose	-	"	1	-
DISACCHARIDES					
Trehalose	α-D-glucopyranosyl-α-D-glucopyranoside	-	"	98	+++
			55°	70	+++
			70°	35	+++
Maltose	4-O-α-D-gluco-pyranosyl-D-glucose	+	"	14	++
			"	7	-
Maltotriose	O-α-D-glucopyranosyl(1→4)-O-α-D-glucopyranosyl-(1→4)-D-glucose	+	"	14	-
Lactose	4-O-ß-D-galacto-pyranosyl-D-glucopyranose	+	"	14	-
Lactulose	4-O-ß-D-galactopyranosyl-D-fructose	+	"	14	+
				35	-
Sucrose	ß-D-fructofuranosyl-α-D-glucopyranoside	-	37°	14	++
			"	35	-
POLYMERS					
Inulin	Polymer of 1-O-ß-D-fructofuranosyl-D-fructose	-	"	7	-
Dextran	Polymer of α-(1→6) -D-Glucopyranose (1→3,1→4 branch)	+	"	7	-
Ficoll	Polymer of ß-D-fructofuranosyl- α-D-glucopyranose	-	"	7	+

Quantitation of activity   Reducing properties
 - no detectable activity   + reducing sugar
 + some activity (10-20% of titre) - non-reducing sugar
++ partial activity (25-40% of titre)
+++ full activity

a residual moisture content of 2.6-3.6% in a number of carbohydrate excipients and stored for one month at either 37°C, 55°C or 70°C. Although some of the carbohydrate excipients stabilised the enzyme during drying and on storage at 37°C (Fig.1, top panel, Table 1), only the samples stabilised using trehalose retained activity when stored at either 55°C or 70°C (Fig.1, middle and bottom panels, Table 1). These results have now been correlated with real-time data and emphasise the finding that, with respect to long-term stability, trehalose is a better stabilising excipient than the other carbohydrates tested under identical drying and storage conditions (Table 1). All the monosaccharides were ineffective, whether reducing or non-reducing, as were polymers such as inulin, ficoll and dextran (Table 1). Reducing sugars, such as lactose and maltose, failed within a month at the lowest temperature studied, 37°C, as did the non-reducing disaccharide sucrose (Table 1). The chemically more stable non-reducing sugars, the sugar alcohols, showed better stabilities than their reducing counterparts, but still failed within a month at 55°C (Table 1).

With regard to the results presented above and in our other studies, it must be emphasised that the stabilisation effects seen with trehalose in our experiments cannot be ascribed to any transition metal effects (4). Previous work has suggested that transition metal ions may in part be responsible for the enhanced stabilisation observed with trehalose (23, 38). However, this does not apply to the studies we have reported as the trehalose used in our experiments was not purified by methods that use deproteination by zinc and barium salts (29), and no trace heavy metals were detectable by atomic absorption spectroscopy. Furthermore, the confusion over the effects of zinc (4, 23) arises from the studies on the enzyme phosphofructokinase as a model for drying, where the effects of the transition metal are mediated largely in the liquid phase (24) and probably reflect the divalent metal ion requirement of this enzyme.

Mechanism of action of trehalose

There are two main hypotheses that have been postulated with respect to the molecular mechanism by which trehalose stabilises biological molecules (30, 31, 36-38). The water replacement theory states that, being a polyol, trehalose can make multiple external hydrogen bonds which could replace the essential structural water molecules that are hydrogen-bonded to biomolecules and thus maintain their molecular structure (30, 38). The glassy state theory postulates that, as the drying trehalose solutions undergo glass transformation, this results in an amorphous continuous phase in which molecular motion, and thus degradative molecular reactions, are kinetically insignificant (31, 36, 37). Our results (11, 14, 15 and below) are not consistent with either hypothesis being a sufficient sole explanation for the mechanism of action of trehalose, but suggest that the chemical inertness of the sugar may be an equally important feature in its mechanism of action.

The water replacement theory suggests that, as polyols, other sugars should also be effective as stabilising excipients, and, if the specific spatial combinations of hydroxyl groups are the crucial feature, then glucose should be as effective as trehalose. However, glucose was in fact among the least effective of the sugars tested (see Table 1), and none of the other polyols tested was found to be as effective as trehalose (see Fig.1 and Table 1). Furthermore, if molecular mimicry of water was important, as might be expected for water replacement, then scyllo-inositol (with all its hydroxyl groups being axial) should, in theory, be the most effective carbohydrate, but it is among the least effective in practice (S.Sen unpublished). The water replacement theory can thus not be a complete explanation for the mechanism of action of trehalose. Similarly, the glassy state theory alone cannot explain the stability conferred by trehalose. In the high temperature storage stability data reported in Fig.1 above, the glass transition temperatures of the samples dried in trehalose to a water content of 2.6-3.6% were all below 37°C as measured by differential scanning calorimetry (Y.Newman *et.al.* manuscript in preparation). Thus, their stability persists at well above their glass transition temperatures, and although the glassy state may be important in other systems, it appears not to be a factor in the long-term high temperature stability of biomolecules dried in trehalose.

It appears that the relative chemical stability and non-reducing nature of trehalose (32, 33) may be significant features in its mechanism of action, especially with regard to the long term stability observed at high temperatures. This was first suggested by an interesting feature observed in the accelerated ageing trial described in Fig.1 above. The development of a brown coloration was noted in a number of the sample wells at all three temperatures after just two weeks of storage of the sample (Fig.2). Furthermore, the extent of the coloration appeared to correlate with the reduction in enzymatic activity in these samples. Increasing coloration was observed in the samples stored at higher temperatures (Fig.2), which also showed the greatest loss of activity (Fig.1). This coloration was highly reminiscent of the non-enzymatic browning seen during the processing and storage of food products. This non-enzymatic browning is the result of the spontaneous reaction between the reducing sugars and proteins that are natural constituents of these foodstuffs, and has been widely studied in food chemistry under the umbrella of the so-called Maillard reaction (Fig.3, 25-27).

The Maillard reaction is actually a cascade of chemical reactions initiated by the spontaneous condensation of reactive carbonyl and amino groups such as those commonly found in reducing sugars and proteins, respectively (Fig.3). The activation energy of the initial condensation to form a Schiff's base is only of the order of 10-15 kcal, is reversible in the presence of water, and the equilibrium is largely in favour of the reactants in aqueous environments (Fig.3, 25-27). The subsequent spontaneous Amadori or Heyns rearrangement of the Schiff's base is irreversible and triggers a complex series of reactions that ultimately result in the production of brown melanoidin pigments and both fragmentation and cross-linking of the proteins involved (Fig.3, 25-27). In the food industry, the Maillard reaction has been widely studied, as it is one of several causes of spoilage, especially of dried food products, during storage. It has even been observed during refrigerated storage of foodstuffs with high protein and sugar contents (25-27). The Maillard reaction is a particular problem with dry foodstuffs, as the equilibrium of the reaction is forced towards the formation of the Schiff's base by the loss of water, and many of the subsequent reactions are accelerated at low water activities (27, 28). We have examined whether the Maillard reaction is a major problem during storage of protein formulations containing carbohydrate excipients, and some of the results are presented below.

Evidence for Maillard reactions in dried protein formulations

Our initial studies attempted to correlate changes in biological activity with the development of brown pigments, as observed in the accelerated ageing studies on restriction enzymes (Fig.2). These studies were carried out on samples of the enzyme alkaline phosphatase which were dried, under the conditions described in the experiments above, from solutions containing glucose, fructose, maltose or trehalose and stored for various periods at 55°C, before enzymatic activity was re-assayed. The results showed that with glucose and fructose, all enzymatic activity was lost within 10 days (Fig.4), with maltose and sucrose, a steady decline in activity was observed, with 15% of the enzymic activity lost within 3 weeks, compared to no detectable loss in the samples dried in trehalose (Fig.4). Perhaps surprisingly, the non-reducing sugar alcohol maltitol showed a much greater loss of activity than its reducing counterpart maltose, with 50% loss of activity within 3 weeks (Fig.4). The most rapid loss of activity was, however, seen with the samples

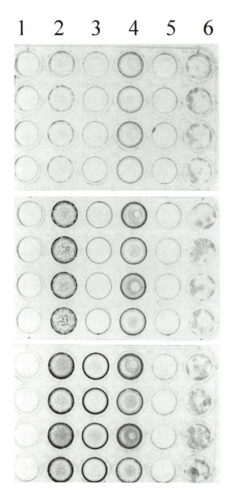

Fig.2. Non-enzymatic browning observed in the samples used in the accelerated ageing study reported in Fig.1 after two weeks storage at 37°C (top panel), 55°C (middle panel) and 70°C (bottom panel). The carbohydrate excipients used were trehalose (row 1), sucrose (row 2), maltose (row 3), reduced isomaltose (row 4), glucopyranosyl-sorbitol (row 5) and glucopyranosyl-mannitol (row 6).

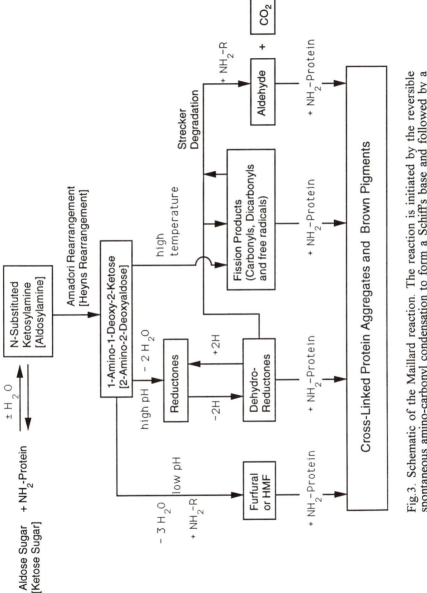

Fig.3. Schematic of the Maillard reaction. The reaction is initiated by the reversible spontaneous amino-carbonyl condensation to form a Schiff's base and followed by a cascade of subsequent reactions resulting in the cross-linking of the proteins involved and the generation of brown melanoid pigments.

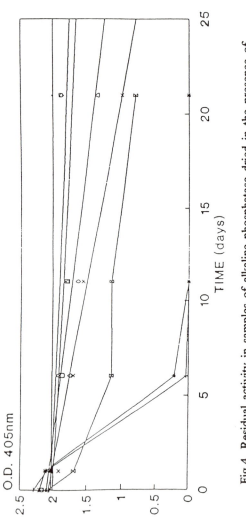

Fig. 4. Residual activity in samples of alkaline phosphatase dried in the presence of various carbohydrate excipients and assayed colorimetrically after storage at 55°C. Enzymic activity (O.D. 405 nm); (·) trehalose, (□) maltose, (◊) sucrose, (×) maltitol, (+) glucose, (*) fructose, (Δ) no additive, (X) wet control.

dried in glucose and fructose (Fig.4), which also showed visual changes with the development of a brown coloration in the sample wells. This development of the brown pigments, as assayed by the increase in absorbance between 277 and 290 nm, did not correlate directly with the rate of loss of enzymatic activity. The production of melanoid pigments occurred later than the loss of enzymic activity assayed colorimetrically (Fig.5). This is consistent with the fact that the generation of brown melanoid pigments occurs in the terminal stages of the Maillard reaction and thus cannot be used to predict enzyme inactivation due to the early reactions of the cascade (Fig.3). Similarly, analysis of the samples by SDS-PAGE showed a complex pattern of protein breakdown and cross-linking in all samples, except those dried in trehalose, and the complexity of these patterns precludes the use of this technique in determining the extent of protein modification by the Maillard reaction.

A surprising result was obtained on analysis of the residual sugar contents of the samples in the studies described above (Fig.4). When dried in glucose or fructose, only the individual sugars were detectable in the samples immediately post-drying (Fig.4). However, on loss of enzymatic activity after high temperature storage, these samples were found to contain mixtures of the two sugars (Table 2). A similar isomerisation was observed in the samples found to contain a mixture of glucose and maltose immediately post-drying, presumably due to partial hydrolysis of the maltose. On loss of activity after high temperature storage, these samples were found to contain a mixture of both glucose and fructose, as well as maltose (Table 2). The absence of any mannose production detected in this non-enzymatic isomerisation is indicative of a chemical reaction pathway involving a common addition compound intermediate, similar to the formation of osazones in the Fisher reaction, by the reversible condensation of phenylhydrazine with reducing sugars (34). Such a reaction could theoretically occur, if the amino acid side chains of the proteins being dried substituted for the amino groups normally contributed by phenylhydrazine in the classical Fisher reaction, and this potential mechanism is currently being tested. It is interesting to note that a recent study of the decomposition of Amadori compounds under physiological conditions has reported the reversal of the Amadori rearrangement of the Schiff's base (39). This results in a similar isomerisation of the aldose or ketose sugar moieties initiating the Maillard reaction.

To enable a more detailed analysis of these chemical modifications of proteins by carbohydrate excipients, we studied a more relevant pharmaceutical model system. The protein modifications of a therapeutic peptide, glucagon, dried for 18 hr under a vacumn of 30 milliTorr, with a shelf-temperature rising from 25 to 42°C, were studied. Formulations containing various carbohydrate excipients were analysed by reverse-phase HPLC analysis, and the comparison of glucose and trehalose is presented in Fig.6. In the samples dried in the presence of glucose, an additional peak that might correspond to Schiff's base derivatives of the peptide was detectable, even immediately post-drying (Fig.6, a), and after just 4 days storage at 60°C, this peak already represented the major fraction of the total peptide detected (Fig.6, b). After storage for 2 weeks at 60°C, this peak represented ~80% of the total protein, and a number of additional peaks corresponding to other reaction products were detectable (Fig.6, c). In the samples dried using trehalose, however, no peaks corresponding to either the addition compounds or other reaction products were detectable, even after 3 weeks storage at 60°C (Fig.6, e). Furthermore, the generation of brown pigments was only observed in the samples dried in glucose. These results suggest that the chemical modification of proteins by carbohydrate excipients such as glucose, can adversely affect their stability during subsequent storage. The results also suggest that the difference between glucose and trehalose as carbohydrate stabilising excipients may be due mainly to their respective chemical reactivities.

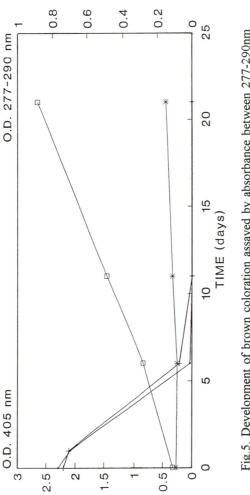

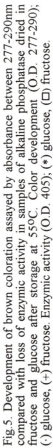

Fig.5. Development of brown coloration assayed by absorbance between 277-290nm compared with loss of enzymic activity in samples of alkaline phosphatase dried in fructose and glucose after storage at 55°C. Color development (O.D. 277-290); (·) glucose, (+) fructose. Enzymic activity (O.D. 405); (*) glucose, (□) fructose.

Table 2. HPLC analysis of carbohydrates in formulations dried in various excipients

Carbohydrate excipient used in two formulations	Post drying activity	Post drying sugar analysis	Activity after 2 weeks	Hplc analysis after 2 weeks storage
1. Glucose Sample A	+	Glucose	-	Glucose + Fructose
2. Glucose Sample B	+	Glucose	-	Glucose + Fructose
3. Sorbitol Sample A	++	Sorbitol	-	Sorbitol
4. Sorbitol Sample B	++	Sorbitol	-	Sorbitol
5. Fructose Sample A	+	Fructose	-	Fructose + Glucose
6. Frucrose Sample B	+	Fructose	-	Fructose + Glucose
7. Maltose Sample A	+++	Maltose + Glucose	-	Maltose + Glucose + Fructose
8. Maltose Sample B	+++	Maltose	+/-	Maltose + Glucose + "Others"
9. Sucrose Sample A	+++++	Sucrose	+++	Sucrose
10. Sucrose Sample B	+++++	Sucrose	++	Sucrose
11. Trehalose Sample A	+++++	Trehalose	+++++	Trehalose
12. Trehalose Sample B	+++++	Trehalose	+++++	Trehalose

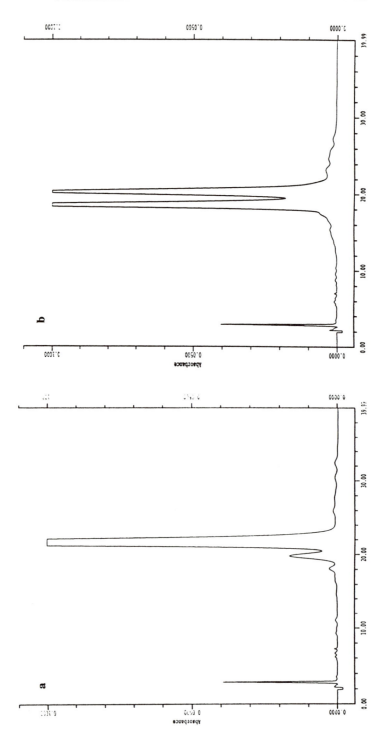

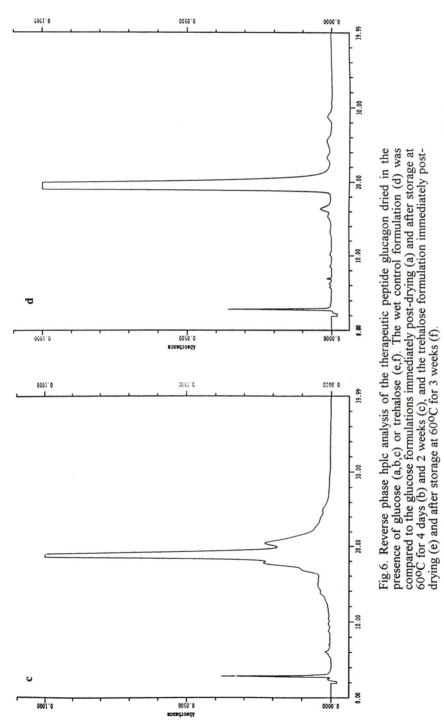

Fig.6. Reverse phase hplc analysis of the therapeutic peptide glucagon dried in the presence of glucose (a,b,c) or trehalose (e,f). The wet control formulation (d) was compared to the glucose formulations immediately post-drying (a) and after storage at 60°C for 4 days (b) and 2 weeks (c), and the trehalose formulation immediately post-drying (e) and after storage at 60°C for 3 weeks (f).

Continued on next page

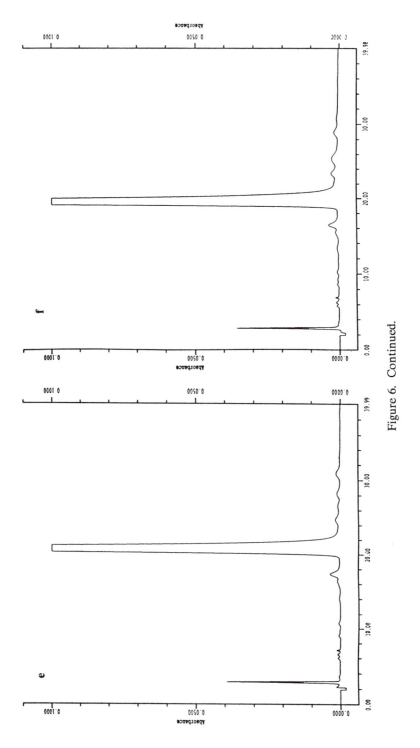

Figure 6. Continued.

The modifications observed in our studies of conventional excipients are consistent with the spontaneous formation of Schiff's bases by amino-carbonyl condensations that initiate the Maillard reaction, as they correlate with published data on the relative chemical reactivity of these sugars in this reaction in food chemistry (25-27), and with the absence of these modifications in samples dried using the non-reducing disaccharide, trehalose (Fig.6, d). Preliminary studies with other non-reducing mono- and disaccharides, mannitol and sucrose, also showed no detectable modifications immediately post-drying, though the appearance of additional peaks was observed in these samples after prolonged storage (results not shown). This might reflect the relative stabilities of the various non-reducing sugars, as compared to trehalose, sucrose is easily hydrolysed (33), and the monosaccharide sugar alcohols are prone to autocatalytic oxidation (35) that yields compounds such as dicarbonyls that are extremely reactive in the Maillard reaction (26, 27). This suggestion is also consistent with the non-enzymatic browning observed in the initial accelerated ageing studies described above, as the more intense browning seen in the sucrose wells, compared to the maltose wells (Fig.2), correlates with the respective reactivities of the hydrolysis products, fructose and glucose (see Fig.5). In the context of the hydrolysis of non-reducing sugars, it is important to also note that some of the reactions in the Maillard cascade may also result in changes in pH, and these changes may in turn accelerate other reactions of the cascade (Fig.3).

Finally, to directly investigate the effect of residual moisture on chemical reactivity, a model system containing lysine with the two non-reducing excipients trehalose and sorbitol, dried and stored at 3 different defined residual water contents, was used. Chemical reactivity was ensured by spiking the drying mixtures with a 5% trace of glucose, and its reaction with the lysine was measured by the quantitation of the glucose remaining after storage at $50^{\circ}C$. In the trehalose samples, chemical reactivity was reduced as the residual water content decreased, and essentially no reactivity was observed at a water content of around 5% (Fig.7a). In contrast, in the sorbitol samples, the chemical reactivity was accentuated under the driest conditions (Fig.7b). These results strongly implicate Maillard-type reactions to yield Schiff bases, as these reactions are driven to completion in systems where water activity is limiting (26, 28).

In conclusion, our results suggest that the Maillard reaction could be an important factor in determining the long term storage stability of dry protein formulations containing carbohydrates as stabilising excipients. Although this preliminary conclusion may be unfamiliar in the context of pharmaceutical formulations, it has long been recognised in the food industry, where the Maillard reaction is known to be a particular problem during the drying of various foodstuffs with high sugar and protein contents (25-28). This is due to both the shift in the equilibrium of the initial amino-carbonyl condensation reaction towards the formation of the Schiff's base by the removal of water during drying and the acceleration of the subsequent reactions at low water activities (27, 28). Most previous work on the stabilisation of dry protein formulations has emphasised the ability of stabilising excipients to replace structural water and/or provide an amorphous or glassy solid matrix in the dry state. This is not entirely incompatible with our findings, as the chemical reactivity of these excipients, documented above, occurs during the storage of the formulations rather than during the drying process itself. It is possibly the unique combination of the properties of water replacement, glass transformation and chemical inertness that might explain the parallel evolutionary selection of trehalose by cryptobionts in many different phyla and the empirically determined superiority of trehalose as a stabilising excipient for dried protein formulations.

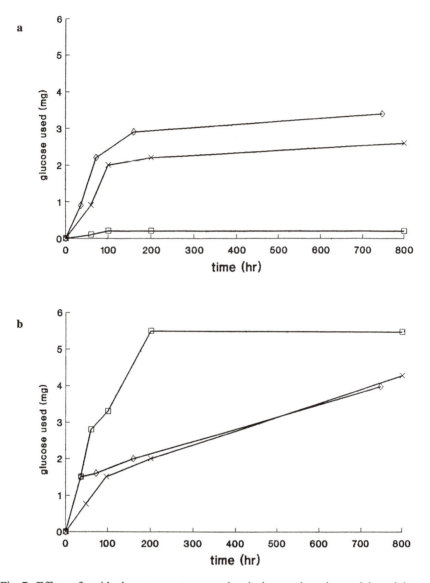

Fig 7. Effect of residual water content on chemical reactions in model excipients. Solutions of 10% w/v lysine and 5% glucose in 85% trehalose (**a**) or 85% sorbitol (**b**) in water were freeze-dried with primary drying at -50°C for 48 hr and secondary drying for a further 24 hr at 20°C. The desired water content was achieved by storage at 20°C over anhydrous P_2O_5 followed by exposure to a saturated water vapour atmosphere for either 0 hr, 8 hr or 25 hr. The actual final water content of the samples was determined by thermogravimetry using a Kahn microbalance.

　　　　　a:- (□) 4.59% water, (◊) 15.09% water, (x) 22.89% water
　　　　　b:- (□) 5.42% water, (◊) 13.12% water, (x) 23.62% water

References

1. Pikal M.J. *Biopharm* **1990**, *3*, 18-27.
2. Pikal M.J. *Biopharm* **1990**, *3*, 26-30.
3. Franks F. *Cryoletters* **1990**, *11*, 93-110.
4. Pikal M.J. 1993 this vol.
5. Carpenter J.F. and Crowe J.H. *Cryobiol.* **1988**, *25*, 244-255.
6. Crowe J.H., Carpenter J.F., Crowe L.M. and Anchordoguy T.J. *Cryobiol.* **1990**, *27*, 219-231.
7. Levine H. and Slade L. *Pure Appl. Chem.* **1988**, *60*, 1841- 1864.
8. Ablett S., Izzard M.J. and Lillford P.J. *J.Chem.Soc. Faraday Trans.* **1992**, *88*, 789-794.
9. Roos Y. *Carbohydrate Res.* **1993**, *238*, 39-48.
10. Carpenter J.C. 1993. this vol.
11. Roser B.J. *Biopharm.* **1991**, *4*, 47-53.
12. Roser B. and Colaco C. *New Scientist* **1993**, *138*, 24-28.
13. Blakeley D., Tolliday B., Colaco C.A.L.S. and Roser B. *The Lancet* **1990**, *336*, 854-855.
14. Colaco C.A.L.S. Blakeley D.,Sen S. and Roser B. *Biotech.Intl.* **1990**, p345-350. Century Press. London
15. Colaco C., Sen.S., Thangavelu M., Pinder S. and Roser B. *Biotech.* **1992**, *10*, 1007-1011.
16. Franks F., Hatley R.H.M. and Mathis S.F. *Biopharm.* **1991**, *14*, 38-55.
17. Franks F. and, Hatley R.H.M. in *"Stability and stabilization of enzymes"* Eds. W.J.J.van den Tweel, Harder A. and Buitelaar. Elsevier, Amsterdam. 1993, p 45-54.
18. Crowe J.H. and Clegg J.S. *Anhydrobiosis.* Dowden, Hutchinson and Ross, Stroudsburg, PA. 1973.
19. Leopold A.C. *Membranes, metabolism and dry organisms.* Cornell Univ. Press, Ithaca. 1986.
20. Coutinho C., Bernardes E., Felix D. and Panek A.D. *J.Biotechnol* **1988**, *7*, 23-32.
21. Gadd G.M., Chalmers K. and Reed R.H. *FEMS Microbiol Letters* **1988**, *48*, 249-254.
22. Crowe J., Crowe L., Carpenter J.F. and Aurell Wistrom C. *Biochem J.* **1987**, *242*, 1-10.
23. Carpenter J.F., Martin B., Crowe L. and Crowe J. *Cryobiology* **1987**, *24*, 455-464.
24. Carpenter J.F. and Crowe J.H. *Cryobiol.* **1988**, *25*, 459-470.
25. Reynolds T.M. *Adv.Food Res.* **1965**, *14*, 167-283.
26. Finot P.A., Aeschbacher H.U., Hurrell R.F. & Liardon R. *The Maillard Reaction in food processing, human nutrition and physiology.* Birkhauser, Basel. 1990.
27. Lendl F. and Schleicher E. (1990). *Ang.Chem.* **1990**, *29*, 565-594.
28. Nursten H.E. in *Maillard browning reactions in dried foods in Concentration and drying of foodstuffs;* Ed. D. Macarthy. Elsevier Applied Science. London. 1986; p53-87.
29. Stewart L.C., Richtmyer N.K. and Hudson C.S. *J.Amer.Chem.Soc.* **1950**, *72*, 2059-2061.
30. Clegg J.S. in *Membranes, metabolism and dry organisms*; Ed. A.C. Leopold Cornell Univ. Press, Ithaca. 1985; p 169-187.
31. Burke M.J. in *Membranes, metabolism and dry organisms;* Ed. A.C.Leopold. Cornell Univ. Press, Ithaca. 1985; p 358-363.
32. Lee C.K. in *Developments in food carbohydrates*; Ed. C.K.Lee. Applied Science Publ. Ltd. London. 1980; vol 2, 1-90.

33. Tewari Y.B and Goldberg R.N. *Biophys. Chem.* **1991** *40*, 59-67.
34. Percival E.G.V. *Advances Carbohyd.Chem.* **1948**. *3*, 23-25.
35. Wolff S.P. et.al. *Free Radicals Biol. and Med.* **1991**. *10*, 339-352.
36. Green J.L. and Angell C.A. *J.Phys.Chem.* **1989**, *93*, 2280-2882.
37. Levine H. and Slade L. *Biopharm.* **1992**, *5*, 36-40.
38. Crowe J.H., Crowe L.M. and Carpenter J.F. *Biopharm.* **1993**, *6*, 28-37.
39. Zyzak D.V., Richardson J.M., Thorpe S.R. and Baynes J.W. *J.Biol.Chem.* **1993**, (in press).

RECEIVED July 5 , 1994

DRUG DELIVERY

Chapter 15

Drug Delivery from Bioerodible Polymers
Systemic and Intravenous Administration

Achim Göpferich[1,2], Ruxandra Gref[1,3], Yoshiharu Minamitake[1],
Lisa Shieh[1], Maria Jose Alonso[1,4], Yashuhiko Tabata[1,5], and
Robert Langer[1]

[1]Department of Chemical Engineering, Massachusetts Institute
of Technology, Building E25, Room 342, Cambridge, MA 02139
[2]Lehrstuhl für Pharmazeutische Technologie, Universität
Erlangen–Nürnberg, Cauerstrasse 4, 91058 Erlangen, Germany
[3]Laboratoire de Chimie Physique Macromoleculaire, Ecole Nationale
Supérieure des Industries Chimiques, 1 rue Grandville, 54001 Nancy
Cedex, France
[4]Laboratorio de Farmacia Galénica, Facultad de Farmacia, Universidade
de Santiago de Compostella, 15706 Santiago de Compostella, Spain
[5]Research Center for Biomedical Engineering, Kyoto University, 53
Kawahara-cho, Shogoin, Sakyo-ku, Kyoto 606, Japan

The recent progress in understanding the erosion of
biodegradable polymers and the manufacturing of controlled
release devices for proteins and peptides is reported. The
erosion mechanism of poly(anhydrides) was investigated as
an example of biodegradable polymers and the erosion behavior is modeled mathematically. The results provide useful
information on the microstructure and chemical environment
inside these polymers during erosion. It is shown how they
might affect the stability and the release of proteins.
Concomitantly proteins were processed in controlled release
devices. Special attention was paid to the stability of the
biomolecules during the manufacturing of dosage forms.
Furthermore, the development of a new type of biodegradable
nanosphere as a future dosage form is shown.

Progress in the field of biology and biochemistry has led to the discovery
of numerous bioactive peptides and proteins in the last few decades. Many
of them, like insulin, as a classical example, are very useful for medical
therapy. The rapid development of biotechnology and progress in peptide
and protein chemistry allowed the mass production of many compounds
and made their broad introduction into medical therapy possible(*1*). The

0097–6156/94/0567–0242$10.88/0
© 1994 American Chemical Society

use of such biomolecules poses, however, severe problems. Some of these compounds have very short half lives in body fluids (under a minute, in some cases(*2*)) and due to degradation in the gastrointestinal tract the majority of them cannot be administered orally. These limitations motivated the development of controlled release dosage forms that improve protein and peptide stability as well as prolong drug activity after application. Processing these substances into dosage forms is not always easily achieved. Many of them have limited chemical and physical stability. Common instabilities are irreversible aggregation(*3*), oxidation(*4*) or conformational changes(*5*) all of which may affect activity. The preparation of controlled release devices for such proteins and peptides has, therefore, become one of the major challenges in the field of controlled release.

Controlled release devices have been prepared for various routes of administration such as oral, parenteral, nasal, rectal, buccal, vaginal and transdermal(*6-7*). They all control the release of peptides and proteins by a few basic principles. Classical systems control release by diffusion through a polymer matrix, in which the drug might be dissolved or suspended(*8*). Closely related to diffusion controlled release systems are those in which swelling controlled release occurs(*9*). The swelling of the polymer, usually an ionic network, increases the permeability of the matrix and allows drugs to be released by diffusion. Parenterally applied dosage forms are the most often used at present as many proteins and peptides are unstable in the gastrointestinal tract resulting in poor bioavailability. The introduction of degradable drug carrier materials brought new progress into this field of controlled release research. Biodegradable polymers as drug carriers control the release of the drugs through diffusion and erosion and have the advantage of dissolving after the application. Instead of simple diffusion-controlled release, drug is release by an erosion-controlled mechanism, which provides the advantage of decreasing the release rate of the proteins and peptides. The search for new biodegradable polymers for controlled release stimulated research in several areas, such as polymer synthesis, polymer processing and the formulation of devices. In the area of polymer synthesis, there have been major efforts to 'design' polymers for the purpose of controlled release. Appropriate examples are poly(ortho esters)(*10*) and poly(anhydrides) (*11*). In providing the biodegradable polymer raw material for controlled release devices for peptides and proteins two fundamental problems arose:

1. In order to efficiently design a controlled release device from biodegradable polymers, the polymer properties and its erosion mechanism should be known.
2. The technology for manufacturing devices ranging from the centimeter to the nanometer scale must be developed. The requirements of stability for such sensitive compounds as peptides and proteins must, thereby, be met.

The successful design of release systems will heavily depend on detailed knowledge in both areas. Unfortunately, a unique theory does not exist for either set of issues. There is not, for example, a basic theory that could predict the erosion of all biodegradable polymers or the stability of proteins and peptides from their structural formulas alone. Due to the expanded research in these areas, however, there is progress towards a broader understanding of polymeric properties and the development of stable protein formulations compatible manufacturing technology. This article intends to show some of the progress that has been made in understanding the release of peptides and proteins from polymers. The focus will be on two major areas:

1. Understanding the erosion mechanism of biodegradable polymers.

2. Providing some examples to illustrate recent advances in the development of controlled release systems. We cover the release of proteins from polymers as well as the development of future dosage forms.

The Erosion of Biodegradable Polymers

Investigating erosion mechanisms of biodegradable polymers might look rather academic and not that beneficial at first glance. The microenvironment of a biodegradable polymer during erosion, however, provides the medium in which peptides and proteins are embedded and from which they have to be released. Factors like pH or the amount of dissolved monomers not only influence release rates, but also affect the stability of proteins and peptides substantially(12) and thus become very important parameters for the stability of a controlled release dosage form. The design of manufacturing systems for the controlled release of proteins and peptides from polymers requires, therefore, a fundamental knowledge the erosion mechanism of a biodegradable polymer.

Diffusion controlled protein release from non-degradable polymers is well-understood(13). The release of such compounds from biodegradable polymers in contrast depends heavily on the erosion of the polymer, a complex process whose mechanism is not entirely understood. The erosion of biodegradable polymers follows mechanisms specific for a certain type of polymer and is influenced by a number of parameters. Most important is the chemical *degradation* of bonds in the polymer chain. The rate at which the bonds are cleaved depends on the type of bonds between the monomers(14), the diffusivity of water in the polymer(15), and the polymer crystallinity(16). The pH of the degradation medium exhibits a catalytic effect on the hydrolysis of bonds(17), influences the solubility of degradation products, and finally regulates their release. Degradation products such as Monomers and oligomers carry functional groups that change pH inside cracks and pores in the eroded polymer and these groups can have feedback on the erosion process(18). It can be concluded that the polymer erosion is a composite process that can be very complex and is specific for each polymer. The erosion mechanisms of poly(lactic-co glycolic acid) and poly(ortho esters), for example, follow completely different mechanisms(19-20).

Investigation of Erosion Mechanisms. We were interested in the erosion mechanism of a class of related poly(anhydrides) consisting of sebacic acid (SA), 1,3-bis(p-carboxyphenoxy)propane (CPP) and a fatty acid dimer (FAD) monomers which are shown in Figure 1. The techniques that were used for these studies may be applied to any class of polymer and the results that we obtained provide a clear outline for the important fundamental characteristics of polymer erosion.

The Structure of Polymers. Prior to the investigation of eroded polymer discs, the polymers were investigated for their internal structure. Figure 2 shows the appearance of various polymers that were taken from thin melted polymer films under polarized light. The Maltese Crosses indicate that these polymers are partially crystalline and consist of spherulites. The crystallinity of the polymers depends on the copolymer composition. In the case of poly(1,3,-bis(p-carboxyphenoxy)propane-co-sebacic acid) abbreviated p(CPP-SA) changing the content of either monomer to a 50:50 composition lowers the crystallinity from more than 60% to almost 10% (*21*). Concomitantly, the glass transition temperature drops substantially. p(FAD) homopolymer is amorphous(*22*), so that the crystallinity of p(FAD-SA) copolymers depends only on the SA monomer content. As shown in Figure 3, increasing the content of FAD monomer above 80% results in amorphous polymers. The effect of microstructure and crystallinity on erosion is observed in investigating eroded polymer discs that were produced from the polymer by melt molding(*11*). Figure 4 a shows a scanning electron photograph of a cross section through a disc of p(CPP-SA) 20:80 that has been eroded in phosphate buffer pH 7.4 at 37°C for 3 days. Eroded and non-eroded polymer sharply separated by the erosion front are readily distinguishable. Observation of this front with time confirms that the erosion of these polymers starts on the surface and moves towards the center of the discs which has been referred to as surface erosion(*23*). Figure 5 shows the kinetics of front movement in p(CPP-SA) 20:80 and p(CPP-SA) 50:50 as measured from SEM pictures. The way erosion affects the polymer can be seen from SEM pictures showing the eroded parts of the polymer in more detail. As shown in Figure 4b the amorphous parts of the spherulites have been eroded, while the large parts of the crystalline skeleton remain in place(*24*). As erosion progresses, a highly porous erosion zone is created. Once the erosion front has reached a certain spot in the matrix, the release of drug begins from that point. The release of compounds is then determined by the diffusion through pores and cracks.

pH Changes during Erosion. When these polymers erode, they release oligomers and monomers that carry carboxylic groups (*25*). We suspected that the pH in the vicinity of the polymer surface or inside the cracks and pores of the polymer is not determined by the pH of the buffer medium, but instead by the released monomers. To support this hypothe-

HOOC-(CH$_2$)$_8$-COOH

sebacic acid (SA)

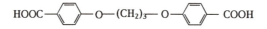

1,3-bis-(p-carboxyphenoxy)propane (CPP)

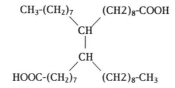

fatty acid dimer (FAD)

Figure 1. Monomers contained in the investigated poly(anhydrides).

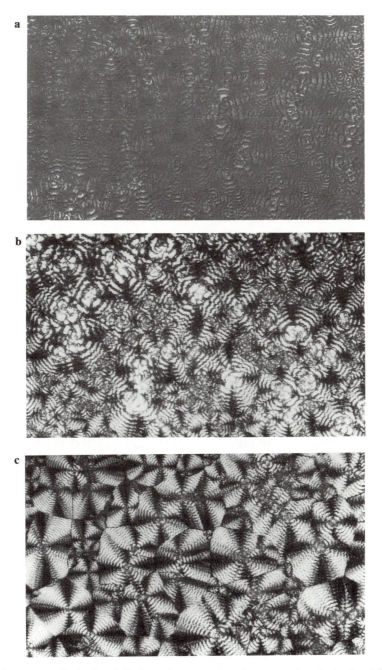

Figure 2. Polarized light microscopic pictures of poly(anhydride) films: a) pSA (250x), b) p-(FAD-SA) 20:80 (400x), c) p-(FAD-SA) 50:50 (400x).

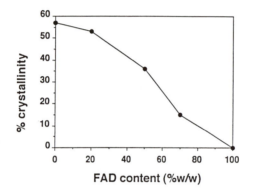

Figure 3. Crystallinity of p(FAD-SA) poly(anhydrides) depending on
FAD content.

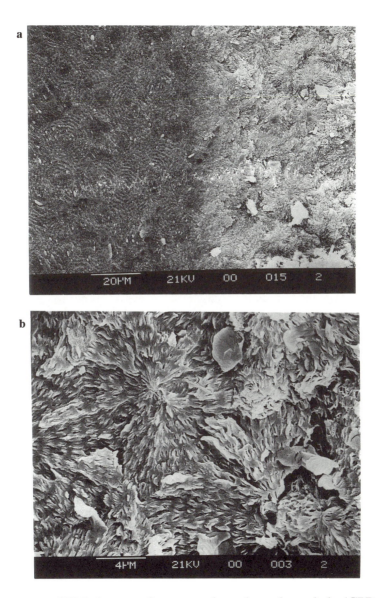

Figure 4. SEM pictures of cross sections through eroded p(CPP-SA) 20:80 dics: a) erosion front separating eroded(right) from non-eroded (left) polymer, b) eroded polymer.

sis, we measured the pH on the surface of p(CPP-SA) 20:80 during erosion using confocal microscopy(26). Figure 6 shows the value for the pH as a function of the distance from the surface of the polymer matrix disc. The pH clearly drops when approaching the surface of the polymer. This effect is due to the high concentration of released monomers upon degradation. Keeping in mind the highly porous structure of the eroded zone of the polymer it can be hypothesized that the pH inside this network is determined by the monomers and is even lower than the pH at the surface(26). This issue is very important with respect to the stability of pH sensitive drugs such as some proteins or peptides.

Changes in Crystallinity. From the structural changes during erosion, changes in polymer crystallinity could be suspected. They were investigated by differential scanning calorimetry and x-ray diffraction. In the case of the p-SA homopolymer, it was found that the crystalline regions are more resistant towards degradation than the amorphous regions. These results agree with the SEM pictures obtained from the degraded part of polymer discs. The monomers SA and CPP were found to crystallize to relatively high extents in all polymers qualitatively shown by wide-angle x-ray diffraction and quantitatively by DSC. Figure 7 shows the calculated amount of crystallized monomers in p(CPP-SA) 20:80, as determined by DSC according to (26). The presence of crystallized material proves indirectly that the pores of eroded polymer are filled with a saturated solution of monomers. FAD monomer is liquid at room temperature and therefore not able to crystallize upon release. Due to its low solubility in aqueous buffer solution, however, it sticks to the eroding discs as a film.

The Release of Monomers during Erosion. The release of acidic monomers indicates the importance of pH during erosion. As an example, the release profile of SA and CPP monomer during the erosion of p(CPP-SA) 20:80 is shown in Figure 8. The release profile of SA is similar to the release of substances from monolithic devices. The release profile of CPP is discontinuous as the release rate increases after SA has completely left the eroded polymer matrix. This spontaneous increase can be explained by the higher solubility of SA compared to CPP. The dissolved SA, therefore, controls the pH inside the pores of the eroded zone limiting, thereby, the solubility of CPP. After SA has been completely released, the pH inside the pores starts to rise, which increases the solubility of CPP(26). This process finally increases the release rate of CPP from the eroded polymer matrix.

Modeling Polymer Erosion. The modeling of polymer erosion is the first step towards the theoretical design of controlled release systems with an a priori adjusted release rate. The lack of knowledge about erosion, however, hindered the development of models substantially. The first attempts

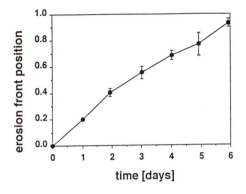

Figure 5. Relative position of the erosion front in p(CPP-SA) 20:80 depending on time.

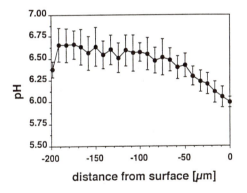

Figure 6. pH in the degradation medium as a function of distance to the surface of a p(CPP-SA) 20:80 polymer disc after 2 days of erosion in daily changed 0.1 M phosphate buffer pH 7.4 at 37°C.
(Reproduced with permission from ref. 26. Copyright 1993 Wiley & Sons.)

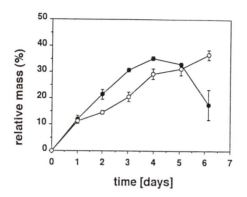

Figure 7. Content of crystallized monomers in a p(CPP-SA) 20:80 matrix during erosion: o CPP, ● SA.
(Reproduced with permission from ref. 26. Copyright 1993 Wiley & Sons.)

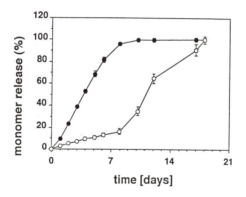

Figure 8. Release of monomers during the erosion of a p(CPP-SA) 20:80 polymer matrix disc: o CPP, ● SA.
(Reproduced with permission from ref. 26. Copyright 1993 Wiley & Sons.):

to describe the erosion of biodegradable polymers stem from the description of the dissolution of erodible tablets(*27*). It was assumed that erosion could be described by surface dissolution kinetics. Later, diffusion processes were taken into account and modeled as a pseudo-steady state process, where erosion was modeled assuming first order kinetics or a constantly moving erosion front(*28-29*). Introducing non-steady state analysis, erosion was described as a moving front problem with constant velocity(*30*). All these models describe erosion in one dimension and do not allow descriptions of the microstructural changes upon erosion. Inspired by the two-dimensional description of the dissolution of matrix tablets(*31-32*), a two-dimensional model able to describe the structural changes of our investigated polymers upon erosion was developed. Rather than taking individual factors of polymer erosion into account, we combined them by regarding the erosion of small parts of polymer matrix as a random event(*33*).

The Basic Concept. A polymer matrix was represented by a two-dimensional computational grid, in which each grid point represents either an amorphous or crystalline part of the polymer matrix. The erosion of such a 'polymer pixel' was assumed to depend on two features: the contact of the polymer with the degradation medium, and the crystalline or amorphous nature of a matrix part which is represented by a pixel. Polymers not in contact with water will not erode. Pixels on the surface of the polymer matrix or next to an eroded neighbor have contact to water. The erosion of crystalline matrix parts occurs at a slower rate than amorphous parts, a characteristic which was observed in the structural studies(*26*). Besides these fundamental assumptions, a general algorithm must be defined to develop a working model: the matrix must be represented by a computational grid divided into individual pixels. A distribution must be chosen for crystallites and amorphous parts. After contact with an eroded neighbor, each pixel is assigned an individual life expectancy after which it is regarded eroded. The life expectancies are distributed according to a first order Erlang distribution(*34*) from which they are chosen randomly. To account for the crystalline or amorphous nature, different constants are chosen for use in the Erlang distribution.

The Representation of a Polymer Matrix. To represent a polymer matrix prior to erosion, it is sufficient to represent only a cutout of the matrix, if the erosion problem is symmetrical. Figure 9 shows such a cutout for a cross section through a cylindrical matrix.

To account for the crystallinity, some of the pixels have to be designated to represent crystalline polymer matrix parts. In the easiest case, a random distribution of them can be assumed:

$$c(x_{i,j}) = \begin{cases} 1-\chi & x_{ij}=0 \\ \chi & x_{ij}=1 \\ 0 & \text{all other values} \end{cases} \quad \begin{array}{l} \text{for:} \\ 1 \le i \le n_x, \\ 1 \le j \le n_y. \end{array} \quad (1)$$

χ is the crystallinity of the polymer. The status $x_{i,j}$ (1 for crystalline and 0 for amorphous) of all pixels is assessed from consecutive Bernoulli trials(35). $c(x_{i,j})$ is the probability that a pixel at location x=i and y=j on the grid represents an amorphous or crystalline part of the matrix. Figure 10a shows the theoretical representation of a polymer matrix prior to erosion. Dark pixels represent crystalline, whereas white pixels represent amorphous parts of the polymer.

The Simulation of Erosion. After having set up the grid, the erosion algorithm can be applied. A first order Erlang distribution(34) was used for the calculation of the life expectancy of a pixel after its contact with water

$$e(t) = \lambda * e^{-t*\lambda} \quad (2)$$

e(t) is the probability that a pixel degrades after a time t. Random variable t is the time that elapsed between the first contact with water and the erosion of the pixels. λ can be considered as a rate constant and is different for amorphous and crystalline pixels. To achieve results independent from the grid size, the grid size n has to be taken into account in equation 2:

$$e_n(t) = \lambda * n * e^{-t*\lambda*n} \quad (3)$$

Using Monte Carlo sampling techniques(36), the life expectancy of an individual pixel can now be calculated from equation 3. Erosion begins at the pixels that represent the surface of the polymer matrix. Their lifetime is calculated and the pixel with the shortest lifetime is determined. Erosion proceeds with the removal of eroded pixels from the grid, now considered to represent aqueous pores. They then initiate the erosion process of non-eroded neighbors, the life expectancies of which are calculated(33). The next pixel is eroded, and so on. As an example, the erosion of the grid shown in Figure 10a can be seen in Figure 10b. In contrast to Figure 10a dark pixels represent non-eroded polymer and white pixels eroded polymer. The appearance of the cross sections of eroded discs shown in Figure 4a agrees with the appearance of the lattice during erosion. In both cases, the erosion front is clearly visible.

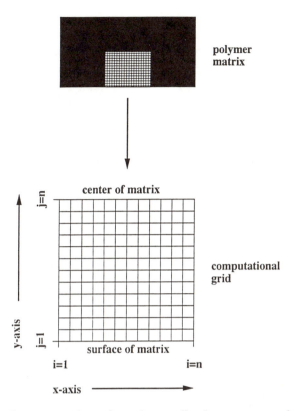

Figure 9. Representation of a polymer disc by a computational grid. (Reproduced with permission from ref. 33. Copyright 1993 ACS.)

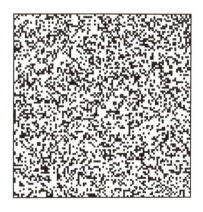

Figure 10a. Theoretical representation of a polymer matrix: prior to erosion (dark pixels = crystalline areas, white pixels = amorphous areas). (Reproduced with permission from ref. 33. Copyright 1993 ACS.)

t * λa = 0.042

t * λa = 0.074

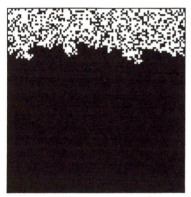

t * λa = 0.104

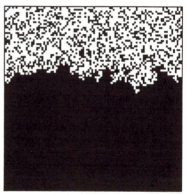

t * λa = 0.157

t * λa = 0.339

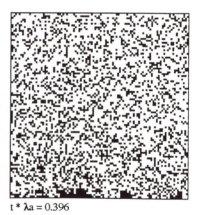

t * λa = 0.396

Figure 10b. Theoretical representation of a polymer matrix: changes during erosion (dark pixels = non-degraded areas, white pixels = degraded areas). (Reproduced with permission from ref. 33. Copyright 1993 ACS.)

The Quantitative Evaluation of Experimental Results. The unknown parameters in this model are the erosion rate constants for amorphous and crystalline pixels. Once these parameters are known the model
becomes a tool to estimate the erosion behavior of polymer matrices and
to predict when the last portion of drug will be released. The computational grid can, thereby, be fit to any desired shape.

In order to determine the erosion rate constants, system parameters
must be defined to develop a model to fit the experimental data. Therefore
two functions were chosen:

$$f(t) = \frac{1}{n} \sum_{i=0}^{n} f_i(t) \tag{4}$$

$$m(t) = \frac{1}{n^2} \sum_{i=1}^{n} \sum_{j=1}^{n} s(x_{i,j}) \tag{5}$$

Equation 4 defines the position of the erosion front $f(t)$ at time t as an average value of the position $f_i(t)$ of the foremost eroded pixel in each column of the grid. Experimentally, the position of the erosion front was determined by SEM. Equation 5 calculates the remaining mass of the matrix
$m(t)$ at time t, $s(x_{i,j})$ being equal to 1 for non-eroded pixels and 0 for
eroded pixels. Experimentally, the remaining mass was determined by
weighing dried eroded polymer matrix discs. Using equations 4 and 5, the
model was fit to experimental data of p(CPP-SA) 20:80 and p(CPP-SA)
50:50, minimizing the squared distances between experimental points and
simulated data calculated from equation 4 and 5 (*37-38*). As shown in
Figures 11 a and b, the movement of the erosion front is almost linear in
both polymers. The larger deviation between experimental and predicted
data for p(CPP-SA) 50:50 (cf. Fig. 11b) is mainly due to problems in the
determination of erosion fronts in this polymer by light microscopy. The
loss of mass after an induction period also follows linear kinetics. From
the good fit to this parameter it can be assumed that the effect of the error
in the determination of the erosion front movement has only moderate effect on the determination of the rate constants. The erosion rate constants
obtained are shown in Table I.

As expected from the structural information on eroded polymer matrices,
the erosion rate constants for crystalline pixels are substantially lower

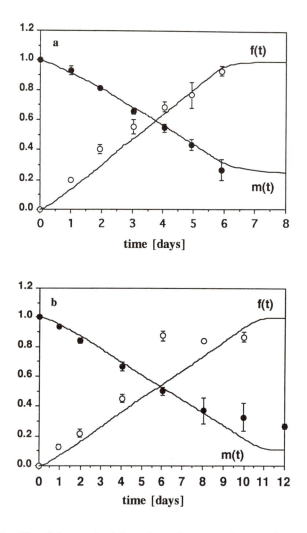

Figure 11. Fit of theoretical functions for mass loss and erosion front movement to experimental data: a) p(CPP-SA) 20:80, b) p(CPP-SA) 20:80.
(Reproduced with permission from ref. 33. Copyright 1993 ACS.)

Table I. Erosion rate constants for crystalline and amorphous areas in two poly(anhydrides)

	λ_a [s^{-1}]	λ_c [s^{-1}]
p(CPP-SA) 20:80	$7.32*10^{-7}$	$8.75*10^{-9}$
p(CPP-SA) 50:50	$2.7*10^{-7}$	$3.85*10^{-11}$

Source: Reprinted with permission from ref. 33. Copyright 1993.

compared to the rate constants for amorphous pixels. The obtained micro-constants allow the simulation of erosion of any kind of polymer matrix cross section under various boundary conditions. A good example is the simulation of erosion of devices partially coated with a water impermeable coating. Such devices could be very useful for the modification of release profiles for slowing down the overall erosion velocity of a biodegradable device.

The Consequences for the Release of Proteins from Polymers. The erosion of the poly(anhydrides) follows a well defined pattern. All polymers degrade from the surface of the polymer matrix discs towards their center. A moving front sharply separates eroded from non-eroded polymer. It could be shown by scanning electron microscopy that crystalline areas of the polymers are more resistant to erosion than amorphous areas. Due to the spherulitic microstructure of the polymers (cf. Figure 2), erosion creates highly porous devices consisting of the crystalline skeleton of eroded spherulites. Due to the high degradation rates, many oligomers and finally monomers are released. Reaching the limit of solubility, these substances start to precipitate out of solution, which was observed by wide angle x-ray diffraction and differential scanning calorimetry. Substantially high amounts of monomers were found to be in a crystalline state(26). Confocal fluorescence microscopy showed that the pH on the surface of the polymers during erosion was lower than in the buffer medium. From the presence of crystalline monomers inside the erosion zone, it can be concluded that the aqueous phase inside the polymers is saturated with monomers. As SA has the highest solubility of all monomers, this compound determines the pH. From the pKa which is about 4.5 and the solubility which is 0.011M, the theoretical pH minimum is about 3.5(39). Due to the diffusion of hydroxide ions from the degradation medium into the erosion zone, however, some of the free acids are neutralized, which buffers the pH probably around the pKa. Buffer salts like HPO_4^{2-} are apparently not able to raise the pH inside the porous zones of the eroded discs. The major reason is the long diffusion pathway into the disc along which they get steadily protonated by dissolved monomers and lose their potential to raise pH. The monomer saturated solution inside polymer pores with a pH of approximately 4.5 is, therefore, the environment that proteins and peptides will be in contact prior to release out of the matrix. Once the stability data of such a biomolecule is known, it can readily be decided whether these types of polymers will make suitable drug carriers. The well defined erosion behavior of the investigated polymers allows the development of mathematical models that describe changes in polymer microstructure during erosion(*33*). Once fit to experimental data, these models are a useful tool to predict changes in polymer microstructure. They describe important parameters like porosity, loss of weight, eroded polymer area and may, therefore, become an important tool to predict the release kinetics of proteins and peptides from such polymers in the future.

Controlled Release Devices for the Delivery of Proteins and Peptides from Polymers

For the delivery of proteins and peptides, non-degradable and biodegradable polymers have been used in controlled release. Due to the high molecular weight of proteins, direct diffusion through a non-degradable polymer matrix is not possible(*40*). By the introduction of a network of pores in manufacturing, however, the release of such large compounds does occur(*41*). Another possibility is the creation of hydrophilic pathways using swellable polymers, or embedding the compounds into gels. In the case of biodegradable polymers, pores are created upon erosion of the polymer matrix enabling the release of proteins from the dosage form. All of these options have certain advantages and disadvantages. Embedding suspended compounds into a non-degradable matrix prevents some of the protein from being released(*42*), or might cause some instabilities due to the intense contact with organic solvents(*43*). Their disadvantage with respect to parenteral application, however, is the need for removing such systems after therapy. By using gels as a carrier, the protein may be released very quickly if not combined with some other sort of material(*44-45*). Degradable polymers change their properties substantially during erosion, a characteristic which may or may not be beneficial for the stability of proteins and peptides. In general, decisions about the suitability of a release device for specific proteins or peptides and appropriate manufacturing technology must be made on a case by case basis. In the following section, we provide a number of examples from the progress we made in the delivery of proteins from polymers.

Biodegradable Polymers for Immunization. The idea of using polymers as antigen releasing carriers for the stimulation of immune responses emerged in the seventies (*46*). Due to their ease of application and the progress in their development, microspheres emerged as a potentially efficient carriers to enhance the immune response (*47*). They have so far been used for a number of vaccines (*48-50*). These microspheres were made from polymers based on poly-lactic acid (PLA) and its copolymers with glycolic acid (PLGA). For these applications, loading a polymer with a high molecular weight protein requires adjusting the desired release rate and preventing the protein's loss of activity.

In an attempt to develop a controlled release system for vaccination against tetanus, we investigated the design of microspheres with a desired release rate and methods to maintain the imunogenicity of the processed tetanus toxoid(*51*). For the manufacturing of microspheres, the solvent evaporation and solvent extraction methods were applied to the double emulsion technique. The solvent evaporation method has been used successfully for the encapsulation of hydrophilic drugs and peptides(*52-53*) and was recently applied to the encapsulation of proteins(*54*). According to the protocol, $50\mu L$ of saline tetanus toxoid solution (3600 Lf/mL) was emulsified by vortex mixing in 1mL organic solution (methylene chloride or ethyl acetate) containing 200 mg of polymer. The emulsion was homogenized by sonication and dispersed into 1mL of a 1% aqueous PVA solution by vortex mixing. The final double emulsion was diluted by adding it to 100 mL of a 0.3% PVA solution while stirring with a magnetic stirrer. The organic solvent was then allowed to evaporate for 3 hours (solvent evaporation) or extracted by adding 200 mL of 2%(V/V) aqueous isopropanol solution (solvent extraction). Both techniques produced microspheres with a high loading capacity (>80%). Critical points in the encapsulation procedure are the protein stability upon contact with moisture(*55*) and organic solvent (*43*) during the formation of the emulsion as well as the subsequent freeze drying.

Protein stability under the manufacturing conditions was investigated varying three parameters. We varied the type of solvent for the dissolution of polymer using methylene chloride and ethyl acetate and investigated the effect of three different types of stabilizers: Pluronic F68, PEG 4,600 and sodium glutamate. To mimic the conditions that prevail during the manufacturing of microspheres, emulsions were prepared from $100\mu L$ aqueous tetanus toxoid solution and dispersed it in 1mL of organic solvent. These emulsions were freeze dried and investigated for aggregation by HPLC(*56*). Table II shows the results from this study.

Table II gives a good survey on the impact of manufacturing conditions on the stability of the protein. Sample one shows that moisture alone causes

Table II. manufacturing conditions and their effect on protein aggregation

Set of conditions	Solvent	Stabilizer	Aggregation (%)
1	none	none	14
2	none	Pluronic F68	0
3	none	PEG 4,600	3.9
4	none	Sodium Glutamate	4.4
5	Methylene chloride	none	13.11
6	Methylene chloride	Pluronic F68	9.6
7	Methylene chloride	PEG 4,600	9
8	Methylene chloride	Sodium Glutamate	3.4
9	Ethyl acetate	none	5.4

the protein to aggregate substantially. The impact of even small amounts of moisture on the structure of lyophilized proteins can be tremendous(55). The aggregation of the protein in methylene chloride (sample 5) is higher than in ethyl acetate (sample 9) which is unfortunately the poorer of both solvents for many biodegradable polymers. The use of stabilizers like Pluronic F68 might be benefical in preventing some of the proteins and peptides from aggregating or from adsorbing to hydrophobic surfaces i.e. polymer forming the microspheres. This has been reported for a number of surfactants(57).

The goal for the release of tetanus toxoid from microspheres was continuous release over a period of weeks. For that purpose two types of polymers were used: L-PLA and D,L-PLGA1:1. The PLGA polymers are amorphous whereas L-PLA are crystalline. Depending on their molecular weight they erode in weeks or months(58). To investigate the effect of erosion velocity the polymers were chosen with a molecular weight of either 100,000 or 3,000 Da. The release was studied by incubation of the microspheres in phosphate buffer pH 7.4 at 37°C. Due to the slow degradation of the polymers it was sufficient to replace the buffer at least every 5 days. The released protein was determined by using a microBCA protein assay. The results for the two types of polymers are shown in Figure 12a and b. The release rate was clearly affected by polymer composition with PLGA faster than from PLA at the same molecular weight. In both cases,

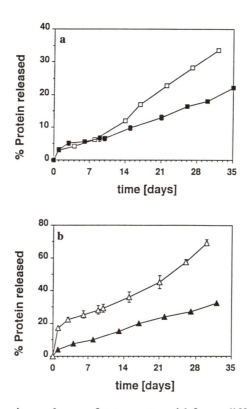

Figure 12. In vitro release of tetanus toxoid from different types of polymer:
a) ■ PLA (Mw 100,000); □ PLGA (Mw 100,000)
b) ▲ PLA (Mw 3,000); △ PLGA (Mw 3,000)
(Reproduced with permission from ref. 56. Copyright 1993 Butterworth-Heinemann Ltd.)

the release from the low molecular weight polymers is faster. This result is due to the smaller size of microspheres prepared from these polymers since the organic polymer solution had a lower viscosity. Other reasons could be their faster degradation due to differences in molecular weight. All preparations release over a period of weeks, as was intended. With the exception of low molecular PLGA, there is no substantial burst effect as reported for other vaccine preparations (*59*).

A general disadvantage of methods that use chemical reactions or physical interactions like chromatographic methods for the determination of protein concentrations is the lack of structural information. We, therefore, tested the *in vivo* response to the encapsulated tetanus toxoid by injecting the microspheres subcutaneously into mice. The sera was assayed for the antitoxin units using the toxin neutralization test(*60-61*) as well as an ELISA test. Figure 13a and b show the results for all microspheres as well as a control injection of fluid toxoid and aluminum adsorbed toxoid. The use of microspheres increases the immune response compared to the fluid toxoid. The use of aluminum adsorbed toxoid produces initially higher antitoxin titers which decrease, however, more rapidly compared to the encapsulated toxoid. The levels resulting from microsphere injection are considerably higher than the estimated minimum protective level which is about 0.01 AU/mL(*62*). The antibodies determined by the neutralization test and the IgG antibodies reach their maximum at different times. This could be explained by the induction of progressive affinity maturation tetanus antibodies induced by prolonged exposure to low concentrations of tetanus toxoid.

Protein Delivery from Poly(anhydride) Microspheres. A major problem associated with many dosage forms for drug release is the release of large amounts of the drug during initial stages of release known as the initial burst-effect. For microspheres, this burst effect tends to be enhanced since microspheres have a larger surface area to volume ratio than more sizable slabs. In addition, the change of biological activity of proteins can occur during microsphere formulation because of their sensitivity to environmental alternation. Our goal in this study was to prepare poly(anhydride) microspheres which permitted the controlled release of biologically active proteins without any initial burst in release. We selected two enzymes, trypsin (Mw24,000) and heparinase (Mw43,000) in addition to BSA (Mw 62,000) which has been used as a model protein. The measurement of enzymatic activity allowed us to investigate the effect of microsphere formulation on protein activity.

We encapsulated proteins into poly(anhydride) microspheres with the solvent evaporation method by using a double emulsion (*54*). However, this technique comprises of a number of formulation processes which may affect protein activity: contact of aqueous protein solution with

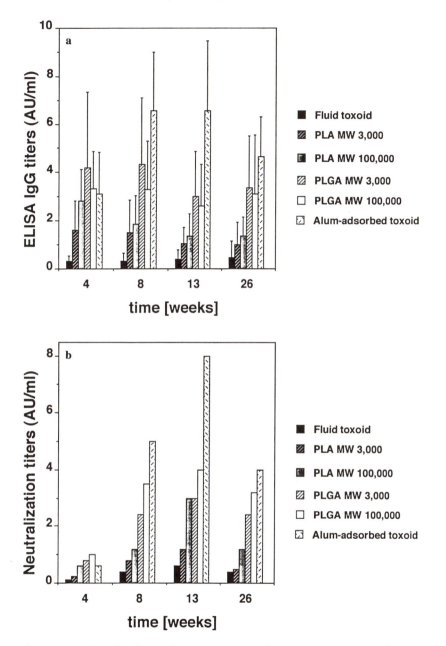

Figure 13. Results from the *in vivo* test of tetanus toxoid loaded microspheres:
a) *In vivo* neutralizing antibody response (antitoxin units/ml serum),
b) IgG antibody response by ELISA.
(Reproduced with permission from ref. 56. Copyright 1993 Butterworth-Heinemann Ltd.)

the organic phase, vortex mixing, exposure to ultrasound and freeze drying.

The activity loss of proteins was measured based on the specific activity of proteins at each stage of the process (63). Table III shows the effect of the formulation process on the enzymatic activity of trypsin.

Table III. Activity loss of trypsin after each step of microsphere preparation

Preparation step	remaining activity (%)
Trypsin solution	100
vortex mixing	100
sonication for 30s	59
3 h solvent evaporation	56
freeze drying	59

The sonication process which was performed in presence of the organic solvent to prepare a primary emulsion, causes the main activity loss of protein, while freeze-drying did not have any addition effect. It is known that ultrasound has an effect on the biological properties of proteins in aqueous solution due to a number of factors such as temperature, mixing and cavitation(64). Table IV shows the effect of ultrasound exposure period on the activity loss of proteins during microsphere preparation together with the additional effect of some protein stabilizers(65).

Table IV. Activity loss of trypsin during microsphere preparation (66)

Period of sonication [s]	Type of stabilizer	Amount of stabilizer (% w/w)	remaining activity
10	none	-	80.5
20	none	-	67.3
30	none	-	58.9
30	BSA	5	58.6
30	BSA	10	58.1
30	Glycine	4	58.7*
30	Glycine	7.5	59.8*
30	Glycine	8	56.6*
30	Glycine	15	57.2*

*Differences within the error margin of the test

The extent of activity loss becomes smaller with a decrease in the exposure period of sonication. In addition, the decreased period did not reduce the size of the primary emulsion. These findings demonstrated that the observed activity loss results from the sonication probe and not from the increase in the interfacial area between aqueous and organic phase. Little effect of the stabilizers used here on activity loss was observed, indicating the poor potential of these substances to protect the investigated proteins against ultrasound exposure. Trypsin and heparinase were encapsulated to study the enzymatic activity of proteins in poly(anhydride) microspheres. Figure 14 shows the comparison of the remaining activity between the proteins encapsulated in microspheres and the proteins in a soluble form after incubation in phosphate buffer pH 7.4 at 37°C. At different stages of the experiment, the remaining % activity of proteins in microspheres was assessed by determining the activity of protein that could be extracted from these systems (66) These results clearly indicate that encapsulation of proteins inside poly(anhydrides) may stabilize their activity for a short period of time (1-2days).

Figure 15 shows release profiles of BSA from p(FAD-SA) 25:75 at various protein loadings. No initial burst is observed irrespective of the protein loading. The protein is released for up to three weeks at a near-constant rate.

The release rate of protein depends on the monomer composition of poly(anhydrides) used as can be seen from Figure 16. Release periods of several days to a couple of weeks are possible for poly(anhydride) microspheres by changing their monomer composition(66).

Biodegradable Injectable Nanospheres Composed of Diblock Poly(ethyleneglycol)-Poly(lactic-co-glycolic acid) Copolymers A major challenge in the area of parenteral administration of drugs is the development of a drug carrier that is small enough for intravenous application and which has a long circulation half-life. There are numerous potential applications for such a system: the suppression of toxic side effects which can occur when the drug is injected in the form of a solution, or the protection of sensitive compounds against degradation in the plasma. Consequently there have been many approaches to develop such systems (67-68). The major obstacle in the use of injectable systems is the rapid clearance from the blood stream by the macrophages of the reticulo-endothelial system (RES). Polystyrene particles as small as 60 nm in diameter, for example, are cleared from the blood within 2 to 3 min(69). By coating these particles with block copolymers based on polyethylene glycol and polypropylene glycol, their half-life was significantly increased (70). Polystyrene particles are, however, not biodegradable and are not therapeutically useful. For this reason, only liposome systems have been developed for intravenous administration (71-76). Due to their small size they were expected

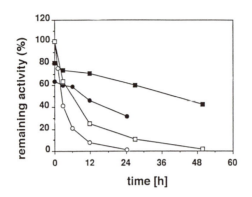

Figure 14. Stability of Heparinase and Trypsin with and without polymer encapsulation in phosphate buffer pH 7.4: ○ Heparinase solution, ● Heparinase encapsulated, □ Trypsin solution, ■ Trypsin encapsulated.
(Reproduced with permission from ref. 66. Copyright 1993 Plenum Press.)

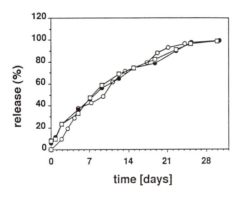

Figure 15. Release of BSA from poly(anhydride) microspheres depending on loading: ○ 1% BSA, ● 2% BSA, □ 4% BSA.
(Reproduced with permission from ref. 66. Copyright 1993 Plenum Press.)

to freely circulate in the blood. It has been found, however, that they are cleared from the blood by uptake through the reticulo-endothelial system (RES). The coating of liposomes with poly(ethyleneglycol) (PEG) increased their half life substantially (71-77). The presence of the flexible and relatively hydrophilic PEG chains induces a steric effect at the liposome's surface, thus reducing protein adsorption and RES uptake (77).

Rather than using liposomes, we designed PEG-coated biodegradable nanospheres that consist of polymer only(78). These are among the first degradable polymer-based systems with potential use for prolonged intravenous administration or drug targeting.

The 'ideal' nanosphere (Figure 17) is biodegradable, biocompatible, has a size of less than 200 nm, and has a rigid biodegradable core. The PEG-coated surface theoretically provides a prolonged half-life in the blood by masking it from the RES. We chose the FDA approved polyesters poly(lactic-co-glycolic acid) (PLGA) to form the core of the particles. These polymers allow us to adjust erosion rates and therefore drug release by changing the monomer ratio. We synthesized a series of diblock copolymers PLGA-PEG, starting with PEG of various chain lengths and progressively increased the chain length of PLGA. Stannous octoate was used as a catalyst for the ring-opening polymerization of lactide and glycolide in the presence of monomethoxy polyethylene glycol (MPEG; Mw 5,000; 12,000 and 20,000). The polymers obtained were characterized by nuclear magnetic resonance spectroscopy (NMR), gel permeation chromatography (GPC), infrared spectroscopy (IR), differential scanning calorimetry (DSC) and X-ray diffraction. The polymerization reaction was followed by GPC from which molecular weight and polydispersity were determined. The consumption of lactide and glycolide can be observed from a decrease of peak D (Figure 18). The shifting of the peak P towards lower retention times (higher molecular weights) indicates that an addition reaction takes place at the hydroxyl end group of the PEG chains. The polydispersity remains low during the polymerization (Mn/Mw less than 1.1). Both the crystallinity and the glass transition temperature (Tg) were determined by using DSC. The diblock copolymers show a single Tg, which shifts to higher temperatures with an increase of the PLGA chain length inside the PLGA-PEG (Figure 19). A complete discussion of the DSC results can be found elsewhere(79). Random lactic acid-ethylene oxide copolymers show two distinct Tg's (80), suggesting a phase separation inside the polymer. We suppose that the single Tg we observed is due to an entanglement effect of long PEG and PLGA chains in the polymers, which cannot easily phase-separate. The exact chemical composition of the polymers was determined by ^{13}C NMR spectroscopy.

Sterically stabilized particles were prepared from diblock PLGA-PEG copolymers or from blends of PLGA and PLGA-PEG. These polymers were dissolved in a common solvent (ethyl acetate or methylene

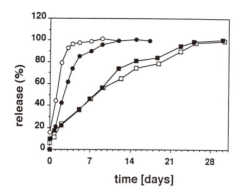

Figure 16. Release of BSA from poly(anhydride) microspheres depending on polymer type: ○ p(SA), ● p(FAD-SA) 8:92, ▢ p(FAD-SA) 25:75, ■ p(FAD-SA) 44:56.
(Reproduced with permission from ref. 66. Copyright 1993 Plenum Press.)

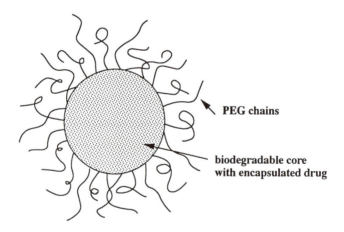

Figure 17. Schematic representation of an 'ideal' nanosphere having a biodegradable drug containing core and a coating of Polyethylene glycol chains.

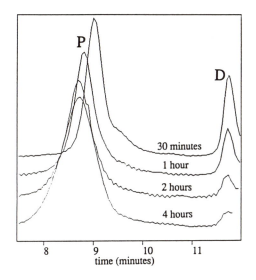

Figure 18. ATime course of the polymerization reaction between PEG and lactide/glycolide followed by GPC: Peak P: polymer PEG-PLGA, peak D: starting material (glycolide+lactide).

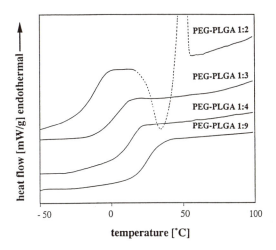

Figure 19. DSC thermograms of PLGA-PEG diblock copolymers (heating rate 10 °C/min, MW PEG=5000, for increasing chain length of PLGA).

chloride). An O/W emulsion was formed by vortex mixing and sonicating for one minute. The organic solvent was then slowly evaporated, at room temperature, under gentle stirring for two hours. Slow removal of the solvent allows the reorganization of the polymer chains inside and on the surface of the droplets. By optimizing the viscosity of the organic phase, the volume ratio of the two phases and the sonication parameters, nanospheres of a mean size of about 150 nm were obtained. After removal of the organic solvent, the nanospheres were isolated from the aqueous phase by centrifugation. They could later be readily redispersed in water, probably due to the presence of PEG chains on their surface.

The nanospheres were characterized by various methods. The surface composition of lyophilized particles was determined by X-ray photoelectron spectroscopy (XPS). Figure 20 shows the carbon 1s envelopes that were observed by analyzing the nanosphere powder. One main carbon environment is observed for PEG. It corresponds to the ether carbon $\underline{C}$-O (81). The spectrum of PLGA polymer (curve a) shows three main carbon environments. In the case of PEG-PLGA nanospheres, the peaks are still present. As the information obtained by XPS corresponds to the first layers on the nanosphere's surface (about 5 nm(82)), this spectrum indicates that PEG is present on the surface of the nanosphere powder. Moreover, we verified that PEG remains on the surface after incubation in distilled water for 24 hours at 37 °C.

To investigate the *in vivo* half life of the systems, the nanospheres were labeled with [111]Indium, which has been used for gamma-scintigraphy studies in humans (83). [111]In was attached directly to the polymer chain by complex formation following a method that has already been used for liposomes (83). The incubation at 37 °C in PBS or horse serum for more than 24 hours showed that there was less than 4 % label loss. This indicated that the label is bound tightly enough to prevent loss during *in vivo* study. Subsequently biodistribution studies were performed by injecting ^{111}In-labeled PLGA and PEG-coated nanospheres into the tail vein of Balb/c mice (18-20g). Five minutes after injection of uncoated PLGA nanospheres, 40% of nanosphere-associated ^{111}In radioactivity was found in the liver and approximately 15% in the blood. In the case of PEG-coated nanospheres, the results were reversed (15% of injected radioactivity in the liver, 60% in the blood). After four hours, 30% of the nanospheres were still circulating in the blood, whereas the PLGA nanospheres had disappeared completely from the blood.

These results are very encouraging and we hope that these nanospheres might be used as drug carriers for intravenous administration in the future. Studies are now underway to achieve the encapsulation of different hydrophobic drugs and small peptides into the core of these nanospheres. We are also determining the effect of PEG molecular weight

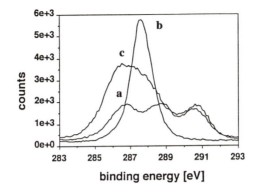

Figure 20. Surface analysis by XPS of PLGA powder: a) PLGA powder, b) PEG crystals, c) PLGA-PEG nanospheres.

and density on the RES uptake and on drug encapsulation efficiency and release.

Summary

Progress in various fields of protein delivery has been illustrated. Using poly(anhydrides) as a model we showed that it is possible to describe erosion mechanisms and to develop mathematical models of the complex processes involved in polymer degradation. Based on such models the design of delivery systems using computers might become possible in the future. Concerning manufacturing technologies for protein delivery, we showed how existing systems, like microspheres, can be improved and provided the development of vaccines as an example for future application of such systems. Finally the recent progress in the development of new dosage forms was illustrated with the example of biodegradable nanospheres with an increased circulation half-life. All these projects aim at a better understanding and an improved design of protein delivery systems.

Acknowledgments

The authors want to thank Dr. R.K. Gupta and Dr. G.R. Sieber for their contribution to the protein stability study and in the vivo immune response studies. Thanks are due to the Deutsche Forschungsgemeinschaft who sponsored the polymer degradation studies with grant No GO 565/1-1 and to the National Institutes of health who sponsored the same project with grant No. GM 26698. We thank Scios Nova for supplying us with polymers.

Literature Cited

1 Sadee, W., *Pharm. Res.* **1986**, 3, pp.3-6.
2 Luft, F.C.; Lang, G.R.; Aronoff, H.; Ruskoaho, M.; Toth, M.; Ganten, D.; Sterzel, R.B.; Unger, T. *J. Pharmacol. Exp. Ther.* **1986**, 236, pp.721-728.
3 Sluzky, V.; Tamada, J.A.; Klibanov, A.M.; Langer, R. *Proc. Natl. Acad. Sci.* **1991**, 88, pp.9377-9381.
4 Witkop B. *Adv. Protein Chem.* **1961**, 16, p.221.
5 Marcritchic *Adv. Protein Chem.* **1978**, 32, p.283.
6 Zhou, X.H.; Li Wan Po, A. *Int. J. Pharm.* **1991**, 75, pp.97-115.
7 Zhou, X.H.; Li Wan Po, A. *Int. J. Pharm.* **1991**, 75, pp.117-130.
8 Bawa, R.; Siegel, R.; Marasca, B.; Karel, M.; Langer, R. *J. Contr. Rel.* **1985**, 1, pp. 259-267.
9 Lee, P.I. In: *Treatise on controlled drug delivery*; Kydonieus, A. Ed.; Marcel Dekker Inc., NY, 1992; pp.155-195.
10 Heller, J.; Sparer, R.V.; Zentner, G.M. In: *Biodegradable Polymers as Drug Delivery Systems*, Chasin, M.; Langer, R. Eds., Drugs and the Pharmaceutical Sciences, Marcel Dekker, NY, 1990; Vol. 45, pp.121-163.
11 Tamada, J.; Langer, R. *J. Biomater. Sci. Polymer Edn.* **1992**, 3.4, pp. 315-353.
12 Samanen, J. M. In: *Peptide and Protein Drug Delivery*; Lee V., Ed.; Advances in parenteral sciences, Marcel Dekker Inc., NY, 1991; Vol. 4, pp137-167.
13 Siegel, R.; Langer R. *J. Contr. Rel.* **1990**, 14, 153-167.
14 Baker, R. W.; Controlled Release of Biologically Active Agents, John Wiley and Sons, NY, 1987; pp.84-131.
15 Joshi, A.; Himmelstein, K. J. *J. Contr. Rel.* **1991**, 15, pp.95-104.
16 Li, S. M.; Garreau, H.; Vert, M. *Journal of Materials Science: Materials in Medicine*, 1990, 1, pp.131-139.
17 Leong, K. W.; Brott, B. C.; Langer, R. *J. Biomed. Mater. Res.*, **1985**, 19, pp.941-955.
18 Li, S. M.; Garreau, H.; Vert, M. *Journal of Materials Science: Materials in Medicine*, **1990**, 1, pp.123-130.
19 Vert, M.; Li, S.; Garreau, H. *J. Contr. Rel.* **1991**, 16, pp.15-26.
20 Heller, J. *J. Contr. Rel.* **1985**, 2, pp.167-177.
21 Mathiowitz, E.; Ron E.; Mathiowitz, G.; Amato, C.; Langer, R. *Macromolecules* **1990**, 23, pp.3212-3218.
22 Tabata, Y.; Domb, A.; Langer, R. *J. Pharm. Sci.* **1993**, in press.
23 Langer, R.; Peppas, N., *J. Macromol. Sci.-Rev. Macromol. Chem. Phys.* **1983**, C23, pp.61-126.
24 Billmeyer, F. W. Textbook of Polymer Science, 3rd Edition, John Wiley & Sons, NY, 1984; pp. 273-281.

25 Tamada, J.A.; Langer R. *Proc. Natl. Acad. Sci. USA* **1993**, 90, pp.552-556.
26 Göpferich, A.; Langer, R. *J. Polymer Sci.* **1993**, 31, pp. 2445-2458.
27 Cooney, D. O. *AIChE Journal* **1972**,18.2, pp.446-449.
28 Baker, R. W.; Lonsdale, H. K. *Am. Chem. Soc. Div. Org. Coat. Plast. Chem. Prepr.* **1976**, 3, pp.229.
29 Heller J.; Baker R. W. In *Controlled Release of Bioactive Materials*; Baker, R. W., Ed., Academic Press, New York, 1980.
30 Thombre, A. G.; Himmelstein, K. J. *AIChE Journal* **1985**, 31.5, pp.759-766.
31 Zygourakis K. *Polym. Prepr. (Am. Chem. Soc., Div. Polym. Chem.* **1989**, 30.1, pp. 456-457.
32 Zygourakis K. *Chem. Eng. Sci.* **1990**, 45.8, pp.2359-2366.
33 Göpferich, A.; Langer, R. *Macromolecules* **1993**, 16, pp. 4105-4112.
34 Drake, A. W., *Fundamentals of Applied Probability Theory*; McGraw Hill Publishing Company, NY, 1988; pp.129-144.
35 Boas, M. *Mathematical Methods in the Physical Sciences*; John Wiley & Sons , NY, 2nd Ed., 1983; pp.685-739.
36 Cashwell, E. D.; Everett, C. J. *A Practical Manual on the Monte Carlo Method for Random Walk Problems*; Pergamon Press, NY, 1959; pp.4-10.
37 Spendley, W.; Hext, G.R.; Himsworth, F.R. *Technometrics* **1962**, 4.4, pp.441-461.
38 Nelder, J.A.; Mead R. *Comp. J* **1964**, pp.308-313.
39 Merck Index, 11th edition, **1989**, pp.1334.
40 Langer, R.; Folkman, J. *Nature* **1976**, 263:5580, pp.797-800.
41 Saltzman, W.M., Langer, R. *Biophys. J.* **1989**, 55, pp.163-171.
42 Siegel, R.; Kost, J.; Langer R. *J. Contr. Rel.* **1989**, 8, pp.223-236.
43 Arakawa, T.; Kita, Y.; Carpenter, F. *Pharm. Res.* **1991**, 8, pp. 285-291.
44 Edelman, E.R.; Mathiowitz, E.; Langer R.; Klagsbrun, M. *Biomaterials* **1991**, 7:12, pp.6119-6126.
45 Andrianov, A.K.; Cohen, S.; Visscher, K.B.; Payne, L.G.; Allcock, H.R.; Langer R. J. Contr. Rel., **1993**, in press.
46 Preis I., Langer R., A single-step immunization by sustained antigen release. *J. Immunol. Methods* 28(1979), p.193-197.
47 Kohn, J.; Niemi, S. M.; Albert, E. C.; Murphy, J. C.; Langer R.; Fox, J. *J. Immunol. Methods* **1986**, 95, 31-38.
48 O'Hagan, D.T.; Rahman, D.; McGee, J.P.; Jeffery, H.; Davies, M.C.; Williams, P.; Davis, S.S.; Challacombe S. *J.Immunology* **1991**, 73, pp. 239-242.
49 Singh, M.; Singh, A.; Talwar, G.P. *Pharm. Res.* **1991**, 8, pp.958-961.
50 Eldridge, J.H.; Staas, J.K.; Meulbroek, J.A.; Tice, T.R.; Gilley, R.M. *Infection and Immunity* **1991**, 59, pp.2978-2986.

51 Alonso, M.J.; Cohen, S.; Park, T.G.; Gupta, R.K.; Siber, G.R.; Langer, R. *Pharm. Res.* **1993**, 51, pp. 945-953.
52 Vranken, M.N.; Claeys, D.A. *U.S. Patent 3,523,906*, 1970.
53 Ogawa, Y.; Yamamoto, M.; Okada, H.; Yashiki, Y.; Shimamoto, T. *Chem. Pharm. Bull.* **1988**, 36, pp.1095-1103.
54 Cohen, S.; Yoshioka, T.; Lucarelli, M.; Hwang, L.H.; Langer R. *Pharm. Res.* **1991**, 8, pp.713-720.
55 Liu, W.R.; Langer, R.; Klibanov, A. *Biotech. Bioeng.* **1991**, 37, pp.177-184.
56 Alonso, M.J.; Gupta, R.K.; Min C.; Siber, G.R.; Langer R. *Vaccine* **1993**, in press.
57 Sluzky, V.; Klibanov, A.M.; Langer R. *Biotech. Bioeng.* **1992**, 40, pp.895-903.
58 Lewis D.H. In: *Biodegradable Polymers as Drug Delivery Systems*, Chasin, M.; Langer, R. Eds., Drugs and the Pharmaceutical Sciences, Marcel Dekker, NY, 1990; Vol. 45, pp.1-41.
59 Esparza, I.; Kissel, T. *Vaccine* **1992**, 10, pp.714-720.
60 Relyveld, E. H.; Mayer, M.M. *Kabat and Mayer's Experimental Immunochemistry*; Thomas, C.C., Ed., Springfield, 1961, pp.22-96.
61 Gupta, R.K.; Maheshwari S.C.; Singh H. *J. Biol. Stand.* **1985**, 13, pp.143-149.
62 Bizzini, B. In: *Bacterial vaccines*; Germanier, R., Ed.; Academic Press, Orlando, 1984, pp. 37-68.
63 Langer, R.; Linhardt, R.; Klein, M.; Flanagan, M.; Galliher, P.; Cooney C. In: *Biomaterials: Interfacial Phenomena and Applications;* Peppas, N.A.; Hoffman, A.; Rather, B.; Cooper, S., Eds.; Advances in Chemistry Series; Washington, D. C., 1982, pp.493-502.
64 Macleod, R. M.; Dunn, F. *J. Acoust. Soc. Am.* **1968**, 44, pp.932-945.
65 Wang, Y.J.; Hanson, M.A. *J. Parent. Sci. Tech.* **1988**, 42(2s), pp.3-26.
66 Tabata, Y.; Langer R. *Pharm. Res.* **1993**, 10.4, pp.487-496.
67 Donbrow, M. *Microcapsules and Nanoparticles in Medicine and Pharmacy*, CRC Press, Boca Raton, Ann Arbor, London, 1992.
68 Müller, R.H. *Colloidal Carriers for Controlled Drug Delivery and Targeting*; CRC Press, Boca Raton, 1991.
69 Davis, S.S; Illum, L., In: *Site-specific Drug Delivery*; E. Tomlinson, E.; Davis, S.S., Eds.; John Willey & Sons, Chichester, 1986, pp. 93.
70 Illum, L.; Davis, S.S. *FEBS Lett.* **1984**, 167, p.79.
71 Allen, T.M.; Hansen, C. *Biochim. Biophys. Acta* **1991**, 1068, 133-141.
72 Allen, T.M.; Hansen, C.; Martin, F.; Redeman, C.; Yau-Young, A. *Biochim. Biophys. Acta* **1991**, 1066, pp.29-36.
73 Torchilin, V.P.; Klibanov, A.L.; Huang, L.; O'Donnel, S.; Nossiff, C.; Khaw, B.A. *FASEB J.* **1992**, 6, pp. 2716-2719.
74 Maruyama, K.; Yuda, T.; Okamoto, A.; Ishikura, C.; Kojima, S.; Iwatsuru, M. *Chem. Pharm. Bull.* **1991**, 39(6), pp.1620-1622.

75 Woodle, M.C.; Matthay, K.K; Hidayat, J.E.; Collins, L.R.; Redemann, C.; Martin, F.J.; Papahadjopoulos, D. *Biochim. Biophys. Acta* **1992**, pp.193-200.
76 Lasic, D.D.; Martin, F.J.; Gabizon, A.; Huang, S.K.; Papahadjopoulos, D.; *BBiochim. Biophys. Acta* **1991**, 1070, pp.187-192.
77 Klibanov, A.; Maruyama, K.; Beckerleg, A.M.; Torchilin, V.P.; Huang, L. *Biochim. Biophys. Acta* **1991**, 1062, pp. 142-148.
78 Gref, R.; Minamitake, Y.; Peraccia, M.T.; Trubetskoy V.; Milshteyn A.; Sinkule, J.; Torchillin, V.; Langer, R. *Proc. Intern. Sympos. Control. Rel. Bioact. Mater.* **1993**, 20, in press.
79 Gref, R.; Minamitake, Y.; Peraccia, M.T.; Langer R. *Macromolecules*, **1993**, pp. 131-132.
80 Zhu, J.E.; Lin, X.; Yang, S.; *J. Polym. Sci., Polym. Lett. Ed.* **1986**, 24, p.331.
81 Brindley, A.;Davies, M.C.; Lynn, R.A.P.; Davis, S.S.; Hearn, J.; Watts, J.F.; *Polymer* **1992**, 33.5, pp. 1112-1115.
82 Miller, D.R.; Peppas, N.A. *J. Macromol. Sci.-Rev. Macromol. Chem. Phys.* **1986**, C26(1), pp.33-66.
83 Torchilin, V.; Klibanov, A. *Critical Reviews in Therapeutic Drug Carrier Systems* **1991**, 7.4, pp. 275-307.

RECEIVED April 19, 1994

Chapter 16

Controlled Intracranial Delivery of Antibodies in the Rat

Samira Salehi-Had and W. Mark Saltzman

Departments of Chemical Engineering and Biomedical Engineering, Johns Hopkins University, Baltimore, MD 21218

To provide long-term delivery of antibodies within the intracranial space, we have developed methods for the controlled release of antibodies from polymer matrices and microspheres. IgG concentrations in the rat brain were measured following injection of IgG in saline or implantation of a polymer matrix containing IgG. IgG (10 to 100,000 ng/g) was detected near the site of administration for 28 days following either injection or implantation. IgG was also detected at lower concentrations in other quadrants of the brain and in the plasma. In all cases, IgG concentrations were highest for animals treated by polymer implantation. Initial rates of elimination from the brain were similar for IgG administered by injection and implantation (half-life ~1 day). The advantage of polymer implantation over direct injection was observed most clearly in the 14 to 28 days following administration. Polymer treated animals continued to have high IgG concentrations, even at brain sites distant from the site of administration, while IgG concentrations in the brain of injection treated animals fell sharply.

Nine thousand new cases of primary brain tumor are reported in the United States each year. Nearly one-half of these cases are malignant gliomas, which are frequently fatal; even with surgical resection and radiation therapy most patients die within two years of diagnosis. Chemotherapy for malignant brain tumors is problematic. Although agents with good activity against human brain tumors are available, the blood-brain barrier limits the entry of most systemically delivered agents into the central nervous system (CNS). To bypass this barrier, improved methods for drug delivery to the CNS have been developed in the last several years (1-4). In addition, novel antitumor strategies have been proposed including antisense-based vaccines (5), gene therapy (6), and immunotherapy with antibody-toxin conjugates (7, 8), growth factor-toxin

0097–6156/94/0567–0278$08.00/0

conjugates (*9*), or activated lymphocytes (*10, 11*). In these new approaches, the limitations of drug delivery to the CNS have become even more apparent, since macromolecules like immunoglobulins, cytokines, and gene therapy vectors must be delivered locally to the brain.

Intracranial controlled release polymers provide sustained levels of active agents directly to a localized brain region. Using animal models, the feasibility of this concept has been demonstrated for anticancer agents (*12, 13*), steroids (*14*), neurotransmitters and their agonists (*15-17*), nerve growth factor (*18-20*), and other macromolecules (*21, 22*). In addition, biodegradable polyanhydride matrices containing 1,3-bis(2-chloroethyl)-1-nitrosurea (BCNU) have been used to treat human patients with recurrent glioblastoma multiforme (*3, 23*). Controlled drug delivery in the CNS can be achieved with a number of agents, but penetration into the tissue surrounding the implant is dependent on biological and physicochemical characteristics of the compound (*24*). Because of their increased retention in the brain extracellular space, compounds that are water-soluble, high molecular weight, and slowly metabolized appear to be the best suited for direct intracranial delivery (*22, 24*).

Antibodies and antibody proteolytic fragments (Fab or F(ab')$_2$) have all of the characteristics that are desirable for an intracranially delivered drug: they are water soluble (>100 mg/mL) and large (50,000 to 1,000,000 daltons) molecules with plasma half-lives of many days (see (*25*) for a review of antibody pharmacokinetics). Different antibody classes (IgG, IgM, or IgA) or antibody proteolytic fragments can be employed, allowing selection of the most desirable elimination and tissue penetration characteristics. Importantly, antibody molecules can be conjugated with toxins to provide intrinsic cytotoxicity (*8, 26, 27*). The specificity of antibody binding to brain-specific antigens can also be selected (*7, 28*), changing the killing properties of the agent as well as its ability to penetrate through brain tissue. Methods for generating "humanized" antibodies or antigen binding proteins with reduced immunogenicity in humans have also been developed (*29, 30*). The kinetics of antibody delivery to brain tissue have been examined following intravenous (*31*), intracarotid (*31*), intraventricular (*32*), and direct intracranial (*33*) injection, but antibody delivery from an intracranial controlled release polymer has never been studied.

In previous reports, different formulations that provide a controlled release of biologically active antibodies have been described (*34-38*). Here, those studies were extended by designing polymer matrices and microspheres appropriate for use in the brain. In addition, one particular antibody delivery system, poly(ethylene-co-vinyl acetate) (EVAc) matrices, was used to study the kinetics of antibody elimination following controlled release in the rat brain.

Methods and Materials

EVAc polymer matrix preparation and characterization The procedure for the preparation of EVAc/antibody matrices has been reported previously (*37*). Briefly, EVAc (ELVAX 40W, Dupont, Wilmington, DE) was washed extensively in water and acetone prior to use. To produce matrices of EVAc and mouse IgG at 40% loading (mass percent IgG), lyophilized mouse IgG (Sigma Chemical Company, St. Louis, MO) was sieved to <180 um particle size, added to EVAc/methylene chloride solution (10% w/v), and vortexed. The mixture was poured into a leveled, pre-chilled mold at

-70°C and allowed to solidify for 15 min. The polymer was removed from the mold and placed in freezer for 48 hr, followed by 48 hr of storage in a vacuum desiccator. Discs (3.5 mm diameter, 0.7 mm thick) were punched from the resulting slab with a cork borer. The average weight of these discs were 7.8±0.5 mg. To produce matrices with less IgG, solid particles containing IgG and ficoll (approximate molecular weight 70,000; Sigma) were first produced. Mouse IgG and ficoll were dissolved separately in doubly de-ionized water, and appropriate amounts of each solution were mixed and lyophilized overnight to give a 100:1 (ficoll:IgG) powder. This powder was sieved and the 40% loaded (mass percent total protein) polymer discs were obtained as described above.

For matrices containing 40% IgG and matrices containing 40% IgG:ficoll, three identical discs were placed in separate polypropylene scintillation vials and 5 mL of phosphate buffered saline (PBS) with 0.02% gentamicin was added to each. The vials were placed in an incubator/shaker at 37°C. At appropriate time intervals, the PBS was removed and assayed for IgG content using an ELISA specific for mouse IgG as described below. The discs were moved to a new vial and fresh PBS was added before further incubation.

Microsphere preparation and characterization Five hundred mg of poly(L-lactic acid) (M_W 50,000; PLA, Polysciences, Warrington, PA) was dissolved in 3 mL methylene chloride and 5 mg bovine γ-globulin (Sigma Chemical) was dissolved in 50 μL double-distilled water (nominal 5% loading). These solutions were combined and sonicated (Tekmar Sonic Disrupter model TM 300) for 10 sec. One mL of aqueous 1% poly(vinyl alcohol) (PVA, 25,000 M_W, Polysciences) and 200 μL L-α-phosphatidylcholine (2 mg/mL in chloroform) were quickly added to this emulsion and vortexed for 15 sec. This double-emulsion was poured into a beaker with 100 mL aqueous 0.3% PVA and stirred for 3 hr to let the microspheres form. The microspheres were collected by centrifuging the solution at 2000 rpm for 10 min. The microspheres were rapidly frozen in liquid nitrogen and lyophilized for 16 to 24 hr. Particle size was determined by suspending a small quantity of microspheres in PBS on a glass slide and measuring particle diameters by computerized image analysis.

The rate of antibody release from the microspheres was determined by incubating 30 mg of microspheres in 4 mL PBS in a centrifuge tube. Periodically, the suspension was centrifuged and the supernatant replaced with fresh PBS. The supernatant was filtered (0.8 μm pore size filter) and the concentration of protein in the filtrate was determined by total protein assay by using Coomassie Blue protein assay reagent (Pierce, Rockford, IL) in a microtiter plate format (38).

Implantation of EVAc matrices into the rat brain Male F-344 rats (6-7 week old , weight ~ 170 g, Harlan, Indianapolis, IN) were used in all experiments. Each rat was anesthetized with an intraperitoneal injection (~3.5 mL/kg) of ketamine/xylazine anesthesia (25 mL Ketamine (100 mg/mL); 2.5 mL Xylazine (100mg/mL); 14.2 mL ethanol (100%); 58.3 mL of 0.9% NaCl) in the lower left quadrant of its abdomen. After removing the hair from the surgical site with electric clippers, the site was wiped with a pad soaked in Betadine solution. The rat was placed in a small animal stereotaxic apparatus (Kopf Instruments, Tujunga, CA). A midline incision into the skin was made with a scalpel using aseptic technique. A line was defined by two points at 2.2 mm rostral and 2 mm lateral to

Bregma, and 1.3 mm caudal and 4-5 mm lateral to Bregma (Figure 1a). The skull was drilled along this line with a dental burr held in a dental drill cap until the scalpel could penetrate through to make a slit for the polymer. An incision was made in the dura and the polymer was inserted into the brain through this incision, after significant bleeding had stopped. The skin incision was closed with surgical staples after implantation.

Injection of IgG solution into the rat brain Rats were anesthetized, placed in the stereotaxic frame, and prepared for surgery, as described above. After the skin incision, a small hole was made in the skull by drilling at a position 1 mm rostral and 2-3 mm lateral to Bregma (Figure 1b). A 5 μL Hamilton syringe (26 gauge needle) mounted on a micro injector (model 5000, Kopf Instruments) was used to deliver 1 μL (20 μg) of mouse IgG (Sigma) over a 30 min interval. Prior to injection, the needle was inserted 6 mm below the surface of the dura and retracted 1 mm. At least 10 minutes were allowed for bleeding to stop before the injection was started. After injection was completed the needle was retracted over an additional 10 min. The skin incision was closed with staples.

Rat sacrifice and sample preparation After appropriate time intervals, three rats were anesthetized and euthanized by exsanguination following direct cardiac puncture. During this procedure the blood was withdrawn into a heparinized syringe. The scalp was opened and the skull was penetrated with bone clippers at the level of the cerebellum. The brain was exposed by further removal of the parietal bone of the skull and the dura covering. The cerebellum was excised from the rest of the brain, and the brain was removed with a spatula. In the case of polymer implants, the position of the polymer was noted and the polymer was removed. The brain was then sectioned into four quadrants using fresh razor blades (Figure 1b). Each section was weighed (average weight of 0.31 g) and extraction buffer (100 mM of tris-HCl; 400 mM NaCl; 2% albumin (w/v); 0.05% Na azide (w/v); 1 mM PMSF (dissolved in dimethyl sulfoxide and diluted 1:1000 to 1 mM in extraction buffer); 7 μg/mL aprotinin; 4 mmol/L EDTA) was added. The volume of the extraction buffer was twice the mass of the section. Samples were sonicated at a 50% duty cycle in 30 sec intervals for 1.5 min and centrifuged for 25 min at 4 °C (14,000 rpm). Blood samples were also centrifuged for 10 min at 3,500 rpm. Supernatants were stored at -70 °C prior to analysis of IgG by ELISA. Brain and blood plasma from control rats were used to make standards for the IgG ELISA. Storage and handling of the samples and standards were identical.

Determination of IgG concentrations in buffered water and tissue samples A Nunc-Immuno Maxisorb flat-bottomed plate (InterMed) was coated overnight at 4°C with 1/500 dilution of affinity purified rat anti-mouse IgG (Jackson ImmunoResearch Laboratories, West Grove, PA) in PBS. The wells were washed twice with PBS/0.05% Tween 20 (wash solution) and blocked with PBS/ 0.05% Tween 20/3% BSA (blocking solution) for 2 hr with orbital rocking. Standards were made from blank brain samples prepared as described above, with known concentrations of mouse IgG added. Standards ranged between 0.6125-320 ng/mL with two fold serial dilution intervals. After washing twice with wash solution, samples and standards were added, followed by one hr of incubation with orbital rocking. The plate was washed three times and antibody conjugate (peroxidase-conjugated affinity purified rat anti-mouse IgG; Jackson

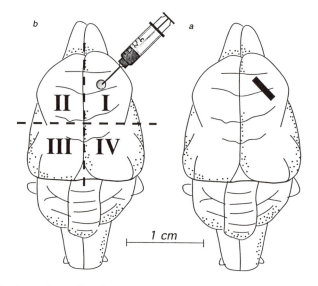

Figure 1: Location of IgG injection and implantation within the adult rat brain. All manipulations were performed with the anesthetized rat mounted in a stereotaxic frame. a) The location for the polymer implant was marked by drilling two small holes; the bone between those points was removed and the dura was carefully incised. b) IgG (20 µg in 1 µL of PBS) was injected from a Hamilton syringe into the right hemisphere. Following sacrifice of the animal, the cerebral hemispheres of every animal were divided into quadrants for analysis as indicated in (a).

ImmunoResearch) was added at a 1/500 dilution with blocking solution. One hour of incubation was followed by three washes with wash solution and one wash with PBS. Biorad peroxidase substrate (ABTS) was added and the plate was read at 405 nm using a microplate reader (Thermomax, Molecular Devices). Softmax version 2.01 was used to analyze the data.

Results

Antibodies were slowly released from several different polymer formulations (Figure 2). When solid particles of IgG were dispersed within an EVAc matrix, the antibody was slowly released (Figure 2a). To ensure reproducible protein release from the EVAc system, the total agent loading must be greater than ~35% (34). For an implant of reasonable size (~10 mg), this loading level usually leads to protein release rates of 10 to 100 µg/day. This high rate of release may not be practical for high-value, potent agents like protein therapeutics. Therefore, we developed methods for releasing very small quantities of IgG, by creating solid particles containing both IgG and a high molecular weight polysaccharide that controls the rate of protein release from the polymer. In the 30 day period of study, the polymer containing IgG particles released ~800 µg of antibody, while the polymer containing IgG:ficoll particles (1:100) released ~5 µg (Figure 2a). In a separate experiment, IgG was shown to be stable in PBS/gentamicin at 37°C for over two weeks (data not shown). We expect gradual release of the remaining IgG from the disc at later time points, as observed in previous studies (34, 37).

An alternate means of antibody delivery was tested with microspheres of PLA. The microspheres were loaded with a bovine antibody. The mean diameter of the spheres was 29 µm with most spheres in the range 16 to 48 µm. This formulation released ~50 µg of protein at a nearly constant rate for 1 month (Figure 2b). Unlike the EVAc matrix, hydration of the ester bonds in the PLA polymer causes gradual degradation of the polymer concurrent with the antibody release. Hence, this biodegradable delivery system leaves no 'matrix after the complete release of the antibody and may be more suitable for clinical situations. However, to get a well defined IgG concentration in these experiments, EVAc polymers were used. It would be difficult to recover a complete ensemble of injected IgG-loaded microspheres at different time intervals after injection, and therefore difficult to estimate local levels of IgG in the brain tissue as a function of time.

To examine the pharmacokinetics of antibody delivery to the brain, IgG was introduced into the rat brain either by direct injection or implantation of a non-biodegradable polymer at the same site in the brain (Figure 1). At various times after administration, the concentration of IgG in brain tissue and plasma was determined by ELISA (Figure 3). In control experiments, we demonstrated that the ELISA was insensitive to rat IgG, so the concentrations reported reflect only mouse IgG delivered by the polymer or injection. In both cases, the brain and plasma contained measurable quantities of IgG for 28 days following administration. Compared to injection, polymer implantation produced higher IgG concentrations in both the brain and plasma. In both cases, the majority of the IgG in the brain was found in quadrant I, the site of administration. In the period 14 to 28 days following administration, the concentration of IgG in the plasma was comparable to the concentration in brain quadrant I.

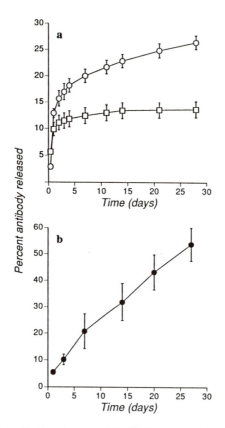

Figure 2: Controlled release of IgG from polymer matrices and microspheres. (a) Solid particles containing either pure mouse IgG (O) or IgG mixed with ficoll (1:100 IgG:ficoll) (□) were dispersed in EVAc matrices at a mass loading of 40%. Using a mouse IgG specific ELISA, the release of IgG from each matrix was measured during continuous incubation in well-stirred PBS. Each symbol represents the mean percent release determined from 3 implants weighing 8-10 mg. The error bars represent the standard deviation. (b) Bovine γ-globulin was encapsulated in spheres of 50,000 M_W PLA (●) at a nominal mass loading of 1%. Thirty mg of spheres were suspended in PBS and the release of antibody was measured using a total protein assay.

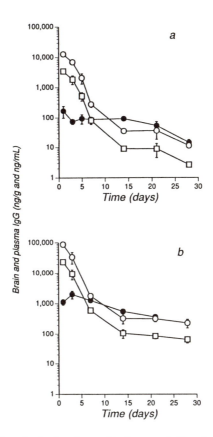

Figure 3: IgG levels in brain and plasma following (a) intracranial injection of 20 µg IgG or (b) intracranial implantation of an EVAc matrix containing ~3 mg IgG (see Figure 1). Concentrations of mouse IgG were determined by a specific ELISA in plasma (●), the cerebral quadrant of Ab administration (O), and both cerebral hemispheres (□). Each symbol represents the mean concentration determined for three rats; error bars indicate the standard deviation.

Both direct intracranial injection and polymer implantation produced the highest levels of IgG at the site of administration. However, significant quantities of IgG were also found in the three other brain quadrants (Figure 4). In every case, polymer implantation produced higher concentrations of IgG. Interestingly, the rate of decrease in IgG concentration within quadrant I was approximately the same for the two forms of administration: concentrations decreased by a factor of 100 over the first week, suggesting a half life of elimination from the brain of approximately 1 day. After that time, however, concentrations stabilized within all quadrants for the polymer implant treated animals (e.g. IgG concentration in quadrant I was in the range 330 to 230 ng/g for days 14 through 28), while concentrations continued to decrease for animals treated by injection.

Discussion

Biologically active antibodies can be released from a variety of polymer formulations (34-38). Here, we have demonstrated controlled antibody release using either matrices or microspheres (Figure 2). To examine the pharmacokinetics of controlled antibody delivery to the brain, we selected implants composed of 40% IgG particles within an EVAc matrix. The implanted matrices were identical to the matrices characterized *in vitro* (Figure 2a, top curve). Although injectable, biodegradable microspheres (Figure 2b) are probably preferable for clinical applications, the EVAc system was selected for this study because i) EVAc is the most well-characterized and reproducible of the available protein release systems, ii) the EVAc matrix is easy to manipulate and can be placed at a specific stereotaxic location in the brain (see Figure 1), and iii) the EVAc matrix does not degrade so the entire matrix can be easily removed from the brain prior to analysis. This last feature is essential for examining the pharmacokinetics of antibody delivery, since the polymer implant is removed prior to analysis, all of the IgG found within the brain tissue must have been previously released from the polymer.

Brain concentrations of IgG were determined by adding an extraction buffer (twice the brain mass) to a brain quadrant, homogenizing and centrifuging the mixture, and assaying the supernatant for the presence of IgG. To calculate IgG concentration within the brain tissue, we assumed that all of the antibody present in the brain sample was extracted into the buffer solution by this procedure. Considering the high solubility of IgG in the extraction buffer this appears to be a reasonable assumption, although it is also probable that some fraction of the IgG in each brain sample remained within the pelleted brain material. In addition, the brain contains a substantial amount of water (~78% by mass in the rat (39)), some fraction of which probably entered the supernatant phase during homogenization. Both of these considerations suggest that our estimates of brain IgG concentration are lower than the actual concentration within the brain tissue. While radiotracer studies of the extraction/homogenization procedure might permit us to quantitate concentrations within the brain samples more accurately, we believe that the present procedure produces a reasonable estimate.

Polymer implantation produced higher IgG concentrations in the brain than direct injection. This result is not surprising, since 20 µg of IgG were injected while larger quantities of IgG were available from the implant. The quantity of IgG released from the polymers into the brain was not

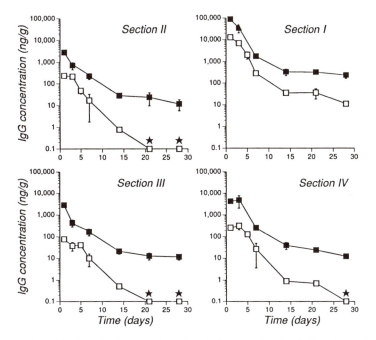

Figure 4: IgG levels in quadrants of the rat brain following either direct injection of 20 μg IgG (□) or intracranial implantation of an EVAc matrix containing ~3 mg IgG (■). Concentrations in each quadrant (see Figure 1) were determined by ELISA. Each symbol represents the mean concentration determined for three rats; error bars indicate the standard deviation. The limit of detection for the ELISA was 0.1 ng/g brain tissue; samples that produced optical densities near this threshold are plotted at 0.1 ng/g and indicated (★).

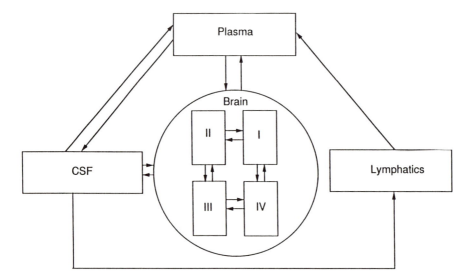

Figure 5: Schematic model of IgG pharmacokinetics following intracranial delivery. Arrows indicate possible pathways for IgG distribution. Abbreviation: CSF, cerebrospinal fluid.

measured directly, however, approximately 400 µg were released during the first 1 day and 800 µg during the first 28 days of incubation in well-stirred PBS (Figure 2). The differences between polymer treated and injection treated animals became more significant during the last 14 days of the experiment. Since a considerable quantity of IgG remained in matrices soaked in PBS for 28 days, and since similar matrices have been demonstrated to release IgG for over 2 years (34), intracranial polymers should continue to provide IgG to the rat brain for a period considerably longer than tested in the present experiment.

During the first 14 days following administration, the rate of clearance of IgG from the brain was similar for injection and polymer implantation. During the second 14 day period, however, the concentration of IgG within the brain following polymer implantation remained nearly constant. For IgG delivered by injection, concentrations decreased more rapidly, particulary in quadrants II, III, and IV. During the second 14 day period, when IgG concentrations are nearly constant for polymer treated animals, the rate of IgG elimination must be balanced by the rate of IgG release from the implant. Injection of IgG results in relatively high concentrations for the first two weeks, but polymer implantation is required to maintain IgG levels beyond this period. Since the rate of elimination of IgG from the brain is reasonably slow, IgG released from the polymer into brain quadrant I can diffuse a considerable distance from the polymer prior to elimination. This observation probably explains the high concentrations of IgG observed within quadrants II, III, and IV, even 28 days following implantation.

The present study provides information on the biodistribution of IgG following intracranial administration. The pathways for IgG transport and distribution following intracranial delivery are undoubtedly complex (Figure 5). IgG administered into quadrant I can diffuse within the tissue extracellular space (40), eventually reaching distant sites in the brain, provided that the rate of elimination is low (24). IgG can enter the systemic circulation, either by permeating through capillary walls, which probably occurs at a very low rate , or via the cerebrospinal fluid (CSF) circulation. In addition, some of the pathways for protein elimination from the brain via the lymphatics have been described (41). In the present case, although IgG was administered into one quadrant of the brain, it was detected within all four quadrants, as well as the plasma, within 24 hr of administration. By coupling experiments in animals, like the ones described here, with mathematical models of antibody biodistribution, as shown schematically in Figure 5, rational strategies for antibody delivery to the CNS can be developed.

Acknowledgements

We thank Richard B. Dause and Amy M. Dodrill for technical assistance. Supported by the National Institutes of Health (CA52857 and GM43873).

Literature Cited

1. Friden, P., L. Walus, G. Musso, M. Taylor, B. Malfroy, and R. Starzyk. 1991. Anti-transferrin receptor antibody and antibody-drug conjugates cross the blood-brain barrier. *Proceedings of the National Academy of Sciences USA.* 88:4771-4775.

2. Neuwelt, E., P. Barnett, I. Hellstrom, K. Hellstrom, P. Beaumier , C. McCormick, and R. Weigel. 1988. Delivery of melanoma-associated immunoglobulin monoclonal antibody and Fab fragments to normal brain utilizing osmotic blood-brain barrier disruption. *Cancer Research.* 48:4725-4729.

3. Brem, H., M. Mahaley, N. Vick, K. Black, S. Schold, P. Burger, A. Friedman, I. Ciric, T. Eller, J. Cozzens, and J. Kenealy. 1991. Interstitial chemotherapy with drug polymer implants for the treatment of recurrent gliomas. *Journal of Neurosurgery.* 74:441-446.

4. Morrison, P. F., D. W. Laske, H. Bobo, E. H. Oldfield, and R. L. Dedrick. 1993. High-flow microinfusion: tissue penetration and pharmacodynamics. *American Journal of Physiology.* in press:

5. Trojan, J., T. R. Johnson, S. D. Rudin, J. Ilan, M. L. Tykocinski, and J. Ilan. 1993. Treatment and prevention of rat glioblastoma by immunogenic C6 cells expressing antisense insulin-like growht factor I RNA. *Science.* 259:94-97.

6. Culver, K. W., Z. Ram, S. Wallbridge, H. Ishii, E. Oldfield, and R. M. Blaese. 1992. In vivo gene transfer with retroviral vector-producer cells for treatment of experimental brain tumors. *Science.* 256:1550-1552.

7. Wikstrand, C. J., P. Fredman, L. Svennerholm, P. A. Humphrey, S. H. Bigner, and D. D. Bigner. 1992. Monoclonal antibodies to malignant human gliomas. *Molecular & Chemical Neuropathology.* 17:137-146.

8. Johnson, V., C. Wrobel, D. Wilson, J. Zovickian, L. Greenfield, E. Oldfield, and R. Youle. 1989. Improved tumor-specific immunotoxins in the treatment of CNS and leptomeningeal neoplasia. *Journal of Neurosurgery.* 70:240-248.

9. Pastan, I., and D. FitzGerald. 1991. Recombinant toxins for cancer treatment. *Science.* 254:1173-1177.

10. Kruse, C. A., K. O. Lillehei, D. H. Mitchell, B. Kleinschmidt-DeMasters, and D. Bellgau. 1990. Analysis of interleukin 2 and various effector cell populations in adoptive immunotherapy of 9L rat gliosarcoma: allogeneic cytotoxic T lymphocytes prevent tumor take. *Proceedings of the National Academy of Sciences USA*. 87:9577-9581.

11. Kruse, C. A., D. H. Mitchell, B. K. Kleinschmidt-DeMasters, D. Bellgrau, J. M. Eule, J. R. Parra, Q. Kong, and K. O. Lillehei. 1993. Systemic chemotherapy combined with local adoptive immunotherapy cures rats bearing 9L gliosarcoma. *Journal of Neuro-Oncology*. 15:97-112.

12. Yang, M., R. Tamargo, and H. Brem. 1989. Controlled delivery of 1,3-Bis(2-chloroethyl)-1-nitrosourea from ethylene-vinyl acetate copolymer. *Cancer Research*. 49:5103-5107.

13. Tamargo, R. J., J. S. Myseros, J. I. Epstein, M. B. Yang, M. Chasin, and H. Brem. 1993. Interstitial chemotherapy of the 9L gliosarcoma: controlled release polymers for drug delivery in the brain. *Cancer Research*. 53:329-333.

14. Reinhard, C., M. L. Radomsky, W. M. Saltzman, J. Hilton, and H. Brem. 1991. Polymeric controlled release of dexamethasone in normal rat brain. *Journal of Controlled Release*. 16:331-340.

15. Howard, M., A. Gross, M. Grady, R. Langer, E. Mathiowitz, R. Winn, and M. Mayberg. 1989. Intracerebral drug delivery in rats with lesion-induced memory deficits. *Journal of Neurosurgery*. 71:105-112.

16. During, M. J., B. A. Sabel, A. Freese, W. M. Saltzman, A. Deutz, R. H. Roth, and R. Langer. 1989. Controlled release of dopamine from a polymeric brain implant: in vivo characterization. *Annals of Neurology*. 25:351-356.

17. Winn, S., L. Wahlberg, P. Tresco, and P. Aebischer. 1989. An encapsulated dopamine-releasing polymer alleviates experimental Parkinsonism in rats. *Experimental Neurology*. 105:244-250.

18. Hoffman, D., L. Wahlberg, and P. Aebischer. 1990. NGF Released from a Polymer Matrix Prevents Loss of ChAT Expression in Basal Forebrain Neurons following a Fimbria-Fornix Lesion. *Experimental Neurology*. 110:39-44.

19. Powell, E. M., M. R. Sobarzo, and W. M. Saltzman. 1990. Controlled release of nerve growth factor from a polymeric implant. *Brain Research*. 515:309-311.

20. Dause, R. B., C. E. Krewson, and W. M. Saltzman. Intracranial nerve growth factor delivery in the rat, in preparation.

21. Mayberg, M., R. Langer, N. Zervas, and M. Moskowitz. 1981. Perivascular Meningeal Projections from Cat Trigeminal Ganglia: Possible Pathway for Vascular Headaches in Man. *Science*. 213:228-230.

22. Dang, W., and W. M. Saltzman. 1992. Dextran retention in the rat brain following controlled release from a polymer. *Biotechnology Progress*. 8:527-532.

23. Chasin, M., G. Hollenbeck, H. Brem, S. Grossman, M. Colvin, and R. Langer. 1990. Interstitial drug therapy for brain tumors: A case study. *Drug Development and Industrial Pharmacy*. 16:2579-2594.

24. Saltzman, W. M., and M. L. Radomsky. 1991. Drugs released from polymers: diffusion and elimination in brain tissue. *Chemical Engineering Science*. 46:2429-2444.

25. Saltzman, W. M. 1993. Antibodies for treating and preventing disease: the potential role of polymeric controlled release. *Critical Reviews in Therapeutic Drug Carrier Systems*. 10:111-142.

26. Vitetta, E. S., R. J. Fulton, R. D. May, M. Till, and J. W. Uhr. 1987. Redesigning nature's poisions to create anti-tumor reagents. *Science.* 238:1098-1104.
27. Kemshead, J. T., and K. Hopkins. 1993. Uses and limitations of monoclonal antibodies (MoAbs) in the treatment of malignant disease: a review. *Journal of the Royal Society of Medicine.* 86:219-224.
28. Schuster, J. M., and D. D. Bigner. 1992. Immunotherapy and monoclonal antibody therapies. *Current Opinion in Oncology.* 4:547-552.
29. Morrison, S. 1985. Transfectomas provide novel chimeric antibodies. *Science.* 229:1202-1207.
30. Riechmann, L., M. Clark, H. Waldmann, and G. Winter. 1988. Reshaping human antibodies for therapy. *Nature.* 332:323-327.
31. Zalutsky, M., R. Moseley, J. Benjamin, E. Colapinto , G. Fuller, H. Coakham, and D. Bigner. 1990. Monoclonal antibody and F(ab')$_2$ fragment delivery to tumor in patients with glioma: Comparison of intracarotid and intravenous administration. *Cancer Research.* 50:4105-4110.
32. Moseley, R. P., A. G. Davies, R. B. Richardson. 1990. Intrathecal administration of ^{131}I radiolabelled monoclonal antibody as a treatment for neoplastic meningitis. *British Journal of Cancer.* 62:637-642.
33. Papanastassiou, V., B. L. Pizer, H. B. Coakham, J. Bullimore, T. Zananiri, and J. T. Kemshead. 1993. Treatment of recurrent and systic malignant glioma by a single intracavity injection of ^{131}I monoclonal antibody: feasibility, pharmacokinetics and dosimetry. *British Journal of Cancer.* 67:144-151.
34. Saltzman, W. M., and R. Langer. 1989. Transport rates of proteins in porous polymers with known microgeometry. *Biophysical Journal.* 55:163-171.
35. Radomsky, M. L., K. J. Whaley, R. A. Cone, and W. M. Saltzman. 1990. Macromolecules released from polymers: diffusion into unstirred fluids. *Biomaterials.* 11:619-624.
36. Radomsky, M. L., K. J. Whaley, R. A. Cone, and W. M. Saltzman. 1992. Controlled vaginal delivery of antibodies in the mouse. *Biology of Reproduction.* 47:133-140.
37. Sherwood, J. K., R. B. Dause, and W. M. Saltzman. 1992. Controlled antibody delivery systems. *Bio/Technology.* 10:1446-1449.
38. Saltzman, W. M., N. F. Sheppard, M. A. McHugh, R. Dause, J. Pratt, and A. M. Dodrill. 1993. Controlled antibody release from a matrix of poly(ethylene-co-vinyl acetate) fractionated with a supercritical fluid. *Journal of Applied Polymer Science.* 48:1493-1500.
39. Katzman, R., and H. M. Pappius. 1973. *Brain Electrolytes and Fluid Metabolism*; Wilkins and Williams, Baltimore.
40. Clauss, M. A., and R. K. Jain. 1990. Interstitial transport of rabbit and sheep antibodies in normal and neoplastic tissues. *Cancer Research.* 50:3487-3492.
41. Cserr, H. F., and P. M. Knopf. 1992. Cervical lymphatics, the blood-brain barrier and the immunoreactivity of the brain. *Immunology Today.* 13:507-512.

RECEIVED June 8, 1994

Chapter 17

Use of Poly(ortho esters) and Polyanhydrides in the Development of Peptide and Protein Delivery Systems

J. Heller

Controlled Release and Biomedical Polymers Department, SRI International, Menlo Park, CA 94025

This brief review summarizes the use of poly(ortho esters) and polyanhydrides in the controlled release of peptides and proteins. Poly(ortho esters) have been used for the controlled release of an LHRH analogue, in the development of an insulin self-regulated drug delivery system and in the development of vaccines, using lysozyme as a model protein. Polyanhydrides based on fatty acid dimers and sebacic acid have been used for the controlled release of a variety of proteins and linear release over a one moth period was achieved. Poly(ortho esters) and polyanhydrides offer a number of advantages over the more commonly used lactide/glycolide copolymers since they offer better control over polymer properties and erosional behavior. Poly(ortho esters) may offer an additional advantage in that initial hydrolysis products are neutral so that liberation of acidic primary hydrolysis products which are inherent in the hydrolysis of both lactide/glycolide copolymers and polyanhydrides may not be a problem.

The need for bioerodible implants for the parenteral administration of peptides and proteins is now well recognized and a significant research effort is currently underway devoted to the development of bioerodible polymers specifically tailored to meet the specialized needs of peptide and protein delivery (1). To this day, most of the work remains focussed on the lactide/glycolide copolymer system, not because it is the best system available, but because it is a system with a proven benign toxicology that degrades to the natural metabolites lactic and glycolic acid (2). However, because these systems undergo bulk hydrolysis and because the ability to vary mechanical properties and the hydrolysis rates is limited, two other polymer systems which do not have such limitations are under active development. These systems are poly(ortho esters) and polyanhydrides. In this review, we will summarize the application of these two polymer systems for the controlled the delivery of peptides and proteins.

[1]Current address: Advanced Polymer Systems, 3696 Haven Avenue, Redwood City, CA 94063

0097–6156/94/0567–0292$08.00/0

Poly(Ortho Esters)

Unlike lactide/glycolide copolymers that undergo a hydrolysis process that occurs throughout the bulk of the material, poly (ortho esters) are capable, under certain conditions, to undergo a hydrolysis process that is largely confined to the surface of the polymer (3). Further, even with systems that do not undergo surface hydrolysis, excellent control over release of incorporated agents can be achieved by controlling polymer hydrolysis rate. Because poly(ortho esters) contain the acid-sensitive ortho ester linkage in the polymer backbone, a wide range of hydrolysis rates can be achieved by the use of excipients incorporated into the polymer matrix.

The development and use of poly(ortho esters) for the delivery of a number of therapeutic agents has recently been exhaustively reviewed (4), and at this time, three structurally distinct families of poly(ortho esters) have been described. The general synthesis and polymer structure of the first family of such polymers is shown in Scheme 1. This polymer was developed at the Alza Corporation (5-8). There is no published information on the use of this polymer system in the release of peptides and proteins.

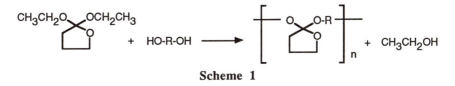

Scheme 1

A second family of poly(ortho esters) is prepared by the addition of polyols to diketene acetals as shown in Scheme 2 (9). Because this condensation proceeds without the evolution of volatile by-products, dense crosslinked materials can be prepared by using reagents that have functionalities greater than two (10). The preparation of a crosslinked material is shown in Scheme 3.

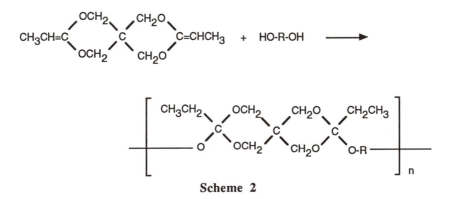

Scheme 2

Poly(ortho esters) have been used to release the LH-RH analogue Nafarelin (11). In this work, a crosslinked poly(ortho ester) prepared as shown in Scheme 3 was used. A semisolid ketene acetal-terminated prepolymer is prepared first and Nafarelin

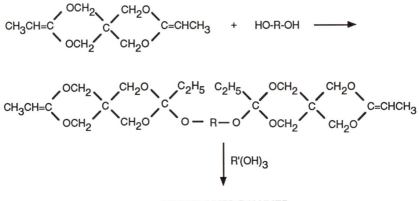

CROSSLINKED POLYMER

Scheme 3

and the crosslinking agent is mixed into the prepolymer at room temperature and without the use of solvents. The viscous paste is then extruded into a suitable mold and the mixture crosslinked at 40°C for 24 hours. The crosslinking can be achieved at room temperature, but many days are required. This procedure provides a very mild means of incorporating proteins into a polymer. However, because the prepolymer contains ketene acetal end-groups, hydroxyl groups of Nafarelin will compete with the triol for those groups and some Nafarelin will become chemically bound to the matrix. However, because the hydrolysis of this family of poly(ortho esters) has been shown to proceed almost exclusively by an exocyclic cleavage of the alkoxy group (*11*), it is expected that hydrolysis of the matrix will liberate unchanged Nafaralenin.

Figure 1 shows daily release of Nafarelin from a crosslinked polymer prepared from a 3-methyl-1,5-pentanediol prepolymer crosslinked with 1,2,6-hexanetriol. The kinetics are those expected from a purely diffusional release. The line crosses the 0.4 μg/day release rate necessary to maintain complete estrus suppression in rats around day 125. These results are in very good agreement with results shown in Figure 2 which shows *in vivo* studies of estrus suppression indicating that the *in vitro* assay is accurate and that Nafarelin is released in its active form from the crosslinked matrix.

However, purely diffusional kinetics are not desirable because during the initial phase of the release profile an excessive amount of Nafarelin is released. For this reason, a more rapidly hydrolyzing polymer was prepared by varying polymer hydrophilicity and by the incorporation of an acidic comonomer, 9,10-dihydroxystearic acid. Results of that study are shown in Figure 3 where a one month release of Nafarelin has been achieved. The change in slope at day 20 is at this time not understood.

A modified version of this polymer has also been used in the construction of a self-modulated insulin delivery system. The device under development is shown in Figure 4 where insulin is dispersed in an acid-sensitive polymer which is surrounded by a hydrogel containing immobilized glucose oxidase (*12*). When glucose diffuses into the hydrogel, it is oxidized by the enzyme glucose oxidase to gluconic acid and the consequent lowered pH accelerates polymer erosion and concomitant insulin release. Such a device requires a bioerodible polymer that can reproducibly and reversibly change erosion rate in response to very small changes in the external pH.

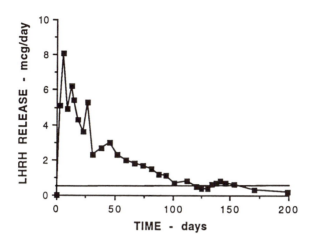

Fig. 1. Daily release of LHRH analogue from a crosslinked poly (ortho ester) prepared from 3-methyl-1,5-pentanediol prepolymer crosslinked with 1,2,6-hexanetriol. Rods, 7.4 x 20 mm in a pH 7.4 buffer at 37°C. Device contains 3 wt% LHRH analogue. (Reproduced with permission from ref. 11. Copyright 1987 Elsevier Science Publishers)

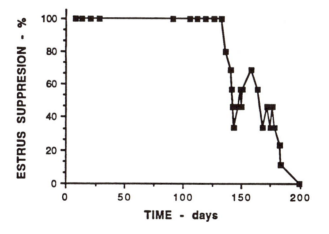

Fig. 2. Percent female rats showing estrus suppression as a function of treatment time. Crosslinked poly (ortho ester) prepared from 3-methyl-1,5-pentanediol prepolymer crosslinked with 1,2,6-hexanetriol. Rods, 7.4 x 20 mm implanted into rats on day 0. Device contains 3 wt% LHRH analogue. (Reproduced with permission from ref. 11. Copyright 1987 Elsevier Science Publishers)

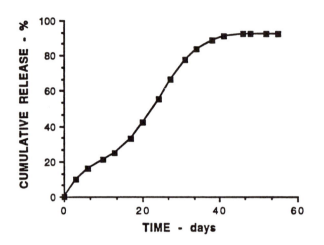

Fig. 3. Cumulative release of LHRH analogue from a crosslinked poly (ortho ester) prepared from triethylene glycol containing 1 wt% 9,10-dihydroxystearic acid prepolymer crosslinked with 1,2,6-hexanetriol. Rods, 7.4 x 20 mm in a pH 7.4 buffer at 37°C. Device contains 3 wt% LHRH analogue. (Adapted from ref. 11. Copyright 1987 Elsevier Science Publishers)

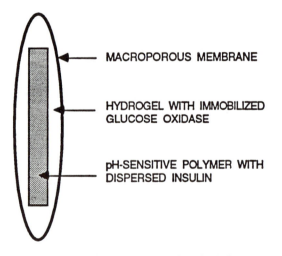

Fig. 4. Schematic representation of proposed insulin delivery system (Reproduced with permission from ref. 12. Copyright 1990 Elsevier Science Publishers)

While poly(ortho esters) are acid-sensitive, their sensitivity is not sufficient to respond to the small pH-changes resulting from glucose oxidation. Therefore, the polymer was modified by the inclusion of tertiary amines in the polymer backbone which enormously increased pH-sensitivity relative to the unmodified polymer. The structure of a polymer from 3,9-bis (ethylidene) 2,4,8,10-tetraoxaspiro [5,5] undecane and N-methyl diethanolamine is shown in Scheme 4.

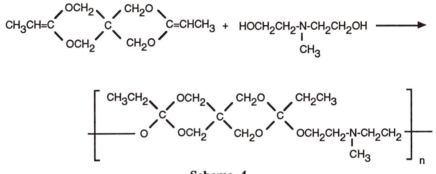

Scheme 4

Disks containing insulin dispersed in the polymer were subjected to well defined low pH pulses and the amount of insulin released during each pulse was determined by radioimmunoassay. As shown in Figure 5, a progressive decrease in insulin levels was achieved with repeated stimulation representing a gradual depletion of insulin from the device.

The data show an excellent control over insulin release and it is encouraging to note that the response of the polymer to a decrease in pH is virtually instantaneous. Further, the release stops as soon as the pH increases, although there is a slight tailing off, perhaps resulting from diffusion of insulin from newly created pores. The rapid response to an increase in pH is important if hypoglicemia is to be avoided. Figure 6 shows the response of the device to pulses of progressively decreasing pH. Insulin release is first detected at pH 6.0 and release increases as the pH decreases, a highly desirable property for an eventually therapeutically useful device.

However, these *in vitro* studies were carried out in a 0.1 M phosphate-citrate buffer. When the studies were repeated in a physiologic buffer, response of the device was only minimal, even at very low pH pulses. These results are consistent with the known general acid catalysis of poly(ortho esters) (*13*). Thus, further work with this polymer was discontinued and a search for another highly acid-sensitive, bioerodible polymer that will undergo specific hydronium ion catalysis is currently underway.

A third family of poly(ortho esters) is prepared as shown in Scheme 5 (*14*).

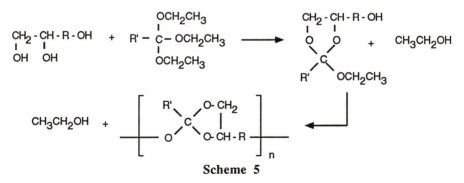

Scheme 5

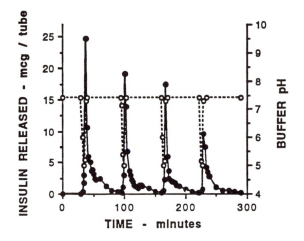

Fig. 5. Release of insulin from a linear polymer prepared from 3,9-bis (ethylidene-2,4,8,10-tetraoxaspiro [5,5] undecane) and N-methyl diethanolamine as a function of external pH variations between pH 7.4 and 5.0 at 37°C. Buffer was continuously perifused at a flow rate of 2 ml/min and total effluent collected at 1 - 10 min. intervals. (o) buffer pH, (●) insulin release (Reproduced with permission from ref. 12. Copyright 1990 Elsevier Science Publishers)

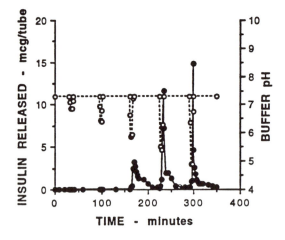

Fig. 6. Release of insulin from a linear polymer prepared from 3,9-bis (ethylidene-2,4,8,10-tetraoxaspiro [5,5] undecane) and N-methyldiethanolamine as a function of external pulses of decreasing pH. Buffer was continuously perifused at a flow rate of 2 ml/min and total effluent collected at 1 - 10 min. intervals. (o) buffer pH, (●) insulin release (Reproduced with permission from ref. 12. Copyright 1990 Elsevier Science Publishers)

When a rigid triol such as 1,1,4-cyclohexanetrimethanol is used in this synthesis, solid materials are obtained (*15*). However, when a flexible triol such as 1,2,6-hexanetriol is used, the polymers are semi-solid, ointment-like materials at room temperature even though polymer molecular weight can be as high as 50 kdaltons. Depending on polymer molecular weight and nature of the R'-group, materials having varying degrees of viscosity can be prepared.

A significant advantage of such an ointment-like material is that therapeutic agents can be readily incorporated into this material by a simple mixing procedure without the need to use solvents or elevated temperatures. This advantage is of particular interest in the delivery of very sensitive proteins that have complex tertiary structures, essential for their biological activity. Biotolerance studies have recently been reported (*16*) and results of these studies show that the polymer is very well tolerated.

In preliminary studies, lysozyme was used as a model protein (*16*). When mixed into an acetate (R=CH₃) polymer, delayed release shown in Figure 7 was achieved. The delay time could be reproducibly controlled by varying molecular weight or, as shown in Figure 8, by varying the nature of the R'-group. Clearly, the effect of the alkyl group is substantial since a 9300 molecular weight valerate polymer has a delay time of about one month while an acetate polymer of similar molecular weight has a delay time of only one week.

When changes in polymer molecular weight were followed concurrently with lysozyme release, data shown in Figure 9 were obtained (*17*). These data are consistent with bulk hydrolysis of the polymer resulting in a continuous decline in molecular weight. Because lysozyme a 14,550 Da protein, it is unable to diffuse from the intact polymer. However, at the point indicated by the arrow, which corresponds to the first appearance of the initial polymer degradation product, the monoester of 1,2,6-hexanetriol, lysozyme is released because the polymer begins to solubilize.

The availability of polymers having different delay times makes possible the construction of devices that can release proteins in well defined and well spaced pulses. To do so, it is only necessary to use a device that contains two or more different polymer formulations in separate domains. Although mechanically dificult, this process could be achieved by placing physically separated polymers in a thin, bioerodible, macroporous cylinder for subsequent implantation, or better, to encapsulate each formulation in a bioerodible, macroporous membrane. In this latter approach, desired release profiles can be achieved by using appropriate mixtures of different capsules.

The initial studies shown in Figures 7 and 8 were generated at room temperature. When these studies were repeated at physiological pH and 37°C, polymer hydrolysis was significantly enhanced and delay times greatly reduced. This is evident in Figure 10 which shows release of lysozyme from the same polymer at three different temperatures. Although details of the release process are still not completely clear, it is apparent that a number of factors combine to accelerate the process. Dominant among these factors is an acceleration of hydrolysis rates with an increase in temperature. When the data in Figure 10 were used to construct an Arrhenius plot, an activation energy of about 24 Kcal/mole can be estimated.

Another important factor is a reduction in polymer viscosity with an increase in temperature. In another study (*18*), we have shown that with this polymer, a change from room temperature to 37°C can result in a 10-fold decrease in polymer viscosity. This decrease in viscosity will not only accelerate diffusion of lysozyme from the polymer, but will also facilitate penetration of water into the polymer.

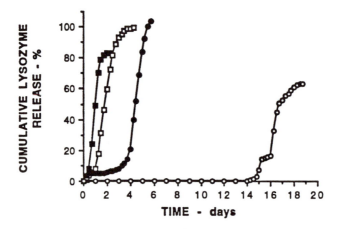

Fig. 7. Release of lysozyme from an acetate polymer (R=CH₃) loaded with 5
wt% lysozyme at pH 7.4 and room temperature. Molecular weights are
(■) 5,350, (□) 6,800, (●) 12,000, (○) 24,400. (Adapted from ref.
16. Copyright 1992 Elsevier Science Publishers)

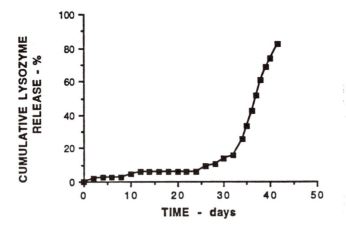

Fig. 8. Release of lysozyme from a 9,300 molecular weight valerate polymer
(R=CH₃CH₂CH₂CH₂) loaded with 5 wt% lysozyme at pH 7.4 and rom
temperature. (Adapted from ref. 16. Copyright 1992 Elsevier Science
Publishers)

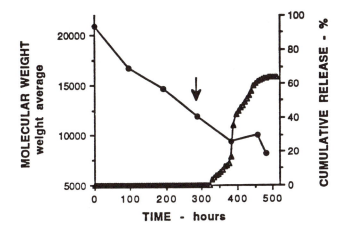

Fig. 9. Change of molecular weight and release of lysozyme from an acetate polymer (R=CH₃) loaded with 5 wt% lysozyme at pH 7.4 and 37°C. Arrow indicates first appearance of 1,2,6-hexanetriol monopropionate.

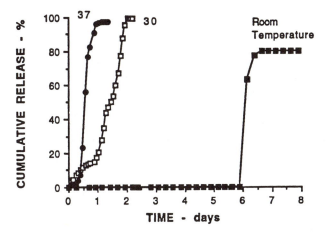

Fig. 10. Effect of temperature on the release of lysozyme from a propionate (R = CH₃CH₂CH₂) 28,000 molecular weight polymer at pH 7.4. Drug loading 5 wt%.

Polyanhydrides

Polyanhydrides represent another class of polymers that are capable of undergoing a hydrolysis process confined largely to the surface of the polymer. Aromatic polyanhydrides were first synthesized in 1909 (*19*) and aliphatic anhydrides in the 1930's (*20*). Subsequent to these early studies, a significant research effort was expended in attempts to improve their poor hydrolytic stability which made them unsuitable as textile fibers (*21,22*). However, it was not until 1983 that it was recognized that this inherent hydrolytic instability renders these materials useful as biodegradable drug delivery systems (*23-25*). Polyanhydrides are synthesized as shown in Scheme 6.

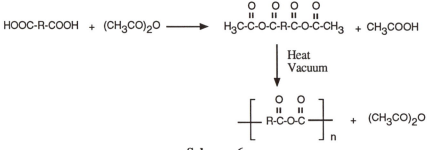

Scheme 6

Polyanhydrides are a highly versatile delivery system which is based on the very large difference in hydrolysis rates between aliphatic and aromatic polyanhydrides. This versatility allows the synthesis of copolymers that have an extremely wide range of erosion rates, as illustrated in Figure 11 for copolymers of sebacic acid and bis(p-carboxyphenoxy) propane. Polyanhydrides undergo a hydrolysis process that is biased towards the outer surface of the polymer because the hydrolysis of anhydride linkages is inhibited by acids. Thus, bulk hydrolysis of polyanhydrides is autosupressed by the generation of diacid hydrolysis products and the polymer undergoes surface erosion.

Because the temperature required to fabricate polyanhydride devices by a hot-melt process is too high for sensitive materials such a proteins, a solvent extraction method must be used [*26*]. In this method, the drug is dispersed or dissolved in a polymer solution containing a volatile organic solvent. This mixture is suspended in an organic oil and the organic solvent is then extracted into the oil, creating microspheres.

When insulin was incorporated into a 50/50 copolymer of sebacic acid and bis(p-carboxyphenoxy) propane, release was considerably faster than polymer erosion, indicating a significant diffusional component to the release mechanism. Studies with streptozotocin-induced diabetic rats showed that insulin retained its activity [*26*].

It has been hypothesized that polyanhydrides based on aromatic diacids become brittle and eventually fragment after exposure to water. This property may cause water-soluble drugs to be released more rapidly than through polymer erosion and diffusion mechanisms. For this reason, a new class of polyanhydrides was prepared from fatty acid dimers derived from naturally occurring oleic and sebacic acids (*27*). The structure of these materials is shown in Scheme 7.

Proteins were incorporated into microspheres prepared from these materials by a solvent evaporation method using a double emulsion [*27-29*]. In this procedure, an aqueous solution of the protein is added to a polymer/methylene chloride solution, the mixture is emulsified and poured into an aqueous solution of 1% poly(vinyl alcohol). This solution was then emulsified to form a double emulsion. Microspheres were obtained by pouring the emulsion into another 0.1% poly(vinyl alcohol) solution and stirring until the methylene chloride was completely evaporated.

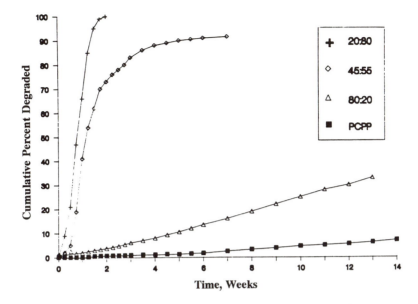

Fig. 11. Degradation profiles of compression molded poly[bis(*p*-carboxy phenoxy) propane anhydride] (PCPP) and its copolymer with sebasic acid (SA) in 0.1 M pH 7.4 phosphate buffer at 37°C. (Reproduced with permission from ref. 24. Copyright 1990, Marcel Dekker.)

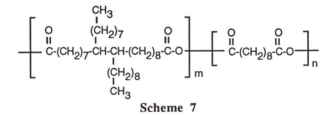

Scheme 7

When lysozyme, trypsin, ovalbumin, bovine serum albumin and immunoglobulin were incorporated into a 25/75 fatty acid/sebacic acid copolymer at a 2 wt% loading, near constant release of about two weeks was achieved (*30*). The fact that they are all released at about the same rates suggests a dominance of an erosion controlled mechanism. Results of that study are shown in Figure 12.

Summary and Conclusions

Poly(ortho esters) and polyanhydrides are currently under active investigation as bioerodible drug delivery systems for a variety of peptides and proteins. Because both systems are based on synthetic schemes that allow considerable variation in mechanical properties and polymer hydrolysis rates, these systems allow a tailoring of erosion behavior to achieve desired release profiles. In this respect, they represent a significant advance over the widely used copolymers of lactic and glycolic acids. However, unlike this latter system that hydrolyzes to the natural metabolites lactic and glycolic acids, neither poly(ortho esters) nor polyanhydrides degrade to natural metabolites so that the

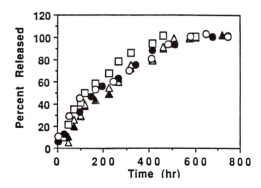

Fig. 12. Release of proteins from a 42,900 molecular weight 25/75 fatty acid
dimer/sebasic polyanhydride in 0.1 M phosphate buffer at 37°C. (○)
lysozyme, (●) trypsin, (△) ovalbumin, (▲) BSA, (□)
immunoglobulin. Protein loading, 2 wt%. (Reproduced with
permission from ref. 30. Copyright 1993 Plenum.)

toxicological burden of assuring complete safety is considerable. Nevertheless,
significant advances in this important field can not be achieved without the availability
of new polymers and the consequent financial burden of toxicological studies.

Literature Cited

1. Heller, J. *Adv. Drug Deliv. Reviews*, **1993**, *10*, 163-204.
2. Heller, J. *CRC Crit. Rev. in Therap. Drug Carrier Syst.* **1984**, *1*, 39-90.
3. Heller, J. *J. Controlled Release*, **1985**, *2*, 167-177.
4. Heller, J. *Adv. in Polymer Sci.*, **1993**, *107*, 41-92.
5. Choi, N. S.; Heller, J. *US Patent*, 4,079,038, **1978**
6. Choi, N. S.; Heller, J. *US Patent*, 4,093,709, **1978**
7. Choi, N. S.; Heller, J. *US Patent*, 4,131,648, **1978**
8. Choi, N. S.; Heller, J. *US Patent*, 4,138,344, **1979**
9. Heller, J.; Penhale, D. W. H.; Helwing, R. F., *J. Polymer Sci., Polymer
    Lett. Ed.*, **1980**, *18*, 82-83.
10. Heller, J.: Fritzinger, B. K.; Ng, S. Y.; Penhale, D. W. H., *J. Controlled
    Release*, **1985**, *1*, 233-238.
11. Heller, J.; Ng, S. Y.; Penhale, D. W. H.; Fritzinger, B. K.; Sanders, L. M.;
    Burns, R. M.; Bhosale, S. S., *J. Controlled Release*, **1987**, *6*, 217-224.
12. Heller, J.; Chang, A. C.; Rodd, G.; Grodsky, G. M., *J. Controlled Release*,
    1990, *13*, 295-302.
13. Fife, T. F., *Accounts of Chem. Res.*, **1972**, *5*, 264-272.
14. Heller, J.; Ng, S. Y.; Fritzinger, B. K.; Roskos, K. V., *Biomaterials*, **1990**,
    11, 235-237.
15. Heller, J.; Ng, S. Y.; Fritzinger, B. K., *Macromolecules*, **1992**, *25*, 3362-
    3364.
16. Wuthrich, P.; Ng, S. Y.; Roskos, K. V.; Heller, J., *J. Controlled Release*,
    1992, *21*, 191-200.
17. Heller, J.; Roskos, K. V.; Duncan, R., *Makromoleculare Chemie*, in press.
18. Merkli, A.; Heller, J.; Tabatabay, C.; Gurny, R., *J. Controlled Release*,
    1994, *29*, 105-112.
19. Bucher, J. E.; Slade, W. C., *J. Am. Chem. Soc.*, **1909**, *31*, 1319-1321.

20. Hill, J.; Carothers, W. H., *J. Am. Chem. Soc.,* **1932,** *54,* 1569-1579.
21. Conix, A. J., *J. Polymer Sci.,* **1958,** *29,* 343-353.
22. Yoda, N., *J. Polymer Sci., Part A,* **1963,** *1,* 1323-1338.
23. Rosen, H. G.; Chang, J.; Wnek, G. E.; Linhardt, R. J.; Langer, R., *Biomaterials,* **1983,** *4,* 131-133.
24. Chasin, M.; Domb, A.; Ron, E.; Mathiowitz, E.; Langer, R.; Leong, K.; Laurencin, C.; Brem, H.; Grossman, S.; *Biodegradable Polymers as Drug Delivery Systems,* Chasin, M.; Langer, R. (Eds.), Marcel Dekker, New York, **1990,** pp. 43-70.
25. Leong, K. W.; D'Amore, P.; Marletta, M.; Langer, R., *J. Biomed. Mater. Res.,* **1986,** *20,* 51-64.
26. Mathiowitz, E.; Saltzman, W. M; Domb, A.; Dor, P.; Langer, R., *J. Appl. Polymer Sci.,* **1988,** *35,* 755-774.
27. Tabata, Y.; Langer, R., *Pharm. Res.,* **1993,** *10,* 391-399.
28. Ogawa, Y.; Yamamoto, M.; Okada, H.; Yashiki, T.; Shimamoto, T., *Chem. Pharm. Bull,* **1988,** *36,* 1095-1103.
29. Sah, H. K.; Chien, Y. W., *Drug. Dev. Ind. Pharm.,* **1993,** *19,* 1243-1263.
30. Tabata, Y.; Gutta, S.; Langer, R., *Pharm. Res.,* **1993,** *10,* 487-496.

RECEIVED April 19, 1994

Chapter 18

Nanoparticles as a Potential Antigen Delivery System

Frank Stieneker and Jörg Kreuter

Institut für Pharmazeutische Technologie, Johann Wolfgang Goethe-Universität, Marie-Curie-Strasse 9, D–60053 Frankfurt am Main, Germany

PMMA nanoparticles showed a significant improvement in the induction of the immune response against different antigens such as influenza and inactivated split whole HIV-1 or HIV-2 when compared to other adjuvants. PMMA induced higher titres, which were stable at consistently high levels over prolonged periods of time. The immunization with PMMA was substantially more reproducible and non-responders were not observed. PMMA was biocompatible without any observable side effects or toxic reactions. The titres showed a clear dependence on the hydrophobicity and on the particle size of the respective adjuvant. Comparison of different adjuvants with HIV-2 as the antigen demonstrated that the specificity of the induced antibodies was dependent on the interaction between the various viral proteins and the adjuvant.

A large number of antigens such as smaller peptides, virus subunits and genetically engineered proteins are weak immunogens and induce little or no protection. For this reason, many vaccines contain not only the antigen but also adjuvants for enhancing the immune reaction.

Until now, a large number of adjuvants have been developed by using mineral compounds (aluminum compounds, bentonite), emulsions (Freund´s complete and incomplete adjuvant), surfactants (non-ionic block polymers), peptides (MDP, T-MDP), lipids (lipid A), liposomes, colloidal carriers (ISCOMs, nanoparticles), and other compounds. Unfortunately, most of these adjuvants cause toxic side effects or cannot be prepared in reproducible manner.

It is known that the particle size of dispersed adjuvants strongly influences the adjuvant effect (1, 2). In particular, slightly different production conditions and aging alter the structure and properties of aluminum compounds (3-6). These production and storage problems may lead to different qualities of alum that show no correlation between the absorption properties of the material and the obtained adjuvant effect (7).

To avoid these problems, many different adjuvants were developed with some success. Nanoparticles were developed as an adjuvant and are defined as solid colloidal particles ranging in size from 10 to 1000 nm. They consist of polymeric materials (polyacrylates, polylactides, denatured proteins) in which the active agent (drug or biologically active material) is dissolved, entrapped, encapsulated and/or to

0097–6156/94/0567–0306$08.00/0

which the active agent is adsorbed or attached (*8*). One attempt was the employment of nanoparticles by Kreuter and Speiser in 1976 (*9*).

One of the most promising polymers for nanoparticulate adjuvants is polymethylmethacrylate (PMMA). This polymer was shown to be slowly biodegradable in the form of nanoparticles without any observable toxic side-effects (*10*). Because of its high safety, it has been used in surgery for 50 years. In addition, PMMA nanoparticles can be polymerized in a physicochemically reproducible manner within narrow limits (*11, 12*). Thirdly, it is rather hydrophobic, which, as it will be shown below, enhances the adjuvant effect for a number of antigens.

Preparation of Nanoparticles

Nanoparticles may be prepared from different materials. For the reasons mentioned already, polymethylmethacrylate (PMMA) holds great promise for use as an antigen delivery system. The preparation of PMMA nanoparticles can be carried out either by heat polymerization or by gamma-irradiation (*12, 13*).

After purifying the monomer (MMA) from polymerization inhibitors (*8, 12*), it is dissolved in distilled water or in a desired buffer solution in amounts of up to 1.5%. Polymerization is initiated by irradiation (500 kRad) or by addition of ammonium peroxodisulfate or potassium peroxodisulfate and heating to 65-85 °C. The resulting dispersion can be stored as an aqueous solution or lyophilized powder.

Antigen may be incorporated into the particles by irradiation-induced polymerization of the monomer in the presence of the antigen. Although this may lead to the denaturation of many proteins, the antigenicity of the influenza antigens was fully retained (*9, 14*). Alternatively, antigens may be adsorbed onto the surface of the particles by adding antigen to the solution after the polymerization and gently agitating the mixture at 4 °C for 24 hours (*14*). Depending on the antigen this may lead to the adsorption of practically all antigen (*14*).

Body Distribution and Elimination of PMMA Adjuvant Nanoparticles

As previously shown, almost all of the radioactivity of ^{14}C-labelled PMMA nanoparticles stays at the injection site after subcutaneous injection (*10, 15*). During the first week after the material was injected, only 6% of the radioactivity was eliminated urinarily and only 5% was eliminated fecally. Between the first week and 70 days post administration, the elimination rate decreased to 0.005% per day. The elimination rate stayed at this low level up to two hundred days post administration. Then the elimination of the radioactive material increased exponentially to 1% of the radioactive dose per day up until day 287. Sixty to seventy per cent of the initial radio labelled nanoparticle dose remained at the injection site until day 287 and was not distributed into the residual body. The shape of the elimination rate curve points to the following scenario which was observed previously with other polymers including polylactic acid, and which may be best explained by biodegradation of the polymer. The initial high excretion rate may be due to the elimination of lower molecular weight components of the nanoparticle polymer. After this, between 70 and 200 days, very little nanoparticle material is excreted. Nevertheless, the biodegradation of the PMMA probably occurs during this time similar to other biodegradable polymers. The transport and elimination of this degraded material, however, seems to take place only when a considerably lower molecular weight is reached, thus explaining the steady exponential elimination rate increase after 200 days up to the sacrifice of the animals after 287 days. Parallel to the elimination rate increasem a comparable 100 to 300-fold increase in radioactivity in the body also was observed.

Important requirements for the use of a protein or antigen delivery system as an adjuvant for vaccines are the biocompatibilty and the biodegradability of the vaccine formulation. Histological examination of the injection site of influenza vaccines containing PMMA nanoparticles showed no abnormal tissue response. The physiological reactions of the tissue were the same as those observed after the injection of the control fluid influenza preparation (16). PMMA, therefore, may be better tolerated than other adjuvants, such as aluminum hydroxide, since aluminum compounds are known to induce granuloma (17, 18).

Physicochemical Characterization

To characterize the physicochemical properties of the nanoparticles different methods were employed. These techniques include scanning and transmission electron microscopy, photon-correlation spectroscopy, BET-surface area analysis, helium pycnometry, X-ray diffraction, measurement of surface charge, determination of wettability (11) and gel permeation chromatography (19). These techniques and their use are listed in Table I.

Influence of Physico-Chemical Properties on the Adjuvant Effect

Particle Size. The influence of the particle size of PMMA and of polystyrene nanoparticles on the immune response to influenza and BSA (bovine serum albumin) was determined after intramuscular injection. The particle sizes ranged from 62 nm to 10 μm. Smaller particles below 350 nm induced a significantly higher immune response (2). The optimal particle size was approximately 100 to 150 nm (2). However, since smaller particles were not thoroughly investigated, it cannot be ruled out that the optimal particle size may be smaller.

Hydrophobicity. The influence of the nanoparticle hydrophobicity on the adjuvant effect was investigated by employing BSA and influenza subunits as antigens (20). The hydrophobicity of the particles was altered by varying the functional groups and the side chains of the monomers. These results indicated that the adjuvant effect of the particles increased with increasing hydrophobicity of the polymers (Figure 1).

Adjuvant Effects of Nanoparticles with Different Antigens

Influenza. In 1976, Kreuter and Speiser (8) described the production of PMMA particles ranging in size from 50 to 300 nm by using γ-irradiation to induce the polymerization of the monomer methylmethacrylate. Adsorption of influenza to the particles led to immune responses which were comparable to those obtained with aluminum hydroxide gel. Polymerization of the particles in the presence of the antigen provided a higher immune response. Similar experiments were performed by using solubilized split influenza antigens, which are weak immunogens. Incorporation, as well as adsorption, of this antigen led to better immune responses than those obtained with aluminum hydroxide adjuvanted vaccines. PMMA induced remarkably high titres with low antigen concentrations (16). These studies also showed an optimal immune response after injection with an adjuvant concentration of 0.5% PMMA nanoparticles. The highest titres were measured 4 weeks after immunization. The greatest differences between aluminum hydroxide and PMMA were observed at low antigen concentrations. In challenge experiments, mice were protected against influenza. Immunization with PMMA provided better protection of mice against the influenza virus than alum immunization. This effect was more pronounced after long time periods (21).

In order to determine the storage and heat stability of influenza vaccines, different preparations (PMMA, alum, fluid) were stored at 40 °C for various time periods up to 10 days and were then administered i.m. to mice. The storage at 40 °C

Table I. Physicochemical characterization methods for nanoparticles

Parameter	Method
Particle size	Photon correlation spectrometry (PCS)
	Transmission electron microscopy (TEM)
	Scanning electron microscopy (SEM)
	SEM combined with energy-dispersive X-ray spectrometry
	Scanned-probe microscopes
	Fraunhofer diffraction
Molecular weight	Gel chromatography
Density	Helium compression pycnometry
Crystallinity	X-ray diffraction
	Differential scanning calorimetry (DSC)
Surface charge	Electrophoresis
	Laser Doppler anemometry
	Amplitude wheighted phase structuration
Hydrophobicity	Hydrophobic interaction chromatography (HIC)
	Contact angle measurement
Surface properties	Static secondary ion mass spectrometry (SSIMS)
Surface element analysis	X-ray photoelectron spectroscopy for chemical analysis (ESCA)

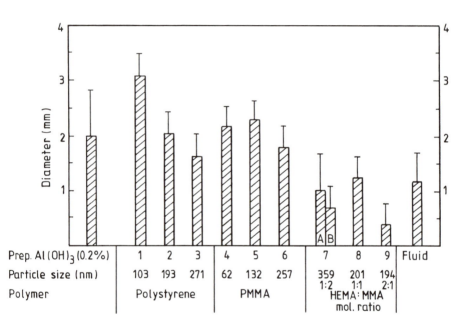

Figure 1.   Precipitation ring diameters (antibody response; mean ± 95% confidence intervals) after immunization of mice with different bovine serum albumine vaccines, particle sizes of the adjuvant particles, and copolymer composition. PMMA, polymethylmethacrylate; HEMA:MMA, 2-hydroxy-ethylmethacrylate/methylmethacrylate copolymer.

resulted in a significant decrease in protection against influenza virus for alum, but there was not a significant decline for PMMA. The protection of mice treated with the stored fluid vaccine was negligible after this preparation was stored for more than 60 hours at 40 °C.

HIV. Three sets of experiments were performed in order to examine the immune response of mice against HIV-1 and HIV-2. In these experiments, the amount of antigen was optimized, the differences in the rate and extent of the immune reactions against both viruses were determined, and some common adjuvants were compared.

Evaluation of the Optimal Antigen Amount. Whole HIV-2 as split antigen was used in three different preparations: antigen dissolved in PBS (fluid), antigen adsorbed to aluminum hydroxide gel (alum), antigen adsorbed onto PMMA nanoparticles (PMMA). Various antigen concentrations ranging from 5 to 50 µg per dose (0.5 ml) were used. The antigen content of each preparation was determined by the Bio-Rad Protein Assay. BSA served as reference solution (22). Antibody titres to the antigen were compared over a time frame of 20 weeks by using an ELISA method (Figure 2a-d) (22, 23).

The titres were determined as follows: Four microtiter plates were treated as one unit. The first six rows per plate contained serial dilutions of the test mouse sera, the last two rows of each plate contained the same dilutions of the negative control. The highest dilution of the negative control was defined as the background. The cut-off was set at zero after subtracting of the mean absorbance of the background and the 3-fold standard deviation from all measured OD_{495} values (optical density at 495 nm) (22).

During the entire observation period, PMMA nanoparticles provided the best adjuvant effect. Between 4 and 20 weeks, the fluid preparation induced significantly lower titres than the other preparations (Figure 2a-d). The alum preparation yielded titres of intermediate levels. They were significantly lower ($p<0.05$) than those obtained with PMMA. After 2 weeks, the titres were relatively low (6,000), but antibody titres increased with increasing antigen amounts. High standard deviations resulted from a lack of reproducibility of the induction of the immune response. At 4 weeks, PMMA induced 2 to 90-fold greater titres (90,000) than those induced by fluid (1000) and alum (50,000). After 10 weeks, titres of 120,000 were measured in the case of PMMA-immunized mice and remained stable up to 20 weeks. In contrast, the titres for fluid-immunized mice increased up to 10 weeks, reaching a maximum level of about 20,000, while alum-immunized mice had lower titres after 10 weeks with a decrease to a minimum of 20,000.

The amount of antigen had little influence on the induced immune response in the case of PMMA, whereas increasing titres with increasing antigen amounts were observed with fluid and alum. After 4 weeks, PMMA-induced immune responses that were largely independent of the antigen concentration resulted in a plateau between 5 and 50 µg antigen per single dose.

Examination of the distribution of non-responders among the mice revealed similar results. With the fluid and the alum preparations, increasing antigen amounts led to a decreasing number of non-responders. In the case of PMMA, the induced immune response was much more reproducible, and during the entire experiment time all of the mice responded to all of the antigen concentrations.

PMMA induced high and reproducible titres with low amounts of antigen. Similar observations with surface antigens of influenza were published earlier (8). The previous study supports the earlier hypothesis that PMMA nanoparticles show good efficiency at low antigen concentrations and with weak immunogens such as HIV antigens.

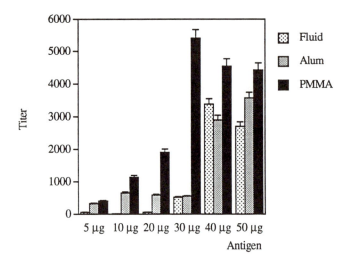

Figure 2a. Serum antibody titres against HIV-2 inactivated split whole virus in mice immunized with 5-50 µg antigen dissolved in PBS (fluid), adsorbed to aluminum hydroxide (alum), or adsorbed to PMMA nanoparticles as measured by ELISA at 2 weeks after immunization.

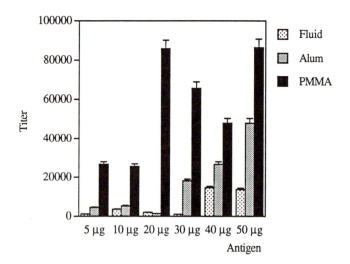

Figure 2b. Serum antibody titres against HIV-2 inactivated split whole virus in mice immunized with 5-50 µg antigen dissolved in PBS (fluid), adsorbed to aluminum hydroxide (alum), or adsorbed to PMMA nanoparticles as measured by ELISA at 4 weeks after immunization.

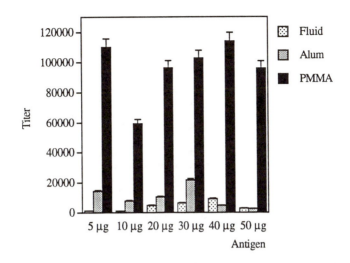

Figure 2c. Serum antibody titres against HIV-2 inactivated split whole virus in mice immunized with 5-50 μg antigen dissolved in PBS (fluid), adsorbed to aluminum hydroxide (alum), or adsorbed to PMMA nanoparticles as measured by ELISA at 10 weeks after immunization.

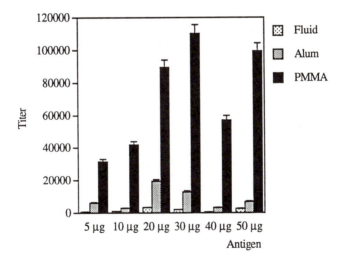

Figure 2d. Serum antibody titres against HIV-2 inactivated split whole virus in mice immunized with 5-50 μg antigen dissolved in PBS (fluid), adsorbed to aluminum hydroxide (alum), or adsorbed to PMMA nanoparticles as measured by ELISA at 20 weeks after immunization.

To avoid undesirable side effects, the amount of the antigen used should be as low as possible but sufficient to induce the desired immune reaction. The described results lead to the conclusion that PMMA seems to fulfill this requirement with split HIV whole virus vaccines. Only 5 µg of antigen that adsorbed to PMMA nanoparticles induced high titres that were stable for 20 weeks and were highly reproducible.

Kinetics of the Immune Response. The kinetics of the immune response against HIV-2 and HIV-1 were determined by immunizing mice with 10 or 20 µg of inactivated HIV-1 or HIV-2 whole virus. The same preparations as described above (fluid, alum and PMMA) were used. Antibody titres between 2 and 40 weeks were determined by ELISA (Figure 3a and 3b) (23, 24).

PMMA induced the highest antibody titres against both HIV-2 and HIV-1. Fluid induced the lowest titres, whereas the titres of alum were of intermediate levels (between 1500 and 5000). The titres induced by alum were significantly lower ($p < 0.05$) than those obtained with PMMA, but at 10 weeks were significantly higher ($p < 0.05$) than those induced by fluid.

With PMMA, the titres against HIV-2 were about 10 times higher than those measured against HIV-1. In the case of the fluid and alum preparations, significant differences between the antigens were not observed. With PMMA, the HIV-1 antibodies generally increased after 10 weeks, whereas the titres of HIV-2 antibodies were significant after 2 weeks and only increased slightly up to 20 weeks. After 4 weeks, PMMA yielded significant levels of antibodies with 20 µg of inactivated HIV-1 whole virus as antigen (20 µg).

Between 10 and 20 weeks, the antibody titres against HIV-1 and HIV-2 decreased and generally diminished to baseline levels after 40 weeks by using either fluid or alum. PMMA, on the other hand, induced antibody titres as great as 120,000 between 4 and 20 weeks. After 40 weeks the titres were still at significant levels (2000) for PMMA.

PMMA possesses the advantage that the titres are highly reproducible as demonstrated by the low standard deviations between the individually determined titres. In addition, non-responders were not observed in PMMA-immunized groups, but fluid or alum-immunized groups sometimes had non-responders.

Although aluminum compounds are widely accepted in humans, inconsistent induction of antibodies often occurs. This problem is well known and is often reported (25, 26). The reason for the inconsistent immune reaction after employment of aluminum compounds may be the pH-dependent adsorption of the antigen to the adjuvant (27).

Remarkably, a more homogenous immune response was obtained when the antigen amount was increased from 10 to 20 µg for the fluid and alum-immunized groups. The use of 20 µg of antigen per dose resulted in lower standard deviations between the individually determined antibody titres and in a reduction of the number of non-responders.

The effectiveness of an immunization can be assessed by measuring the kinetics of the appearance of antibodies. A fast seroconversion followed by long-term titres of specific antibodies is desirable. Ideally, a single dose of vaccine preparation should lead to lifelong protection, making boosters unnecessary.

PMMA seems to be able to change the kinetics in the appearance of binding antibodies and to reduce the lag time until the seroconversion is established. Furthermore, PMMA prolongs the time during which specific antibodies are detectable.

Comparison of Different Adjuvants. The adjuvant effect of 24 different vaccine preparations was compared by measuring the immune response 10 weeks

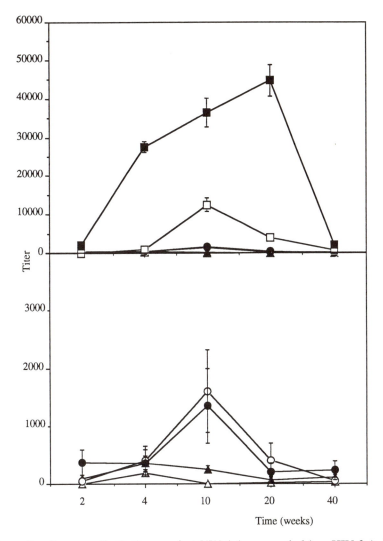

Figure 3a. Serum antibody titres against HIV-1 (open symbols) or HIV-2 (solid symbols) inactivated split whole virus in mice immunized with 10 µg antigen dissolved in PBS (Δ, fluid), adsorbed to aluminum hydroxide (O, alum), or adsorbed to PMMA nanoparticles (□) as measured by ELISA between 2 and 40 weeks after immunization. The upper part of the figure shows all obtained results, whereas the lower part of the figure only shows the titres obtained with alum and fluid.

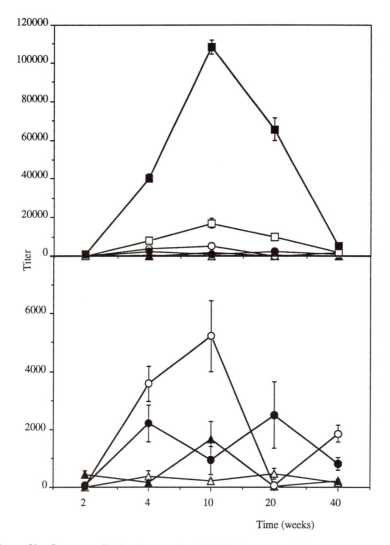

Figure 3b. Serum antibody titres against HIV-1 (open symbols) or HIV-2 (solid symbols) inactivated split whole virus in mice immunized with 20 μg antigen dissolved in PBS (Δ, fluid), adsorbed to aluminum hydroxide (O, alum), or adsorbed to PMMA nanoparticles (□) as measured by ELISA between 2 and 40 weeks after immunization. The upper part of the figure shows all obtained results, whereas the lower part of the figure only shows the titres obtained with alum and fluid.

after administration of 5 µg inactivated HIV-2 whole virus with different adjuvants (*23*). Adjuvants of different origins such as aluminum compounds, liposomes, ISCOMs, SAF-1, different nanoparticles, surfactants, Freund´s complete and incomplete adjuvant were employed. The antibody titres were determined by ELISA (Figure 4) and the quality of the immune response was assessed by Western blot analysis (Table II).

These investigations also showed that the best adjuvant effects were induced by PMMA, with titres of approximately 50,000. High titres between 30,000 to 40,000 were achieved with preparations containing Synperonic PE L121 in pure form or in combination with aluminum phosphate or PMMA nanoparticles (*23*). Intermediate titre levels were induced by aluminum compounds (aluminum hydroxide gel Behring, aluminum hydroxide gel Hem, aluminum phosphate), Freund´s complete (FCA) and incomplete adjuvant (FIA), bentonite, SAF-1, Aerosil 200, and R972 (*23*). Lower titres were observed with the fluid preparation (standard control) than with Adjuvant 65, three liposomal preparations (MLV with incorporated antigen, SUV with the antigen integrated into the bilayer, SUV with the antigen adsorbed onto the surface of the vesicle), muramyl dipeptide (MDP), Pluronic F68, two nanoparticle preparations consisting of either polyhexylcyanoacrylate (PHCA) or polybutylcyanoacrylate (PBCA), and ISCOMs, where only 0.5 µg of the antigen was attached to the complex, because an antigen dose of 5 µg led to lethal side effects (*23*).

Non-responders were not observed with PMMA, alum Behring, alum Hem, aluminum phosphate, FCA, FIA, SUV with adsorbed antigen, bentonite, Aerosil 200, Aerosil R972, SAF-1, PHCA, PBCA, Pluronic F68, Synperonic PE L121, and the combinations of PMMA/Synperonic PE L121, PMMA/Pluronic F68 or aluminum phosphate/Synperonic PE L121. The other preparations showed some non-responders.

The titres increased with increasing hydrophobicity of the adjuvant as previously observed (*26*). PMMA induced higher titres than PHCA or PBCA. A similar effect was observed when comparing aluminum phosphate/Synperonic PE L121 to aluminum phosphate, SAF-1 to MDP, Synperonic PE L121 to Pluronic F68 or PMMA to its combinations with Synperonic PE L121 and Pluronic F68. Increasing the hydrophobicity of the adjuvant seems to lead to a better interaction of the antigen-adjuvant-complex with immune competent cells (*20*). Increasing the hydrophobicity of adjuvants also intensifies their cell-adhesion and phagocytosis (*28*). Similar observations were made after covalent linking of different antigens to fatty acids (*29*, *30*).

Besides the hydrophobicity of the employed adjuvants, the particle size also has a strong influence on the immune response. In the case of PMMA nanoparticles, good titres against different antigens were achieved if the particle sizes ranged between 100 to 200 nm. As mentioned earlier, particle sizes from 0.5 to 1 µm resulted in lower titres or no response (*2, 8, 26, 31, 32*). The investigations with 5 µg HIV-2 confirm these former results. PMMA (120 nm) induced higher titres than PBCA (443 nm) or PHCA (961 nm), although the influence of the particle size can not be separated from the influence of the hydrophobicity of the employed polymers on the adjuvant effect.

The liposomal presentation of the antigen showed a dependence on the kind of association of the antigen with the liposome. The highest titres were induced if the antigen was adsorbed onto the surface of the bilayer. Integration of the antigen into the bilayer resulted in intermediate titres, whereas incorporated antigen induced lower and weaker titres than those induced by the fluid control preparation.

Literature data are contradictory as to whether association of the antigen with the liposome influences the immune response. In some cases, entrapped antigen seemed to inhibit the immune response (*33*), and only surface-associated antigen induced a significant immune response (*34*). Other investigations revealed no difference between incorporated, adsorbed, or bilayer integrated antigen (*35*).

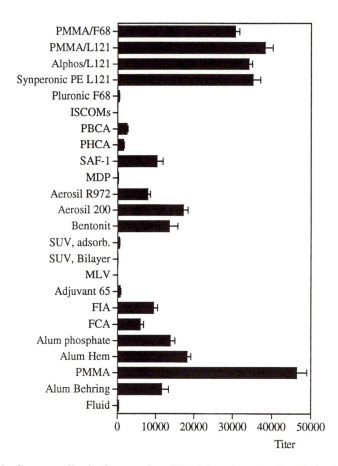

Figure 4. Serum antibody titres against HIV-2 inactivated split whole virus in mice immunized with 5 µg antigen in combination with different adjuvants. Mice immunized with ISCOMs received 0.5 µg antigen. The following abbreviations are used: PMMA (polymethylmethacrylate nanoparticles), Alum Hem (aluminum hydroxycarbonate produced by S. L. Hem, Purdue University), FCA (Freund's complete adjuvant), FIA (Freund's incomplete adjuvant), MLV (multilamellar large vesicles, incorporated antigen), SUV, bilayer (small unilamellar vesicles, antigen integrated into the bilayer), SUV, adsorb. (small unilamellar vesicles, antigen adsorbed onto the surface), MDP (N-acetyl-muramyl-L-alanyl-D-isoglutamine), SAF-1 (Syntex Adjuvant Formulation-1), PHCA (polyhexylcyanoacrylate nanoparticles), PBCA (polybutylcyanoacrylate naoparticles), ISCOMs (immunostimulating complexes), Alphos/L121 (combination of aluminum phosphate and Synperonic PE L121), PMMA/L121 (combination of PMMA and Synperonic PE L121), PMMA/F68 (combination of PMMA and Pluronic F68).

Table II. Antibody responses induced by different adjuvants against various antigens of 5 μg HIV-2 as determined in Western Blot

Adjuvant	viral antigens				
	p17	p24	RT	gp41	gp120
Fluid	(+)	(+)	-	-	-
Alum (Behring)	++	++	+	-	+
Alum (Hem)	++	++	+	-	+
Aluminium phosphate	++	++	+	-	(+)
FCA	+++	++	+	+	+
FIA	++	++	+	-	+
Adjuvant 65	+	++	-	-	-
MLV	-	-	-	-	-
SUV, bilayer	-	+	-	-	-
SUV, adsorbed	+	+	-	-	-
Bentonite	++	++	+	-	-
Aerosil 200	++	++	+	-	+
Aerosil R972	++	++	+	-	(+)
MDP	-	+	-	-	-
SAF-1	++	++	+	-	-
PMMA	++	++	+	+	-
PHCA	(+)	(+)	-	-	-
PBCA	+	+	-	-	-
ISCOMs (0.5 μg antigen)	(+)	-	-	-	-
Pluronic F68	+	++	-	-	-
Synperonic PE L121	++	++	+	-	-
Aluminium phosphate/L121	++	++	+	-	-
PMMA/L121	++	++	++	(+)	-
PMMA/F68	++	++	+	-	-

+++ very strong, ++ strong, + medium, (+) weak, - no reaction

To assess the quality of the immune response, the collected sera were individually examined by Western blot (Table II). Only FCA induced antibodies against all viral antigens. FIA induced a similar antibody pattern with the except that antibodies against the transmembrane glycoprotein gp41 were not raised. PMMA also showed good responses in the Western blot. However, PMMA did not induce antibodies against the outer envelope glycoprotein gp120.

Clear responses against gp120 were detected after immunization with both aluminum hydroxide as well as FCA, FIA and Aerosil 200. A weaker immune reaction was observed in the case of aluminum phosphate and Aerosil R972. A positive response against the transmembrane glycoprotein gp41 was only observed with the sera of mice immunized with FCA, PMMA, and the combination of PMMA and Synperonic PE L121. The most prominent immune responses were generated against p24 and p17 independent of the adjuvant.

A clear relationship between the hydrophobic properties of the adjuvant and the pattern of antibodies against the different antigens of the inactivated HIV-2 split whole virus was observed. Hydrophobic adjuvants induced antibodies against more hydrophobic antigens such as p17, p24, RT or gp41, whereas only hydrophilic adjuvants such as alum preparations or Aerosil 200, and those containing a hydrophilic phase, i.e. the emulsions, generated antibodies against the hydrophilic antigen gp120. Particularly, significant titres of antibodies against the hydrophobic glycoprotein gp41 were only induced after administration of PMMA and FCA. A weaker response was obtained after injection of the combination of PMMA and Synperonic PE L121. This observation underlines the importance of hydrophobic interactions between adjuvant and antigen. Therefore, the evaluation of an adjuvant can only be undertaken in combination with the antigen of interest.

As shown in the previous section, PMMA provided the most rapid seroconversion. The in vivo distribution of PMMA nanoparticles indicates that there may be two modes of action for these adjuvant particles. The first mechanism seems to be a depot effect induced by particles residing at the injection site and a resulting controlled release or delayed desorption of the antigen. Secondly, the particles may act as transport vehicles to the regional lymph nodes, as described by Kreuter et al. (10).

PMMA may accomplish, at least partly, the requirements of the World Health Organization for the design of new vaccine preparations. A vaccine that fulfills these requirements should stimulate the immune system for longer time periods and would avoid the need for multiple administrations of the antigen (36). One of the disadvantages of PMMA is that it did not induce antibodies against the whole spectrum of antigens, but this problem may be overcome by the combination with another adjuvant.

Conclusions

PMMA showed a significant improvement of the induction of the immune response against different antigens such as influenza, inactivated split whole HIV-1 or HIV-2 compared to other adjuvants. PMMA induced higher titres that remained at consistently high levels over prolonged periods of time.

Immunization with PMMA as the adjuvant was substantially more reproducible than the immunization with other adjuvant preparations. With PMMA, animals without significant specific antibody titres (non-responders) were not observed, whereas non-responders were sometimes observed after administration of other adjuvants, especially alum.

During the entire observation period, PMMA showed exceptional biocompatibility. Side effects or toxic reactions were not observed, whereas lethality was recorded in some cases after the injection of ISCOMs or adjuvants containing Synperonic PE L121.

The titres showed a clear dependence on the hydrophobicity and the particle size of the respective adjuvant. The levels of the titres increased with increasing hydrophobicity of the preparation and with decreasing particle size.

Comparison of different adjuvants with HIV-2 inactivated split whole virus as the antigen also demonstrated a dependence of the specificity of the induced antibodies on the interaction of the various viral proteins with the adjuvant. Antibodies against hydrophobic antigens were preferably induced with hydrophobic adjuvants, whereas antibodies against hydrophilic proteins were preferably observed after injection in combination with hydrophilic adjuvants. These findings demonstrate that the quality of an adjuvant can only be evaluated in direct combination with a specific antigen, and not generally for all antigens. Along these lines, it has to be concluded that an optimized immunogenic preparation should always be envisaged as an antigen-adjuvant-complex.

Literature Cited

1. Grafe, A. Arzneim. Forsch. 1971, 21, 903.
2. Kreuter, J.; Berg, U.; Liehl, E.; Soliva, M.; Speiser, P. P. Vaccine 1986, 4, 125.
3. Kerkhof, N.J.; White, J. L.; Hem, S. L. J. Pharm. Sci. 1975, 64, 940.
4. Nail, S.L.; White, J. L.; Hem, S. L. J. Pharm. Sci. 1976, 65, 1188.

5. Nail, S.L.; White, J. L.; Hem, S. L. *J. Pharm. Sci.* **1976**, *65*, 1192.
6. Nail, S.L.; White, J. L.; Hem, S. L. *J. Pharm. Sci.* **1976**, *65*, 1195.
7. Pyl, G. *Arch. Exp. Veterinärmed.* **1953**, *7*, 9.
8. Kreuter, J. *Pharm. Acta Helv.* **1983**, *58*, 196.
9. Kreuter, J.; Speiser, P. P. *Infect. Immunol.* **1976**, *13*, 204.
10. Kreuter, J.; Nefzger, M.; Liehl, E.; Czok, R.; Vogels, R. *J. Pharm. Sci.* **1983**, *72*, 1146.
11. Kreuter, J. *Int. J. Pharm..* **1983**, *14*, 43.
12. Berg, U. E.; Kreuter, J.; Speiser, P. P.; Soliva, M.; *Pharm. Ind.* **1986**, *48*, 75.
13. Kreuter, J.; Zehnder, H. J. *Radiation Effects* **1978**, *35*, 161.
14. Kreuter, J.; Speiser, P. P. *J. Pharm. Sci.* **1978**; *65*, 1624.
15. Kreuter, J.; Täuber, U.; Illi, V. *J. Pharm. Sci..* **1979**, *68*, 1443.
16. Kreuter, J.; Mauler, R.; Gruschkau, H.; Speiser, P.P. *Expl. Cell Biol.* **1976**, *44*, 12.
17. Barr, M.; Glenny, A. T.; Butler, N. R. *Br. Med. J.* **1955**, *2*, 635.
18. Turk, J. L.; Parker, D. *J. invest. Dermatol.* **1977**, *68*, 336.
19. Bentele, V.; Berg, U. E.; Kreuter, J. *Int. J. Pharm.* **1983**, *13*, 109
20. Kreuter, J.; Liehl, E.; Berg, U.; Soliva, M.; Speiser, P. P. *Vaccine* **1988**, *6*, 253.
21. Kreuter, J; Liehl, E. *J. Pharm. Sci.* **1981**, *70*, 367.
22. Stieneker, F.; Kreuter, J.; Löwer, J. *AIDS* **1991**, *5*, 431.
23. Stieneker, F. *Entwicklung und Prüfung von Adjuvantien für HIV Impfstoffe.* VAS-Verlag für Akademische Schriften oHG: Frankfurt am Main, 1992; 80-140.
24. Stieneker, F., Löwer, J.; Kreuter, J.; *Vaccine Research* **1993**, *2*, 111.
25. Edelman, R. *Rev. Infect. Dis.* **1980**, *2*, 370.
26. Kreuter, J.; Haenzel, I. *Infect. Immun.* **1978**, *19*, 667.
27. Seeber, S. J.; White, J. L.; Hem, S. L. *Vaccine* **1991**, *9*, 201.
28. Oss, C. J. van; Gillman, C. F.; Neumann, A. W. *Phagocytic Engulfment and Cell Adhesivness.*; Marcel Decker: New York, Basel, 1975.
29. Wiesmüller, K.-H.; Jung, G.; Hess, G. *Vaccine* **1988**, *7*, 29.
30. Krug, M.; Folkers, G.; Haas, B.; Hess, G.; Wiesmüller, K.-H.; Freund, S.; Jung, G. *Biopolymers* **1989**, *28*, 499.
31. Torrigani, H.; Roitt, I. M. *J. Exp. Med.* **1965**, *122*, 181.
32. Freeman, M.J. *Immunology* **1968**, *15*, 481.
33. Rooijen, N. van; Nieuwmengen, R. van In *Targeting of Drugs;* Gregoriadis, G.; Senior, J.; Trouet, A. Eds.; Plenum Publishing Corporation: New York, 1982; 301-326.
34. Thérien, H.-M.; Lair, D.; Shahum, E. *Vaccine* **1990**, *8*, 558.
35. Gregoriadis, G.; Davis, D.; Davies, A. *Vaccine* **1987**, *5*, 145.
36. Ada, G. *Mol. Immunol.* **1991**, *28*, 225.

RECEIVED April 19, 1994

Chapter 19

Formulation and Aerosol Delivery of Recombinant Deoxyribonucleic Acid Derived Human Deoxyribonuclease I

D. C. Cipolla, I. Gonda, K. C. Meserve, S. Weck, and S. J. Shire

Pharmaceutical Research and Development, Genentech, Inc., South San Francisco, CA 94080

Purulent lung secretions from most patients with Cystic Fibrosis (CF) have high concentrations of DNA (3-14 mg/mL). Human DNase produced by recombinant DNA technology (rhDNase) can improve the rheological properties of CF sputum. We report here the formulation and aerosol delivery by jet nebulizers of rhDNase. Although the developed formulation must provide 1-2 years of stability on storage it also must meet additional requirements that are unique to its delivery as an aerosol. First, the formulation must not cause adverse pulmonary reactions such as cough or bronchoconstriction. Secondly, the formulation components should not interfere with the nebulizer performance. For example, the amount of product that is generated as an aerosol should not be decreased, and the desired size distribution (~1 to 6 µm) should not be altered by formulation components. Finally, the formulation also must stabilize the rhDNase sufficiently to ensure that the protein survives the rigors of the nebulization process. We present here the characterization of the physical properties of rhDNase aerosols (droplet size distribution, composition and quantity of rhDNase solution converted to an aerosol), and the chemical and physical stability of rhDNase before and after the nebulization process.

Most protein drugs that are currently marketed are delivered by the parenteral route. The lung has been cited as an ideal candidate for an alternate route of administration because the pulmonary route is relatively non-invasive and the lung has a large surface area available for drug targeting. However, systemic absorption of proteins delivered to the lung is usually low and can be variable (1). Therefore, protein drugs with narrow therapeutic windows may be difficult to administer by the pulmonary route. From another perspective, protein drugs that are targeted for specific local action in the lung should be administered directly to the lung since the same barriers that prevent systemic absorption can prevent direct targeting to the lung when administered by the parenteral route. We present here a case study of the development of a protein pharmaceutical designed for topical pulmonary administration for the treatment of cystic fibrosis.

Cystic fibrosis (CF) is a genetic lung disease caused by mutations in the cystic fibrosis transmembrane conductance regulator (CFTR) gene which regulates the

0097–6156/94/0567–0322$08.18/0

synthesis of a chloride ion transfer protein, in the lung (*2*). The abnormal ion transport may result in dehydrated viscous mucous in the lung that contributes to decreased mucocilliary clearance and persistent bacterial infections. It has been shown that DNA derived from the neutrophils which infiltrate the airways in response to the bacterial infections is present at high concentrations (3-14 mg/mL) in CF purulent lung secretions (*3-5*). This high concentration of DNA results in an increased viscosity and elasticity of the lung mucus which leads to occlusion of the pulmonary airways and respiratory dysfunction. Bovine deoxyribonuclease I (bDNase) was initially used to treat cystic fibrosis patients but was discontinued because of adverse effects caused by the immune response to a non-human protein, and DNase preparations that were also contaminated with proteases (*6, 7*). The problems associated with the bovine DNase preparations have now been circumvented by the use of recombinant DNA technology to produce human DNase I, rhDNase, (*8*). The viscosity and elasticity of purulent sputum from CF patients was decreased after incubation of the sputum with rhDNase, and suggested that CF patients might benefit from an administration of rhDNase directly into the obstructed airways (*8*). In order to deliver rhDNase directly into the airways the protein was formulated for delivery as an aerosol using so-called "jet nebulizers". Although the formulation must provide the requisite 1-2 years of shelf life, it must also meet additional requirements that are unique to delivery as an aerosol. In this paper we discuss the development of a liquid formulation for rhDNase that provides the requisite stability as well as compatibility with the pulmonary route of administration.

Choice of Formulation Components and Stability of rhDNase Liquid Formulation.

Requirement for an Isotonic Formulation Without Buffer Salts. This drug is administered as an aerosol for local pulmonary delivery, and requires a formulation that is compatible with its delivery to the upper airways of the lung. It has been shown that the osmolality and pH of nebulizer solutions are critical variables that effect bronchoconstriction and subsequent adverse reactions during pulmonary delivery of drugs (*9-14*). In particular, it has been recommended that whenever possible nebulizer solutions should be formulated as isotonic solutions at pH >5 (*11*). Moreover, recent studies have shown that the droplet size distribution of an aerosol will be altered during delivery as the result of a loss or uptake of water vapor from the airways if the formulation is not isotonic (*15, 16*). It is also not uncommon to find that buffer components can cause adverse reactions such as cough (*17, 18*) and, therefore, we have attempted to formulate rhDNase as an unbuffered isotonic solution. Since protein solubility and many protein degradation pathways are highly pH dependent, the control of the pH in an unbuffered formulation is a major concern. Fortunately, although the rhDNase is formulated without a buffer, the protein at 1 mg/mL itself provides sufficient buffering capacity so that the pH of the formulated drug product is quite stable over the recommended storage life of the drug (Figure 1).

Requirement for Calcium in the Formulation. It is well known that another nuclease, bDNase requires calcium for activity as well as structural stability (*19-23*). The requirement for calcium in the formulation of rhDNase is supported by at least two lines of evidence. First, we have shown that after treatment with EDTA, there are 1-1.5 calcium ions that remain bound to rhDNase (data not shown), and subsequent formulation into a phosphate buffer results in a loss of activity compared to an untreated sample (Figure 2). Deamidation is a major degradation route as discussed in more detail in the next section. It has been established that phosphate catalyzes deamidation in many

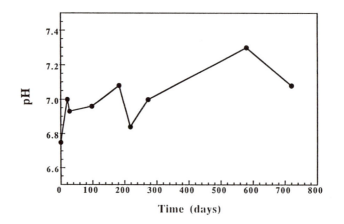

Figure 1. The pH as a function of time of unbuffered rhDNase in 150 mM NaCl, 1 mM CaCl$_2$. In this example, rhDNase was stored in stoppered glass vials at 2-8°C.

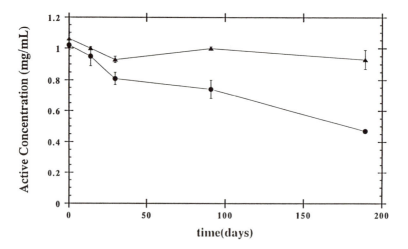

Figure 2. Effect of calcium on activity of rhDNase stored at 25°C at ~pH 6. rhDNase was either formulated in 150 mM NaCl and 1 mM CaCl$_2$ (solid triangles) or treated with EDTA and formulated in isotonic 10 mM PO$_4$ (solid circles). The active concentration was determined by the methyl green activity assay.

peptides *(24)*, however, the rate of deamidation was similar with and without the phosphate buffer (data not shown). Presumably the phosphate buffer effectively competes with any remaining calcium bound to the protein, and removal of this residual calcium results in a loss of activity. Secondly, trypsin is unable to digest rhDNase unless sufficient calcium is removed by treatment with EDTA (Frenz, J., Genentech, Inc., personal communication, 1992). This observation strongly suggests that the binding of calcium stabilizes the conformation of rhDNase. The addition of 1 mM calcium (~33 fold molar excess compared to rhDNase) was shown to be sufficient to maintain stability at the recommended storage temperature of 2-8°C (Figure 3).

Stability and Major Degradation Route of rhDNase. In order to determine a pH rate profile for degradation of the protein, the stability of rhDNase at 2-8, 15, 25 and 37° C as a function of pH was studied by formulating the protein with 5 mM isotonic buffers from pH 5 to 8. A variety of assays including size exclusion chromatography, SDS polyacrylamide gel electrophoresis (SDS PAGE), tentacle ion exchange chromatography, and an activity assay were used to monitor product stability. The ion exchange chromatography detects a specific deamidation at Asn 74 *(25)* that can be analyzed as a pseudo-first order rate reaction (Figure 4, Table 1). The activity assay is a modification of the methyl green DNA binding assay (methyl green actvivity assay) *(26)*. The methyl green dye binds to DNA by intercalating between the stacked base pairs. After hydrolysis of the DNA by rhDNase the dye is released into solution at pH ~ 7.5 where it undergoes an intramolecular rearrangement that results in a fading of the green color *(27)*. The binding of the dye to DNA apparently prevents the structural conversion of the methyl green and the subsequent fading of the color. A standard curve is generated using an rhDNase standard and the resulting activity of a rhDNase sample is expressed as a concentration. The measured concentration from the activity assay can then be compared to the expected concentration based on a UV spectrophotometric determination. The deamidation at Asn 74 is the major route of degradation of rhDNase as detected by the methyl green activity assay. This degradation route, however, does not lead to a completely inactive molecule but rather a protein with ~50% of the activity of the non-deamidated protein. The correlation between deamidation and activity as assessed by the methyl green activity assay is shown by comparing the pseudo-first order kinetics for deamidation and activity loss. When the differences in methyl green activity between deamidated and non-deamidated rhDNase at pH ~7.7 and 37°C are taken into account , the first order plots (Figure 5 a and b) yield similar pseudo first order rate constants, 0.078 ±.005 days^{-1} and 0.069±006 days^{-1} for deamidation and methyl green activity respectively.

The pseudo-first order rate constants for deamidation were obtained using real time data after 88 days of storage and are presented along with the standard errors from a linear regression analysis (Table 1). The errors in the determined rate constants of deamidation at 2-8° and 15° C at pH values below 7 were large (±100%) because of the small changes in deamidation. In some cases the experimental error was larger than the observed change in deamidation resulting in apparent positive slopes. As expected, the pseudo first order rate constants for deamidation (Table 1) are highly pH dependent and decrease with a reduction in pH. Although the rate constant for deamidation at pH 5 is smaller than at pH 8 (~ 0.004 vs 0.1 days^{-1} at 37°C storage), precipitation occurs at 37°C at pH 5. These data suggested that the formulation pH should be kept low enough to effectively control the rate of deamidation but not too low since precipitation could occur upon storage. Although the unbuffered formulation in glass vials was usually in the pH range of 6-7, stability studies of rhDNase in glass vials showed that the unbuffered product at pH ~7 had a fractional activity above 0.8 after ~ 1 year at 2-8° C (Figure 3).

These data show that deamidation is correlated with the loss of rhDNase activity as monitored by the methyl green assay. Although the methyl green assay is a convenient assay for monitoring the protein activity, it does not provide information regarding the

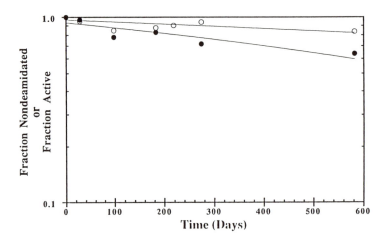

Figure 3. Kinetics of deamidation (solid circles) and activity loss (open circles) for rhDNase in 150 mM NaCl, 1 mM CaCl$_2$, pH 7, in glass vials at 2-8°C. Fraction non-deamidated is computed as $C_{nondeam,t}/C_{nondeam,t=0}$ where $C_{nondeam,t}$ and $C_{nondeam,t=0}$ are concentrations of non-deamidated rhDNase at time t and 0, respectively.

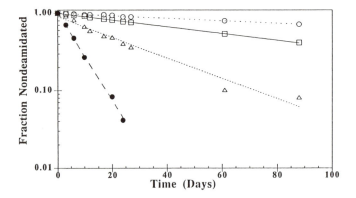

Figure 4. Kinetics of deamidation of rhDNase as determined by tentacle ion exchange chromatography at 37°C in 1 mM CaCl$_2$, 150 mM NaCl and the following 5 mM buffers: acetate, pH 5 (open circles); succinate, pH 6 (open squares); tris, pH 7 (open triangles); and tris, pH 8 (solid circles). Fraction non-deamidated is computed as $C_{nondeam,t}/C_{nondeam,t=0}$ where $C_{nondeam,t}$ and $C_{nondeam,t=0}$ are concentrations of non-deamidated rhDNase at time t and 0, respectively.

Table 1. First Order Rate Constants (days^{-1}) for Deamidation of rhDNase as Assessed by Tentacle Ion Exchange Chromatography[†]

Formulation*	2-8°C	15°C	25°C	37°C
acetate, pH 5	+	~0	$7 \pm 1 \times 10^{-4}$	$3.7 \pm 0.15 \times 10^{-3}$
succinate, pH 5	$3.6 \pm 10 \times 10^{-4}$	$7 \pm 7 \times 10^{-5}$	$5.6 \pm 1.5 \times 10^{-4}$	$3.8 \pm 01.5 \times 10^{-3}$
citrate, pH 5	+	+	$7.3 \pm 1.4 \times 10^{-4}$	$4.0 \pm 01.3 \times 10^{-3}$
histidine, pH 6	$3 \pm 3 \times 10^{-4}$	$1.3 \pm 0.2 \times 10^{-3}$	$3.6 \pm 0.15 \times 10^{-3}$	$1.4 \pm 0.03 \times 10^{-2}$
succinate, pH 6	+	$7 \pm 7 \times 10^{-5}$	$1.6 \pm 0.1 \times 10^{-3}$	$9.96 \pm 0.16 \times 10^{-3}$
maleate, pH 6	+	$2.4 \pm 1 \times 10^{-4}$	$1.3 \pm 0.08 \times 10^{-3}$	$9.66 \pm 0.65 \times 10^{-3}$
tris, pH 7	$1.9 \pm 0.5 \times 10^{-3}$	$5.7 \pm 0.8 \times 10^{-3}$	$1.3 \pm 0.08 \times 10^{-2}$	$3 \pm 0.15 \times d10^{-2}$
tris, pH 8	$1.1 \pm 0.1 \times 10^{-2}$	$2.3 \pm 0.13 \times 10^{-2}$	$4.8 \pm 0.3 \times 10^{-2}$	$1.3 \pm 0.03 \times 10^{-1}$

* : buffers consist of 5 mM buffer salt, 150 mM NaCl and 1 mM CaCl$_2$

† : + indicates that slope was positive.

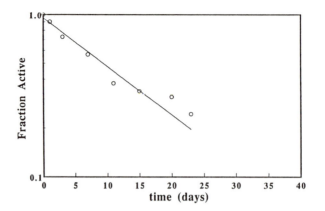

Figure 5a. Kinetics of activity loss of rhDNase as determined by the methyl green activity assay at pH ~7.7 and 37° C. The solid line is the result of a fit to psuedo-first order kinetics (k= 0.069± 0.006 days^{-1} and R^2= 0.986).

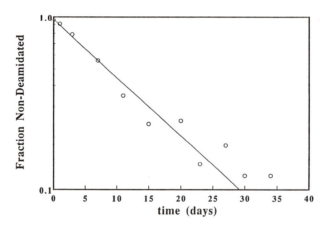

Figure 5b. Kinetics of deamidation of rhDNase as determined by tentacle ion exchange chromatography at pH ~7.7 and 37° C. Fraction non-deamidated is computed as $C_{nondeam,t}/C_{nondeam,t=0}$ where $C_{nondeam,t}$ and $C_{nondeam,t=0}$ are concentrations of non-deamidated rhDNase at time t and 0, respectively. The solid line is the result of a fit to first order kinetics (k= 0.078± 0.005 days-1 and R2= 0.992).

details of the kinetics of DNA hydrolysis. The classical hyperchromicity method for quantitation of DNase bioactivity was used to study the kinetics of DNA hydrolysis (*28*) by non-deamidated rhDNase and rhDNase that was deamidated at Asn 74. This assay measures the increase of absorbance of the polynucleotide at 260 nm upon hydrolysis. The initial rates as a function of DNA concentration were fitted by commercially available non-linear regression software [Kaleidagraph, Synergy Software, Reading, PA] to the Michaelis-Menten equation. The kinetics of DNA hydrolysis by rhDNase are undoubtedly more complex than Michaelis-Menten kinetics since there are multiple hydrolysis sites on the DNA substrate. However, if the hydrolysis rates at each site are independent of each other (i.e. no cooperativity between the individual reaction sites on the substrate) then the analysis may be appropriate. The K_m values for deamidated and non-deamidated rhDNase were 296 ± 52 and 36 ± 7, µg/mL respectively, whereas V_{max} was ~ $2x10^5$ U/mg and was essentially independent of the extent of deamidation (Figure 6). The deamidation site at Asn 74 is very close to the binding pocket for DNA in the crystal structure of bDNase (*29*). These data suggest that the deamidation that results in a negatively charged aspartate residue affects the binding of DNA (i.e., K_m) but not the turnover rate of the enzyme.

The choice of formulation components and pH were discussed in the previous sections. Although the formulation is stable for at least one year, it is important to demonstrate that the rhDNase formulation can be successfully aerosolized for local pulmonary delivery. The following sections describe the generation and the physical and chemical characterization of rhDNase aerosols.

Generation of rhDNase Aerosols by Jet Nebulizers.

Aerosols of rhDNase were generated with jet nebulizers using an air compressor (Pulmo-Aide Model # 5610D, DeVilbiss, Somerset, PA) at a nominal flow rate of 7 L/min. The experimental setup for generation of the aerosols was designed to duplicate the conditions used in the human clinical trials as closely as possible. In the human clinical trials, the initial Phase I and II studies were done with a Respirgard II model #124030 nebulizer (Marquest, Englewood, CA) modified by removal of the expiratory one-way valve. In the later Phase III trials, a T Up-Draft II model #1734 nebulizer was used to generate the rhDNase aerosols (Hudson RCI, Temecula, CA). In addition to the two nebulizers that were used in the clinical trials, we also investigated two other jet nebulizers with similar performance specifications, the Acorn II model #124014 (Marquest, Englewood, CA), and the Airlife Misty with Tee Adapter model # 0020308 (Baxter-American Pharmaseal Company, Valencia, CA). In each experiment, 2.5 mL of rhDNase was placed in the nebulizer bowl and nebulized for ten minutes. In all experiments, the nebulization was complete within ten minutes as observed by cessation of aerosol generation.

Jet nebulizers generate an aerosol from a liquid by introducing a fast stream of air through the solution. Over 97% of the aerosol generated is of large non-respirable droplets that impact on the surfaces and baffle system of the nebulizer, and are recirculated (*30*). Proteins are generally surface active molecules and are susceptible to surface denaturation at air water interfaces (*31*). Denaturation of a protein may also expose hydrophobic amino acid residues normally found in the interior of the protein. The exposure of hydrophobic residues can lead to aggregation of the protein which may result in reduced activity. An additional possibility is the generation of potentially immunogenic species due to the unfolding and/or aggregation of the protein. Immunogenicity, of course, is a major concern with proteins that must be administered chronically such as rhDNase. Denaturation of a globular protein may also effect water structure which, in turn, might alter the surface tension properties of the rhDNase solution. Alteration of the surface tension could effect the droplet size distribution which would alter the amount of rhDNase that reaches the airways. Thus, it is important to determine the effect of the nebulization process on rhDNase .

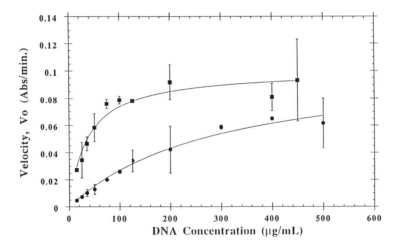

Figure 6. Hyperchromicity assay of non-deamidated rhDNase (solid squares) and rhDNase deamidated at Asn 74 (solid circles). The lines are the fit to the Michaelis-Menten equation. The K_m values were 296 ± 52 and 36 ± 7 µg/mL for deamidated and non-deamidated rhDNase, respectively, whereas both proteins had similar V_{max} values of ~ 2 x 10^5 U/mg (1 unit is defined as the activity which causes an increase in absorbency of 0.001 per minute per milliliter at 25°C).

Effect of Nebulization Process on rhDNase Stability.

Many proteins have been formulated for delivery by aerosol, but most of the studies have not ascertained what happens to the protein during nebulization. In this study, we have collected nebulized rhDNase and determined the effect of the nebulization process on activity, protein size distribution, deamidation, and protein conformation.

Collection of rhDNase Aerosols. Aerosols of rhDNase were collected by impaction into a test tube placed in an ice-filled 250 mL side arm flask. The aerosol was delivered to the test tube via a narrow flexible piece of tygon tubing (3/16 inch id) attached to a 2.0 mL plastic pipet. The narrow diameter tip of the pipet and relatively high aerosol flow rate ensured that a majority, ~70%, of the rhDNase that exited from the nebulizer mouthpiece was collected by impaction. The remaining aerosol presumably either collected on surfaces prior to collection in the test tube, or failed to collect by impaction because the droplet size was sufficiently small to allow the droplets to remain in the air stream that was pulled out of the side-arm flask by the applied vacuum. An improved collection procedure described elsewhere resulted in recovery of over 90% of the nebulized rhDNase, and showed that the additional material recovered did not differ in its composition from the rest of the aerosol (*32, 33*).

Not all of the rhDNase aerosol exits from the mouthpiece since there is a holdup volume [ranging from 0.6 to 0.85 mL] which is dependent on the nebulizer. After ten minutes of nebulization, the protein left in the dead volume of the device has undergone the nebulization process throughout the time of delivery. The protein remaining in the nebulizer bowl has been subjected to the air stream and the accompanying stress longer than the collected aerosol, and was also analyzed at the completion of the experiment.

Stability of Nebulized rhDNase. The assays used for monitoring stability (i.e., SDS PAGE, sizing chromatography, tentacle ion exchange chromatography, and the methyl green activity assay) were also used to assess the effects of nebulization on rhDNase. The results for rhDNase formulated at 1 and 4 mg/mL are shown in Table 2.

Protein Aggregation. As mentioned previously, aggregation could lead to either a decrease in protein activity or to an immunogenic protein. One of the few thorough studies investigating the effect of nebulization on a protein was done with human growth hormone (*34*). Formulations of rhGH without surfactant were highly aggregated after nebulization. It was also noted that the protein remaining in the nebulizer bowl was even more highly aggregated than the collected aerosol. This result is not surprising since the residual protein has been recirculated and subjected to continuous nebulization throughout the delivery period. Inclusion of a surfactant in the formulation greatly decreased the amount of particulates and soluble aggregate that formed as a result of the nebulization process. In contrast, even without inclusion of a surfactant in the formulation, SDS PAGE and size exclusion chromatography revealed that essentially 100% of the rhDNase eluted as monomer after nebulization (Table 2).

Protein Activity. The enzyme fully retained its activity during the nebulization process, as well as in the residue in the nebulizer after cessation of the aerosol generation as evidenced by the methyl green assay of the collected material (Table 2). The active concentration (determined by methyl green activity assay) of rhDNase when divided by its concentration in solution (as determined by UV spectroscopy) was the same in the collected nebulized samples and in the reservoir residua. The increase in deamidation as a result of nebulization was usually less than 1 % which is completely consistent with the observed retention of activity as measured by the methyl green activity assay. The increase in the rhDNase concentrations in the reservoir solutions

Table 2. rhDNase Formulation Before and After Delivery By Jet Nebulization

rhDNase Sample	4 mg/mL rhDNase			1 mg/mL rhDNase		
	Activity* (SD = ± 0.05)	% monomer (SD = ± 0.1)	% change deamidation† (SD = ± 0.01)	Activity* (SD = ± 0.05)	% monomer (SD = ± 0.1)	% change deamidation† (SD = ± 0.01)
Before Nebulization	1.13	99.9	-	1.00	100	-
Hudson T Up-Draft II						
Residua After Nebulization	1.14	99.9	0.7	1.00	100	0.1
Collected Aerosol	1.13	99.6	0.3	0.90	100	1.3
Baxter Airlife Misty						
Residua After Nebulization	1.10	99.9	0	1.01	100	0.4
Collected Aerosol	1.09	99.9	0.1	0.96	100	0.6
Marquest Acorn II						
Residua After Nebulization	ND	ND	ND	0.94	100	0.4
Collected Aerosol	ND	ND	ND	0.93	98.5	0.1
Marquest Customized Respirgard II						
Residua After Nebulization	ND	ND	ND	1.01	100	0.3
Collected Aerosol	ND	ND	ND	0.96	100	0.1

* Normalized activity expressed as ratio of active concentration of rhDNase as determined by methyl green activity assay to concentration determined by ultraviolet absorption spectroscopy.

† % change deamidation expressed as [% deamidation after nebulization/ % deamidation before nebulization)-1] x 100.

(Table 2) was proportionately matched by an increase in the active concentrations. The air used to generate the aerosol is generally not saturated with water vapor, and thus water evaporates from the formed droplets during the nebulization. Such evaporative losses from jet nebulizers have been well documented in the past (*30, 35-37*).

Protein Conformation. The effect of nebulization on the secondary and tertiary conformation of rhDNase was evaluated by circular dichroism (CD). Changes in protein tertiary structure as well as aggregation of protein can be monitored by near ultraviolet (UV) CD if the aromatic chromophores are exposed to different environments as a result of protein denaturation or protein association (*38, 39*). The near UV CD spectrum of rhDNase at 1 mg/mL was unaffected by nebulization in the Acorn II (Marquest), T Up-Draft II (Hudson), and Airlife Misty (Baxter) jet nebulizers as shown by representative CD spectra in Figure 7. This result suggests that the overall tertiary structure of rhDNase is unaffected by nebulization in these devices. Optical activity in the far UV region of the CD spectrum is dominated by the peptide backbone, i.e., the secondary structure of the protein (*40*). Nebulization in the Acorn II (Marquest), T Up-Draft II (Hudson), and Airlife Misty (Baxter) jet nebulizers resulted in no change in the far UV circular dichroism spectrum of rhDNase (see for example Figure 8). This suggests that the secondary structure of rhDNase was unperturbed by nebulization in these devices. The samples nebulized in the Customized Respirgard II (Marquest) were not analyzed and are assumed to yield spectra similar to those for the Acorn II nebulized samples as the nebulizer bowl and baffling system are identical to that of the Acorn II device.

Characterization of rhDNase Aerosols.

The droplet size distribution of an aerosol is an important variable in defining the site of droplet or particle deposition in the patient (*30*). A distribution of particles between 1 and 6 μm in size should give a relatively uniform deposition of rhDNase in the airways (*41-43*). Droplets larger than 6 μm will deposit mainly in the oropharynx whereas droplets less than 1 μm are likely to be exhaled during normal tidal breathing (*30*). Although the choice is somewhat arbitrary, this portion of the size spectrum (1-6 μm) will be referred to as the "respirable fraction", RF.

Jet nebulizers cannot effectively aerosolize all of the 2.5 mL of rhDNase solution because some of it remains in the 'dead volume' of the equipment. The fraction of the dose initially placed in the nebulizer that is actually aerosolized and delivered to the exit from the mouthpiece during the nebulization time (10 minutes) is termed the "nebulizer efficiency", E. The "delivery efficiency", DE, is the product of the respirable fraction times the nebulizer efficiency:

$$DE = E \times RF$$

This parameter estimates the potential of delivery of the aerosol to the target airways. The actual delivery will vary depending on the severity of the disease as well as pulmonary function of the individual patient.

In this section, we discuss the evaluation of aerosols generated by jet nebulizers with similar performance specifications. In order to assure consistent delivery of rhDNase, the various nebulizers tested should have similar values for the measured delivery. Moreover, since most proteins are surface active agents the effect of protein concentration on droplet size distribution and nebulizer efficiency was also investigated.

Experimental Method for Determination of Aerosol Droplet Size Distribution. The droplet size distribution for rhDNase aerosols was determined with a seven-stage cascade impactor (In-Tox Products, Albuquerque, NM). This device separates particles in an airstream as a result of differences in inertial mass (*15, 44, 45*).

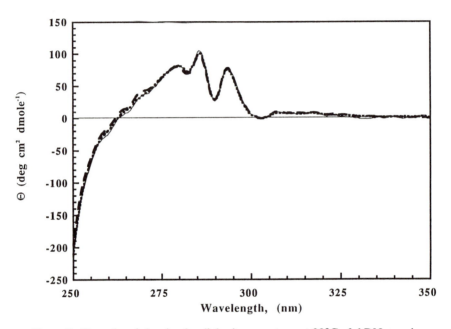

Figure 7. Near ultraviolet circular dichroism spectrum at 20°C of rhDNase prior to nebulization (——), collected aerosol (— — —), and reservoir residua (---). All spectra were obtained with an Aviv model 62DS spectropolarimeter (Aviv Inc., Lakewood, N.J.) using rhDNase diluted to between 0.2 and 0.4 mg/mL in a 1 cm cylindrical cuvette. Data were obtained every 0.2 nm with a 5 sec averaging time at a 1.5 nm bandwidth. Triplicate scans were averaged for each sample. Reproduced by permission from Cipolla et al. (50).

The inertia of a particle in the airstream will tend to cause it to travel in its original path when the airstream is deflected. In the cascade impactor, the velocity of the airstream is increased as it passes through the different stages of the impactor by successively decreasing the size of the nozzles above each of the stages. In this manner, the inertia of the particles is increased on their passage through the impactor. The larger particles with greater mass impact on the upper stages and the smaller particles, once they have been sufficiently accelerated impact on the lower stages.

The individual stages of the seven-stage cascade impactor were cleaned, and weighed before assembly. The assembled impactor was connected to the mouthpiece of the jet nebulizer by a short flexible 6 inch piece of tubing and a vacuum source controlled by an inline flow meter (Model #821, Sierra Instruments, Monterey, CA). A 0.2 micron nylon filter (MSI Calyx, Westboro, MA) was placed in line to protect the flow meter from any liquid residue. On average, less than 5% of the generated aerosol impacted on the walls of the connecting tubing, resulting in >95% of the aerosol output reaching the cascade impactor. This indicates that the tubing did not significantly alter the aerosol flow profile or particle size distribution. The aerosol reaching the cascade impactor was thus representative of the aerosol that would reach the patient's mouth. The nebulizer was connected to the compressor with the tubing provided by the nebulizer manufacturer. The entire assembly (Figure 9) was placed in a laminar flow hood to insure that the air drawn through the cascade impactor was free of interfering particulates. In a control experiment, air was passed through a dry cascade impactor for 24 hours and the amount of material collected in the impactor was found to be insignificant. The higher flow rate through the impactor (8 L/min) ensured that all the rhDNase aerosol (generated at 7 L/min) was continuously drawn through the cascade impactor.

The cascade impactor used in these studies has effective cutoff diameters (ECDs) from 8 to 0.5 µm at 5 L/min. However, in our work, it was used exclusively at 8.0 L/min. The ECD for each stage at 8.0 L/min was calculated by multiplying the manufacturer specified ECDs at 5 L/min by $(5/8)^{1/2}$ based on the theory that the ECD is inversely proportional to the square root of the flow rate (*44*). The new ECDs were 6.11, 3.84, 2.42, 1.51, 0.96, 0.60 and 0.37 µm for stages 1-7, respectively. As the main aspect of the studies reported here was to inquire about the *in vitro* equivalence of the nebulizers tested, the absolute accuracy of the sizing procedure was not essential.

Determination of Aerosol Droplet Size Distribution. After 10 minutes of nebulization, the nebulizer was disconnected from the cascade impactor and the stages were dried by passing air through the impactor at 8.0 L/min for ~ 20 hours. The dry weight of the excipients and rhDNase collected on each stage and the filter was determined. The dry weight of the undelivered excipients and rhDNase was also determined by transferring the rhDNase solution remaining in the nebulizer and on the walls of the tubing to a weight boat and drying in a dessicant-filled jar. The dry weights of both the delivered and undelivered materials were compared to the dry weight of the excipients and rhDNase in the initial 2.5 mL solution. The total recovery of dry rhDNase plus excipients at the end of each cascade impaction run (i.e., the sum of dry masses of deposits from the cascade impactor, the nebulizer and the connecting tubing) was on average >89% for all nebulizers at both protein concentrations. Thus, the majority of the rhDNase used in these experiments could be accounted for by mass balance, indicating that the size analysis is representative of the aerosol that would be delivered to patients.

The aerosol droplet size distribution was essentially independent of the jet nebulizer used to generate the rhDNase aerosol or the concentration of the rhDNase in the formulation (Figures 10 a and b). In all cases, the respirable fraction (i.e. the fraction of dry weight accumulated on stages 2 - 5, approximately 1 - 6 µm) was about 46 - 53% for each nebulizer (Table 3). The nebulizer efficiency for all four devices tested was ~ 50%, and the overall delivery efficiency (DE) of rhDNase by these

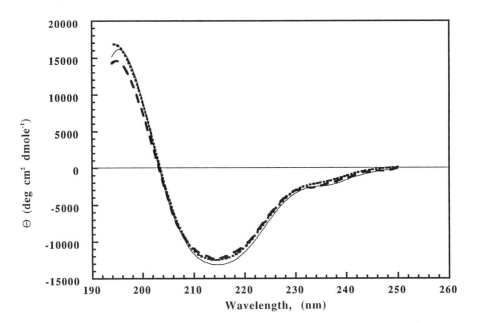

Figure 8. Far ultraviolet circular dichroism spectrum at 20°C of rhDNase prior
to nebulization (———), collected aerosol (— — —), and reservoir residua (---).
All spectra were obtained with an Aviv model 62DS spectropolarimeter (Aviv
Inc., Lakewood, N.J.) using rhDNase diluted to between 0.2 and 0.4 mg/mL
in a 0.01 cm cylindrical cuvette. Data were obtained every 0.2 nm with a 5 sec
averaging time at a 1.5 nm bandwidth. Triplicate scans were averaged for each
sample. Reproduced by permission from Cipolla et al. (50).

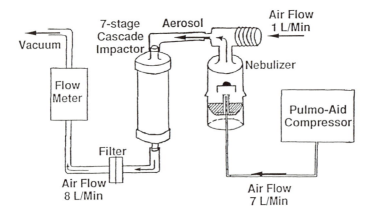

Figure 9. Experimental set-up for characterization of nebulizer performance.

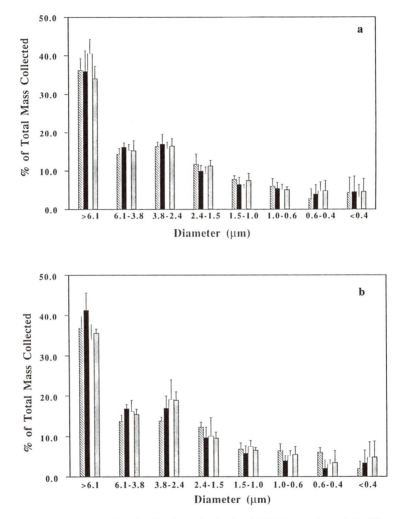

Figure 10. Droplet size distribution of nebulized rhDNase at 1 mg/mL (Figure 10a) and 4 mg/mL (Figure 10b). Nebulizers used were Respirgard II ▧ , Acorn II ■, Baxter ☐ , and Hudson ▨. Adapted by permission from Cipolla et al. (*50*).

Table 3. rhDNase Delivery By Jet Nebulizers

Nebulizer	n[†]	4 mg/ml rhDNase			n[†]	1 mg/mL rhDNase		
		RF*	E*	DE*		RF*	E*	DE*
Marquest Customized Respirgard II	7	46.7±2.4%	45.9±6.9%	21.6±4.1%	8	50.5±2.9%	48.0±6.0%	24.4±4.1%
Marquest Acorn II	8	49.2±4.3%	53.1±8.7%	25.9±3.7%	8	49.7±3.6%	54.7±6.4%	27.2±3.2%
Hudson T Up-Draft II	6	50.6±3.5%	52.4±5.2%	26.6±3.7%	8	50.7±3.7%	49.6±4.9%	25.1±1.9%
Baxter Airlife Misty	5	52.8±2.2%	43.9±2.8%	23.2±2.3%	7	46.3±3.3%	44.4±2.2%	20.5±1.3%

† table values are a result of n independent measurements using n different nebulizers and are given as the mean value and standard deviation.

* RF=Respirable Fraction, E= Efficiency, DE=Delivery

nebulizers ranged from 21-27% which was independent of rhDNase concentration. Thus, the reduction in concentration from 4 to 1 mg/mL led to a proportional (fourfold) reduction in the respirable dose delivered to the mouthpiece.

An assumption inherent in our analysis is that rhDNase and excipients distribute equally in all aerosol particles, and thus, that the percent dry weight of either species on each stage and the filter is identical to the percent of combined weight on that location. The proportion of rhDNase to excipients in an aerosol particle should be independent of the particle size and only dependent upon the concentration in the reservoir solution. The concentration of rhDNase before and after nebulization was determined by UV absorption spectroscopy using an absorptivity of 1.7 $cm^{-1}(mg/mL)^{-1}$. The concentration of rhDNase remaining in the nebulizer bowl was greater than the concentration of rhDNase before nebulization (Table 4, average 29% greater for 1 mg/mL and 15% greater for the 4 mg/mL formulation). The increase in concentration of the protein is due to evaporation in the dry air stream during nebulization. The concentration of sodium ions before and after nebulization was determined by flame photometry (Model IL943, Instrumentation Laboratories, Lexington, MA). The ratio of rhDNase to sodium ions was the same before and after nebulization suggesting that rhDNase and excipients do distribute equally in all aerosol droplets (Table 4).

CONCLUSIONS

Most proteins have been formulated and delivered by the parenteral route of administration. rhDNase is one of the first protein drugs to be formulated for local delivery to the lung. We have successfully formulated rhDNase as a liquid that is stable for 1.5 years at 2-8°C. In addition the formulated protein can be delivered as an aerosol by using current jet nebulizer technology. Although the recirculation of protein solutions under high shear rates in the nebulizer bowl could potentially denature the protein (*30, 46*), we have demonstrated that rhDNase is unaltered during the nebulization process.

This work also shows that the four nebulizers tested are able to deliver 1 mg/mL and 4 mg/mL aqueous solutions of rhDNase in respirable size particles. In particular, the droplet size distribution, and delivery efficiency was similar for the four nebulizers. The analyses indicate a delivery to the mouthpiece of approximately 25% of the rhDNase in the respirable range. These results are comparable to the maximum delivery of other solutes by jet nebulizers (*47*). Nebulizers that have similar in vitro performance characteristics can theoretically deliver similar amounts of rhDNase to the airways. Most jet nebulizers with similar *in vitro* performance characteristics could potentially be used to deliver rhDNase, provided that the rhDNase is not altered by the nebulizer. This conclusion is further supported by the human clinical trials (*48, 49*) where delivery of rhDNase over a wide range of concentrations (0.25 mg/mL to 4 mg/mL) yielded similar clinical efficacy.

Acknowledgements

The authors gratefully acknowledge Ms. Nancy Dasovich for her excellent laboratory support, and Mr. Hans Christinger and Genentech Bioassay Services for assay support. We wish to thank Dr. Marjorie Winkler for sharing with us her preliminary rhDNase hyperchromicity experiments. Our experiments verified her observation that deamidation of rhDNase results in alteration of K_m rather than V_{max}. We also thank Dr. Tom Bewley for critical reading of the manuscript. The authors also wish to acknowledge Dr. Rodney Pearlman for continued support and discussions.

Table 4. rhDNase and Sodium Concentrations Before and After Jet Nebulization

rhDNase Sample	4 mg/mL rhDNase			1 mg/mL rhDNase		
	rhDNase Conc. (mg/mL)*	Na Conc. (mM)	Na/rhDNase (mole ratio)†	rhDNase Conc. (mg/mL)*	Na Conc. (mM)	Na/rhDNase (mole ratio)†
Before Nebulization	3.75	127.1 ± 1.0	990	1.06	155.8 ± 5.8	4300
Hudson T Up-Draft II						
Residua after Nebulization	4.25	150.3 ± 0.4	1040	1.28	184.1 ± 2.3	4200
Collected Aerosol	4.01	141.4 ± 2.0	1030	1.11	150.9 ± 2.5	4000
Baxter Airlife Misty						
Residua after Nebulization	4.38	151.4 ± 2.3	1010	1.38	195.4 ± 2.5	4200
Collected Aerosol	3.96	137.5 ± 1.8	1020	1.05	144.6 ± 2.5	4000
Marquest Acorn II						
Residua after Nebulization	ND	ND	ND	1.45	208.3 ± 7.5	4200
Collected Aerosol	ND	ND	ND	1.22††	162.0 ± 4.2	3900
Marquest Customized Respirgard II						
Residua after Nebulization	ND	ND	ND	1.35	193.1 ± 4.4	4200
Collected Aerosol	ND	ND	ND	1.20	163.0 ± 6.1	4000

* Concentration determined by UV absorption spectroscopy using an absorptivity of 1.7 $(mg/mL)^{-1} cm^{-1}$ that was determined by quantitative amino acid composition analysis.

† Weight concentrations of rhDNase were converted to moles of rhDNase using a protein molecular weight of 29339 based on expected molecular weight from protein sequence as coded by cDNA sequence.

†† This sample was corrected for light scattering (≈10%).

Literature Cited

1. Niven, R. W. *Pharm. Tech.*, **1993**, 17, 72-82.
2. Riordan, J. R. M., R. J.; Kerem, B. S.; Alon, N.; Rozmahel, R.; Grzelczak, Z. *Science*, **1989**, 245, 1066-73.
3. Chernick, W. S.; Barbero, G. J. *Pediatrics*, **1959**, 24, 739-745.
4. Potter, J.; Matthews, L. W.; Lemm, J.; Spector, J. S. *Am. J. Dis. Child.*, **1960**, 100, 493-5.
5. Matthews, L. W.; Spector, S.; Lemm, J.; Potter, J. L. *Am. Rev. Respir. Dis.*, **1963**,88, 199-204.
6. Raskin, P. *Am. Rev. Respir. Dis.*, **1968**, 98, 597-8.
7. Lieberman, J. J. *Am. Med. Assoc.*, **1968**, 205, 312-313.
8. Shak, S.; Capon, D. J.; Hellmiss, R.; Marsters, S.; Baker, C. L. *Proc. Natl. Acad. Sci. USA*, **1990**, 87, 9188-9192.
9. Fine, J. M.; Gordon, T.; Thompson, J. E.; Sheppard, D. *Am. Rev. Respir. Dis.*, **1987**, 135, 826-830.
10. Balmes, J. R.; Fine, J. M.; Christian, D.; Gordon, T.; Sheppard, D. *Am. Rev. Respir. Dis.*, **1988**, 138, 35-39.
11. Beasley, R.; Rafferty, P.; Holgate, S. T. *Br. J. Clin. Pharmacol.*, **1988**, 25, 283-287.
12. Desager, K. N.; Van Bever, H. P.; Stevens, W. J. *Agents and Actions*, **1990**, 31, 225-228.
13. Snell, N. J. C. *Resp. Med.*, **1990**, 84, 345-348.
14. Sant'Ambrogio, G.; Anderson, J. W.; Sant'Ambrogio, F. B.; Mathew, O. P. *Resp. Med.*, **1991**, 85 (Supplement A), 57-60.
15. Gonda, I.; Kayes, J. B.; Groom, C. V.; Fildes, F. J. T. In *Particle Size Analysis* ; Stanley-Wood, N. G., Eds. ; Wiley Heyden Ltd.: N.Y., 1982, pp.52P
16. Gonda, I.; Phipps, P. R. In *Aerosols* ; Masuda, S., Takahashi, K., Eds.; Pergamon Press: N.Y., 1991, pp. 227-230.
17. Godden, D. J.; Borland, C.; Lowry, R.; Higenbottam, T. W. *Clinical Sci.*, **1986**, 70, 301-306.
18. Auffarth, B.; de Monchy, J. G. R.; van der Mark, T. W.; Postma, D. S.; Koeter, G. H. *Thorax*, **1991**, 46, 638-642.
19. Poulos, T. L.; Price, P. A. *J. Biol. Chem.*, **1972**, 247, 2900-2904.
20. Price, P. A. *J. Biol. Chem.*, **1975**, 250, 1981-1986.
21. Douvas, A.; Price, P. A. *Biochem. et Biophys. Acta.*, **1975**, 395, 201-212.
22. Lizarraga, B.; Sancchez-Romero, D.; Gil, A.; Melgar, E. *J. Biol. Chem.*, **1978**, 253, 3191-3195.
23. Lizarraga, B.; Bustamante, C.; Gil, A.; Melgar, E. *Biochim. et Biophys. Acta*, **1979**, 579, 298-302.
24. Capasso, S.; Mazzarella, L.; Zagari, A. *Peptide Research*, **1991**, 4, 234-238.
25. Cacia, J.; Quan, C. P.; Vasser, M.; Sliwkowski, M. B.; Frenz, J. *J. Chromatogr.*, **1993**, 634, 229-239.
26. Kurnick, N. B. *Arch. Biochem.*, **1950**, 29, 41-53.
27. Kurnick, N. B.; Foster, M. *J. Gen. Physiol.*, **1950**, 34, 147-159.
28. Kunitz, M. *J. Gen . Physiol.*, **1950**, 33, 349-362.
29. Suck, D.; Oefner, C.; Kabsch, W. *The EMBO J.*, **1984**, 3, 2423-2430.
30. Byron, P. R. In *Respiratory drug delivery* ; Byron, P. R., Eds.; CRC Press: 1990.
31. Horbett, T. A. In *Stability of protein pharmaceuticals, Part A: Chemical and physical pathways of degradation* ; Ahern, T. J., Manning, M. C., Eds. ;Plenum Press: N.Y., **1992**, pp. 195-210.
32. Cipolla, D.; Gonda, I. In *Protein Formulations and Delivery* ; Cleland, J. L., Langer, R., Eds.; ACS Symposium Series; N.Y., **1994**.

33. Cipolla, D.; Gonda, I., In *Abstract, ISAM 9th Congress* Garmisch-Partenkirchen, Germany, March 30-April 4, **1993**

34. Oeswein, J. Q.; Daugherty, A. L.; Cornavaca, E. J.; Moore, J. A.; Gustafson, H. M.; Eckhardt, B. M., In *Proceedings of the Second Respiratory Drug Delivery Symposium* Dalby, R. N., Evans, R., Eds.; Continuing Pharmacy Education, University of Kentucky, **1991**; pp. 14-49.

35. Mercer, T. T.; Tillery, M. I.; Chow, H. Y. *Am. Ind. Hyg. Assoc. J.*, **1968**, 29, 66-78.

36. Newman, S. P.; Pellow, P. G. D.; Clay, M. M.; Clarke, S. W. *Thorax*, **1985**, 40, 671-676.

37. Phipps, P. R.; Gonda, I. *Chest*, **1990**, 97, 1327-1332.

38. Strickland, H. E. *CRC Crit. Rev. Biochem.*, **1974**, 2, 113-175.

39. Shire, S. J.; Holladay, L. A.; Rinderknecht, E. *Biochemistry*, **1991**,30, 7703-7711.

40. Manning, M. *J. Pharm. Biomed. Analysis*, **1989**, 7, 1103-1119.

41. Gonda, I. *J. Pharm. Pharmacol.*, **1981**, 33, 692-696.

42. Gonda, I. *J. Pharm. Pharmacol.*, **1981**, 33 (Suppl 1), 52P.

43. Rudolph, G.; Gebhart, J.; Heyder, J.; Scheuch, G.; Stahlhofen, W. *Ann. Occup. Hyg.*, **1988**, 32 (Suppl 1), 919-938.

44. Hinds, W. C., *Aerosol technology: Properties, behavior, and measurement of airborne particles.* ; John Wiley and Sons: N. Y.,N.Y., **1982**.

45. Milosovich, S. M. *Pharm. Tech.*, **1992**, 16, 82-86.

46.Niven, R. W.; Butler, J. P.; Brain, J. D., In *Abstracts 11th annual Meeting of the American Association for Aerosol Research, Oct. 12-16, 1992* San Franccisco, CA, **1992**; pp. p160.

47. Lewis, R. A. In *Drugs and the Lung* ; Cumming, G., Bonsignore, G., Eds.; Plenum Press: N. Y., N. Y., **1984**, pp. 63-83.

48. Hubbard, R. C.; McElvaney, N. G.; Birrer, P.; Shak, S.; Robinson, W. W.; Jolley, C. *New Engl. J. Med*, **1992**, 326, 812-815.

49. Fuchs, H. J. *Pediatr. Pulmonol.*, **1992**, Suppl. 8, pp. 149.

50. Cipolla, D.; Gonda, I.; Shire, S. J. *J. Pharm. Res.*, in press.

RECEIVED April 19, 1994

Chapter 20

Method for Collection of Nebulized Proteins

D. C. Cipolla and I. Gonda

Pharmaceutical Research and Development, Genentech, Inc., South San Francisco, CA 94080

Fine particle size fractions of nebulized solutions of small molecular weight drugs are typically captured dried on microfilters for subsequent drug characterization. This procedure normally does not lead to chemical modification of small molecules. However, it may cause adhesion and denaturation of proteins. To quantitate active, unmodified protein drugs in the fine size fractions of nebulized aerosols, we developed a method to overcome these problems. A sintered glass filter funnel was introduced downstream from the aerosol generator which produced a cloud containing droplets of an aqueous solution of protein at a rate of 7 L/min. The aerosol was drawn into the sintered glass filter together with 16-19 L/min of air pre-humidified at 47° C and collected in an ice-cooled filter flask. The humid dilution air was mixed with the aerosol in an attempt to induce condensation growth of the droplets. By increasing the droplet size, an increase in impaction collection efficiency should result. To test whether this arrangement reduced the fraction of droplets that escaped impaction, a glass fiber filter was placed at the exhaust from the aerosol line. The weight of the dry material collected on this filter was only 2% of the initial load when the humid air was used. In addition, the presence of the humid air resulted in an overall aerosol recovery of 98%. Analysis of the collected protein indicated full retention of its activity and structural integrity. This setup will most likely enable quantitative collection of other aqueous protein aerosols that can then be subjected to biochemical analysis.

Until recently, delivery of therapeutics to the lung by inhalation has been limited to small molecular weight drugs. The major concern when these drugs are delivered is how much drug will reach the airways. However, it has been suggested that proteins may be more prone to degradation during aerosol generation (1). Thus, an additional concern is the delivery of active, essentially unaltered protein to the lung.

With the recent increase in the availability of recombinant human proteins, delivery to the lung of these substances for either local action or systemic absorption has been extensively studied (2). However, there are very few reports of the effect of nebulization on protein integrity. Although the presence of a small percentage of denatured or altered protein in the aerosol may have a minimal impact on the effectiveness of the aerosol treatment, altered protein may cause adverse reactions such as an immune response in the patients. For these reasons, the aerosol must be collected and the protein analyzed for the retention of both its activity and integrity.

0097–6156/94/0567–0343$08.00/0

© 1994 American Chemical Society

Aerosolized small molecular weight drugs are typically captured and dried for subsequent drug characterization (3). This procedure is not recommended for capturing proteins which may adhere to the collection surface and denature. Reports of methods that have been used for collecting aerosolized proteins [human growth hormone collected by impaction in a test tube (4), and α1-antitrypsin (5) and secretory leukoprotease inhibitor (6) collected by bubbling through an impinger] evaluate protein activity and integrity but do not report recoveries. Our experiences collecting recombinant human deoxyribonuclease I (rhDNase) by impaction in a test tube resulted in quite variable and low recovery efficiencies of between 60 and 80% (7). Similarly, collecting the aerosol by bubbling through buffer in an impinger may also suffer from low collection efficiencies because a significant fraction of the fine aerosol droplets will be likely to remain entrained in the airstream and escape collection. Thus, the aerosolized protein collected by these methods, although fully active and intact, may not be representative of the protein that is in the fine aerosol droplets escaping collection.

Alternatively, studies using a multistage liquid impinger to collect nebulized insulin reported a collection efficiency of 92.5%. However, detailed analysis of protein integrity and activity was unavailable (8). Although this method yields additional information concerning the aerosol particle size distribution, one drawback is the requirement for collection and analysis of multiple fractions to determine the effect of nebulization on protein integrity. Additionally, activity and integrity analysis of the collected protein on the terminal filter may not be possible due to surface adsorption or denaturation, drying, and difficulty in resolubilizing the entrapped protein. Thus, it is desirable to devise a method to collect the protein aerosol more efficiently to allow for analysis of its activity and integrity.

We are interested currently in aerosol administration of rhDNase for treating cystic fibrosis, and as such, rhDNase has been our model protein throughout these studies. Patients with cystic fibrosis (CF) often suffer from thick, viscous secretions in the airways of their lungs which are difficult to expectorate and contribute to reduced lung volumes and expiratory flow rates (9). The presence of high concentrations of DNA in these secretions is associated with its viscous nature (10). rhDNase is a protein which specifically cleaves DNA. *In vitro* experiments indicate that rhDNase cleaves DNA in the sputum from cystic fibrosis patients and reduces sputum viscosity (11). Clinical trials in CF patients show that short-term treatment with rhDNase administered by inhalation improves lung function (12, 13) and long term therapy additionally reduces the need for parenteral antibiotics (14).

We report here the development of a simple and efficient collection method for recovering aerosolized rhDNase and possibly other proteins by impaction in a coarse sintered glass filter apparatus with prehumidified aerosol dilution air.

Materials and Methods

Recombinant human rhDNase (dornase alpha, rhDNase) was purified from cell culture supernatants and supplied in a bulk solution at 4.7 mg/mL. This rhDNase, formulated in 150 mM NaCl and 1 mM $CaCl_2$, pH 7.0 ± 1.0, was diluted to 4.0 and 1.0 mg/mL with formulation vehicle for use in these experiments.

Generation of the Aerosol. In each experiment, 2.5 mL of a 4.0 or 1.0 mg/mL rhDNase solution was placed in the reservoir of the Hudson RCI T Up-Draft II, Neb-U-Mist, disposable jet nebulizer (Model #1734; Temecula, CA). A vacuum was attached and adjusted so that the flow rate of air through the collection device was either 8.0 L/min (for the collection device in Figure 1) or between 23-26 L/min (for the collection device in Figure 2). This flow rate was kept steady throughout the experiment and monitored with a Sierra Instruments 820 Mass Flow Meter (Model #821; Carmel Valley, CA), range 0-30 standard liters per min. The DeVilbiss

Pulmo-Aide Compressor (Model #5610D; Somerset, PA), which generated an approximate flowrate of 7 L/min at a pressure of 43 psi, was connected to the nebulizer, turned on and the nebulization was allowed to proceed for ten minutes. In all experiments, the nebulization was complete during this time frame as defined by a cessation of aerosol generation. However, even when nebulization was complete, residual rhDNase coated the surface of the nebulizer and lowered the efficiency of the nebulization process.

Collection of the Aerosol with the Test Tube Impaction Apparatus (Figure 1). The rhDNase aerosol (generated at 7 L/min) was diluted with 1 L/min room air and fed into a narrow piece of flexible tygon tubing (3/16 inch i.d.) and through a narrow 2.0 mL plastic pipet. The larger aerosol particles exited the pipet and impacted in the ice-cooled test tube. The fine aerosol particles remained entrained in the air stream and were drawn away uncollected by the vacuum at 8 L/min. After the ten minute nebulization was complete, the walls of the tubing and pipet were rinsed with formulation vehicle and this solution was combined with the rhDNase collected in the bottom of the test tube. The rhDNase collected as well as that remaining in the nebulizer reservoir were then analyzed for activity and integrity and a percent recovery was determined.

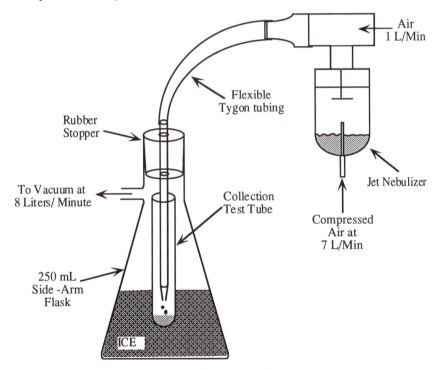

Figure 1. Test Tube Impaction Collection Method. The aerosol is drawn through the flexible tygon tubing and directed into a test tube. The majority of the droplets impact in the test tube allowing for subsequent analysis of the protein in these collected droplets.

Collection of the Aerosol with the Sintered Glass Filter Collection Apparatus (Figure 2). The nebulized rhDNase aerosol (generated at 7 L/min) was diluted with either room air or prehumidified dilution air (at 16-19 L/min) and drawn at 23-26 L/min into the prewetted collection device. The dilution air was humidified by passing it through a bubbler filled with water and maintained at $47 \pm 2°$ C by using a thermostated Fisher Scientific Isotemp Immersion Circulator, Model 730 (Pittsburgh, PA). Parafilm laboratory film (American Can Co., Greenwich, CT) was used to seal the connection between the nebulizer mouthpiece and the collection device. The aerosol collection device consisted of a 150 mL capacity vacuum filter funnel (coarse porosity) with a Buchner joint and a 50 mL round bottom flask, obtained from ChemGlass (Vineland, NJ). The aerosol stream passed through the coarse sintered glass filter (approximately 40-60 micron maximum pore size) where the majority of the aerosol particles impacted on and inside the sintered glass filter.

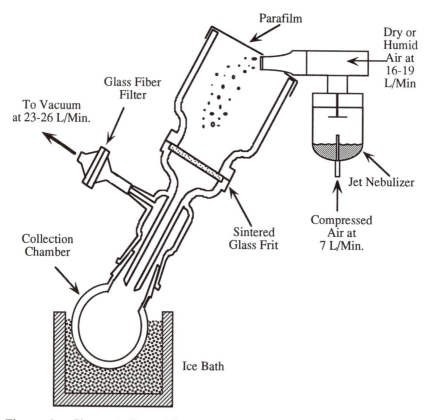

Figure 2. Sintered Glass Filter Collection Apparatus. The aerosol is directed into a prewetted collection chamber. The bulk of the droplets accumulate on the sintered glass frit without drying, allowing for subsequent analysis of protein integrity in the collected droplets.

The air stream was then drawn through a 25 mm glass fiber filter, type A/E, Gelman Sciences, Inc. (Ann Arbor, MI) to determine the percent of rhDNase which escaped impaction. Following aerosol generation and collection, air was drawn through the filter for an additional twenty minutes to ensure complete loss of moisture from the filter. Twenty minutes was deemed adequate because no further loss of moisture was observed as measured gravimetrically when drying for longer periods of time. The liquid from the coalesced aerosol droplets in the sintered glass filter drained into the 50 mL ice-cooled collection flask. After the ten minute nebulization was complete, the walls of the collection funnel and the sintered glass filter were rinsed with formulation vehicle and the liquid was drained into the collection flask. The rhDNase collected and that remaining in the nebulizer reservoir were then analyzed for activity and integrity, and the percent recovery was calculated.

Protein Concentration Determinations. Solutions of rhDNase were diluted to approximately 0.5 mg/mL. The collected aerosol fractions that were less than 0.5 mg/mL were not diluted further. Each rhDNase sample was loaded into a 1 cm quartz cuvette, and the absorbance was read in a Hewlett Packard 8451 Diode Array Spectrophotometer (Mountain View, CA). The concentration of rhDNase was then determined by using an absorptivity of 1.6 cm^{-1} $(mg/mL)^{-1}$ at 280 nm without correcting for light scattering. Light scattering at 320 to 400 nm, indicative of high molecular weight aggregates, was not appreciable for any of the rhDNase solutions. The absorptivity was previously determined by a variety of techniques (Hoff, E., Genentech, Inc., unpublished data.).

Determination of Protein Aggregation. The amount of rhDNase monomer and aggregated rhDNase were determined by size exclusion chromatography with a 30 cm TSK 2000SWXL column (HP, Mountain View, CA). The mobile phase was 5 mM HEPES, 150 mM NaCl, 1 mM $CaCl_2$, titrated to pH 7.0 with NaOH. The flow was 1.0 mL/min for 15 min. Absorbance was monitored at 214 nm. Peak areas and retention times were recorded. The elution times of the low molecular weight Bio-Rad (Hercules, CA) gel filtration standards, consisting of thyroglobulin (670 kD), gamma-globulin (158 kD), ovalbumin (44 kD), myoglobin (17 kD), and cyanocobalamin (1.35 kD) were used to estimate an apparent molecular size for the rhDNase species. The values for percent monomer were determined from duplicate injections.

Determination of Protein Activity. The 1 mg/mL rhDNase samples were diluted sequentially into assay diluent to 0.8, 0.4, and 0.2 μg/mL and assayed in the methyl green activity assay (*15*). Assay diluent was 25 mM HEPES, 4 mM $CaCl_2$, 4 mM $MgCl_2$, 0.1% BSA, 0.01% Thimerosol, 0.05% Polysorbate 20, pH 7.55 ± 0.05. In this assay, active rhDNase hydrolyzes the DNA, releasing the intercalated methyl green dye. The release of the dye into solution and the subsequent fading of its green color at pH 7.5 was followed spectrophotometrically. The active fraction of rhDNase was determined by dividing the active concentration by the concentration determined by UV spectroscopy. The active fractions were normalized by dividing by the active fraction of the 1 mg/mL rhDNase control.

Determination of Protein Deamidation. The deamidation of Asn 74 results in an approximately 50% drop in the protein's bioactivity (*16*). The amount of deamidated rhDNase (Asn 74 to Asp 74) was determined by ion exchange chromatography with a E. Merck (Gibbstown, NJ) Separations Lichrosphere 1000 SO3 column at a flow rate of 0.5 mL/min (*17*). Mobile phase consisted of 10 mM acetate, 1 mM $CaCl_2$, and 2 mM $MgCl_2$, pH 4.5. A linear gradient of 0 to 0.7 M NaCl was achieved in 30 minutes. Absorbance was monitored at 214 nm. Peak areas and retention times were recorded. The relative values for deamidation were determined as:

$$\text{Percent Deamidation} = \left[\frac{A_{df}/(A_{df}+A_{nf})}{A_{di}/(A_{di}+A_{ni})} - 1\right] \times 100$$

where A stands for area under the peak, and the letters in the subscripts are: f and i for final (after nebulization) and initial (before nebulization), d and n for deamidated and nondeamidated rhDNase respectively. Thus, a value of zero means that there has been no increase in the percent deamidation relative to the control sample.

Mass Balance Calculations. A Sartorius (McGaw Park, IL) BA 4100S balance with an accuracy of 0.01g was used to determine weights of nebulized solutions. Volumes of rhDNase were calculated from weights using a density of 1.00 g/mL (*18*). The mass of rhDNase was calculated from the volumes and from the concentration as determined by UV spectroscopy.

(a) **Initial rhDNase Load.** Approximately 2.5 mL of either the 1.0 or 4.0 mg/mL rhDNase solution was placed in the nebulizer reservoir. The actual volume was determined by weighing the nebulizer before and after addition of rhDNase. The initial mass load of rhDNase was calculated from the volume and the concentration of rhDNase in solution.

(b) **Percent Collected as Aerosol.** The volume of the collected aerosol was determined by weighing the solution. The rhDNase concentration in solution was measured. The mass of rhDNase in solution was then calculated from the volume and concentration of rhDNase. The fraction of rhDNase collected as aerosol is the recovered mass of rhDNase divided by the initial mass load of rhDNase.

(c) **Percent Remaining in the Nebulizer.** The volume of rhDNase solution remaining in the nebulizer following termination of nebulization was determined by weight difference. The mass of rhDNase remaining in the nebulizer was determined by multiplying the rhDNase concentration by the volume of solution. The fraction of rhDNase remaining in the nebulizer is thus the recovered mass divided by the initial rhDNase mass load.

(d) **Percent Recovery of rhDNase.** The total amount of rhDNase recovered is the summation of the percent rhDNase by mass collected as aerosol and the percent rhDNase by mass remaining in the nebulizer. The activity and integrity of the rhDNase in these samples were evaluated.

(e) **Percent on the Filter.** The dry weight of the rhDNase and salts collected on the glass fiber filter was measured with a Sartorius (McGaw Park, IL) Research R200D semi-microbalance, with an accuracy of 0.01 mg. Knowing the proportion of rhDNase to salts in the formulation (Appendix 1), the amount of the dry weight attributed to rhDNase was calculated. The fraction of rhDNase on the filter was determined by dividing the amount of rhDNase on the filter by the initial mass of rhDNase loaded into the nebulizer. rhDNase on the filter was not assayed for activity or structural integrity. It only served to determine the quantity of rhDNase in aerosol particles that were uncollected by the sintered glass filter collection apparatus. It also allowed a determination of mass balance.

(f) **Total rhDNase Recovery.** The overall recovery of rhDNase is the summation of the masses of rhDNase in the collected aerosol, remaining in the nebulizer and on the glass fiber filter.

Results and Discussion

Collection and Recovery of rhDNase. The mass of rhDNase from the collected aerosol and that remaining in the nebulizer reservoir was approximately 73% of the initial nebulizer load when the test tube impaction method (Figure 1) was used (Table I, Expt. 1). The recovery was quite variable ranging from 60 to 80%. The remaining rhDNase which was unaccounted for (27% of the initial load) was either in small aerosol droplets which escaped impaction, or impacted on the walls of the collection tubing and apparatus and was inefficiently recovered.

By using the sintered glass filter collection apparatus (Figure 2) with room air as the aerosol dilution air, the recovery of rhDNase was increased to $92.1 \pm 1.3\%$

Appendix 1

The percent of the dry weight which is rhDNase or salt is calculated for the 1.0 and 4.0 mg/mL rhDNase solutions.

4.0 mg/mL rhDNase Solution:

(4.0 mg/mL rhDNase)(2.5 mL)	= 10.00 mg
(150 mM NaCl)(2.5 mL)(58.44 mg/mmole)(1 L/1000 mL)	= 21.92 mg
(1 mM CaCl$_2$)(2.5 mL)(110.99 mg/mmole)(1 L/1000 mL)	= 0.28 mg
Total Dry weight	= 32.20 mg

Percent rhDNase by weight = 10mg/32.2 mg = 31.1%

1.0 mg/mL rhDNase Solution:

(1.0 mg/mL rhDNase)(2.5 mL)	= 2.50 mg
(150 mM NaCl)(2.5 mL)(58.44 mg/mmole)(1 L/1000 mL)	= 21.92 mg
(1 mM CaCl$_2$)(2.5 mL)(110.99 mg/mmole)(1 L/1000 mL)	= 0.28 mg
Total Dry weight	= 24.70 mg

Percent rhDNase by weight = 2.5mg/24.7 mg = 10.1%

Table I. Collection and Recovery of Nebulized rhDNase Solutions[a]

Experiment Number	Percent Collected as Aerosol	Percent Remaining in Nebulizer	Percent Recovery of rhDNase	Percent on the Filter	Total rhDNase Recovery
Experiment 1	36.1 ± 5.1	37.0 ± 5.1	73.0 ± 7.1	N.D.	N.D.
Experiment 2	43.4 ± 1.7	48.7 ± 2.8	92.1 ± 1.3	3.7 ± 0.6	95.9 ± 1.3
Experiment 3	45.2 ± 3.3	50.7 ± 3.3	95.8 ± 1.3	2.3 ± 0.4	98.1 ± 1.7
Experiment 4	51.6 ± 6.6	43.0 ± 6.6	94.6 ± 0.1	2.0 ± 0.9	96.9 ± 0.9

[a]*Explanation of the experimental set-up*

Experiment Number, n	Collection Method	Preheated Humid Air	rhDNase Concentration
Expt. 1, n=4	*Test Tube Impaction*	*no*	*1.0 mg/mL*
Expt. 2, n=3	*Sintered Glass Filter*	*no*	*4.0 mg/mL*
Expt. 3, n=5	*Sintered Glass Filter*	*yes*	*4.0 mg/mL*
Expt. 4, n=3	*Sintered Glass Filter*	*yes*	*1.0 mg/mL*

(Table I, Expt. 2). In addition, $3.7 \pm 0.6\%$ of the initial rhDNase load was collected on the glass fiber filter downstream from the collection apparatus. The rhDNase on the filter was due to the presence of aerosol droplets which escaped impaction. The total recovery was thus $95.9 \pm 1.3\%$. Therefore, $4.1 \pm 1.3\%$ of the initial rhDNase load was unaccounted for, presumably residing in collected aerosol droplets that were not recovered in the collection apparatus. The droplets may have dried on the collection apparatus leaving the rhDNase irreversibly bound to the collection surface.

Attempts were made to increase the recovery of rhDNase by reducing both the number of droplets of rhDNase which escaped impaction and the amount of rhDNase that was not recovered and possibly irreversibly bound to the apparatus. By preheating and humidifying the dilution air, we expected that the condensation of water vapor on the aerosol droplets would result in an increase in the droplet size. Thus, there would be fewer small droplets which usually pass through the collection chamber. Furthermore, the putative loss of rhDNase through drying and irreversible adsorption on the collection surface would also be decreased as a result of the humid air which decreases the rate of droplet evaporation.

To ensure that rhDNase was not denatured by the dilution air, the temperature of the water in the bubblers was equilibrated at 47° C, well below that of the thermal transition temperature of approximately 60° C for rhDNase (Chan, H. K., Genentech, Inc., unpublished data.). By using this preheated, humid dilution air, there was indeed an increase in the recovery of rhDNase from $92.1 \pm 1.3\%$ to $95.8 \pm 1.3\%$ (Table I, Expt. 3). The difference in the recoveries is significant at the $p < 0.01$ level using the two way t-test. The percent of rhDNase in droplets that escaped impaction decreased from $3.7 \pm 0.6\%$ to $2.3 \pm 0.4\%$. Thus, summing the rhDNase in these two fractions results in an increased total recovery of rhDNase from $95.9 \pm 1.3\%$ to $98.1 \pm 1.7\%$.

Finally, it is expected that it may be more difficult to fully recover the nebulized rhDNase during nebulizations with lower concentrations of rhDNase. In particular, if a fixed amount of protein adsorbed on the surface of the nebulizer, or was lost in a similar manner in the collection device, the protein recovery would decrease. This lost protein would be a greater percent of the initial load for smaller initial loads of rhDNase. To test whether a significant portion of rhDNase is lost in this manner, an experiment was performed in which the concentration of rhDNase was decreased from 4 to 1 mg/mL. This was the concentration administered to patients in the phase 3 clinical trials of rhDNase for cystic fibrosis. There was a small but statistically insignificant drop in the recovery of rhDNase from $95.8 \pm 1.3\%$ to $94.6 \pm 0.1\%$. As expected, the percent of rhDNase in droplets escaping impaction and collected in the glass fiber filter changed only slightly from 2.3% to 2.0%. The total rhDNase recovery decreased insignificantly from $98.1 \pm 1.7\%$ for 4 mg/mL rhDNase to $96.9 \pm 0.9\%$ for 1 mg/mL rhDNase.

Characterization of the Collected rhDNase. Previous experiments suggested that nebulization alone does not alter the activity or integrity of the rhDNase solution remaining in the Hudson jet nebulizer (7,16,18). rhDNase can withstand the physical rigors of repeated nebulization and impaction within the nebulizer. Thus, the presence of altered rhDNase in the collected aerosol would be likely to indicate that either rhDNase is modified during the generation and subsequent delivery of only the very fine aerosol particles which escape through the mouthpiece, or that the collection procedure itself causes rhDNase degradation. If a small but constant amount of rhDNase was modified during nebulization or during the collection procedure, this denatured protein would be more apparent when the lowest rhDNase concentration was used because the altered rhDNase would make up the largest percent of the recovered protein. Thus, the activity and integrity of the rhDNase in the collected aerosol and that remaining in the nebulizer reservoir were analyzed only for the experiments with the 1 mg/mL rhDNase.

The absence of particulates during visual inspections and the lack of light scattering during UV spectroscopic concentration determinations suggests that protein precipitation did not occur in the nebulizer reservoir residual or in the collected aerosol samples. Aggregation in the rhDNase samples was also not observed by SEC as the soluble rhDNase eluted as 100% monomer (Table II). The area of the monomer peak correlated with the rhDNase mass load indicating that rhDNase was not lost to the column (data not shown). The percent of deamidated rhDNase did not change for either the nebulizer reservoir or the collected aerosol components when compared to that of the initial sample (Table II). Finally, the methyl green activity assay indicated that the rhDNase samples were fully active within assay error when compared to the unnebulized control rhDNase sample (Table II). Clearly, the absence of altered or inactive rhDNase suggests that rhDNase in the aerosol droplets is fully active and intact, and the collection method does not alter or inactivate the rhDNase in the collected aerosol droplets.

Conclusions

The challenge when collecting nebulized proteins is not only to efficiently collect the protein droplets, but also to ensure that the protein is unaltered by the collection method. The advantage of the sintered glass filter collection method (Figure 2) for collecting protein aerosols is that the protein droplets remain wet during collection, reducing the possibility of protein denaturation and unfolding likely to occur during indiscriminate drying on the collection surface. Additionally, this collection method is also highly efficient for collecting protein aerosols. The use of the sintered glass frit collection method results in an increase in the recovery of rhDNase to approximately 92% compared to recoveries of 60-80% when the test tube impaction method (Figure 1) was used. The sintered glass method also results in consistent and reproducible recoveries. A further increase in rhDNase recovery from 92.1% to 95.8% for 4 mg/mL rhDNase (94.6% for 1 mg/mL rhDNase) occurs when the dilution air is humidified and preheated to 47° C. The use of this dilution air results in a significant reduction in the fine rhDNase aerosol particles which would otherwise escape impaction and subsequent analysis. This result is consistent with the hypothesis that condensation growth of aerosol particles leads to an increase in their impaction and collection efficiency. The rhDNase in the collected aerosol and the rhDNase remaining in the reservoir are analytically and biochemically identical to the unnebulized control rhDNase. Thus, this novel aerosol collection configuration increases the recovery of rhDNase aerosols without altering the protein. This aerosol

Table II. Characterization of rhDNase Solutions

Sample[a]	Percent Change in Deamidation	Percent Monomer	Fraction Active
Initial rhDNase Solution	0.0 ± 1.0	100.0 ± 1.0	1.00 ± 0.10
Reservoir Residual Solution	0.0 ± 1.0	100.0 ± 1.0	0.98 ± 0.10
Collected rhDNase Aerosol	0.2 ± 1.0	99.3 ± 1.0	0.94 ± 0.10

[a]*For all three solutions, the values for the percent change in deamidation, percent monomer, and fraction active are the average and standard deviation of three experiments using the sintered glass filter collection apparatus with preheated, humidified dilution air.*

collection configuration may be generally applicable to the efficient collection of other aerosolized proteins without altering protein integrity. Note that proteins which are found to be perturbed by a jet nebulizer or other aerosol generating device, may also be more likely to be perturbed during the collection process. However, our experience so far with other proteins (unpublished results) indicates that this collection device does not cause their denaturation or alteration.

Acknowledgements

The authors are grateful to assay services for determining the methyl green activity of the rhDNase samples. Many thanks to Milianne Chin for technical editing of this manuscript.

Literature Cited

1. Byron, P. R., *Adv. Drug Del. Rev.* **1990**, *5*, pp. 107-132.
2. Niven, R. W., *Pharm Tech.* **1993**, *17*, pp. 72-82.
3. Mercer, T. T.; Tillery, M. I.; Chow, H. Y., *Am. Ind. Hyg. Assoc. J.* **1968**, *29*, pp. 66-78.
4. Oeswein, J. Q.; et al. In *Proceedings of the Second Respiratory Drug Delivery Symposium*; Dalby, R. N.; Evans, R., Eds.; Continuing Pharmacy Education, University of Kentucky: KY, **1991**, pp. 14-49.
5. Hubbard, R. C.; et al., *Pro. Nat. Acad. Sci.* **1989**, *86*, pp. 680-684.
6. Vogelmeier, C.; et al., *J. Appl. Physiol.* **1990**, *69*, pp. 1843-1848.
7. Cipolla, D. C.; Gonda, I.; Shire, S. J., *Pharm. Res.* **1994**, (in press).
8. Colthorpe, P.; et al., *Pharm. Res.* **1992**, *9*, pp. 764-768.
9. Welsh, M. J.; Fick, R. B., *J. Clin. Invest.* **1987**, *80*, pp. 1523-6.
10. Chernick, W. S.; Barbero, G. J., *Thorax* **1953**, *8*, pp. 295-300.
11. Shak, S.; et al., *Proc. Nat. Acad. Sci.* **1990**, *87*, pp. 9188-92.
12. Aitken, M. A.; et al., *Jama* **1992**, *267*, pp. 1947-1951.
13. Hubbard, R. C.; et al., *New. Eng. J. Med.* **1992**, *326*, pp. 812-815.
14. Fuchs, H. J.; et al., *Abstr. 36th Annual Conf. Chest Disease, Intermountain Thoracic Society* **1993**.
15. Sinicropi, D.; Baker, D.; Shak, S., *Pediatr. Pulmonol. Suppl.* **1992**, *8*, pp. 302.
16. Cipolla, D.; Gonda, I.; Meserve, K.; Weck, S.; and Shire, S., In *Protein Formulations and Delivery* , Cleland, J. L.; Langer, R. Eds.; ACS Symposium Series, **1994**, (in press).
17. Cacia, J.; Quan, C. P.; Vasser, M.; Sliwkowski, M. B.; and Frenz, J., *J. Chrom.* **1993**, *634*, pp 229-239.
18. Cipolla, D.; Clark, A.; Chan, H.; Gonda, I.; and Shire, S., *STP Pharma Sciences*, **1994**, 4, (in press).

RECEIVED April 19, 1994

Author Index

Affiliation Index

Subject Index

A

Absorption spectroscopy, protein
formulation and delivery system
assessment, 36–37
Activity, determination, 347
Additives, role in structure of dried
proteins, 159–165
Adrenocorticotrophic hormone, stability,
109–110
Aerosol delivery, recombinant DNA
derived human deoxyribonuclease I,
322–340
Aerosol droplet size distribution,
determination, 333,335–340
Aerosolized small molecular weight
drugs, collection, 344
Aggregation, determination, 347
Alkyl hydroperoxide, role in protein
oxidation, 64–69
Analytical methods for protein
formulation and delivery system
assessment
chemical degradation, 31–32
chromatography, 28–31
electrophoresis, 23,25–28
MS, 33–35
primary structure analysis, 32–33
protein alterations during formulation
development, 23,24t
requirements, 23
secondary and tertiary structure
analysis, 35–40
Analytical ultracentrifugation, protein
formulation and delivery system
assessment, 38–39
Anhydrobiotic organisms, function, 223

Antibodies
controlled intracranial delivery in rat,
279–289
use for brain tumor treatment, 279
Antibody proteolytic fragments, use for
brain tumor treatment, 279
Antigen delivery system, nanoparticles,
306–320
Aqueous parenteral formulations, peptide
stability, 100–112
Asparagine residues, instability, 53–56
Aspartic acid residues, instability,
47–52

B

Basic fibroblast growth factor
description, 85
disulfide nature of soluble multimers,
87,91f
fate in solution, 96–98
heparin binding site, 94,95f
lyophilization effect on structure,
149,150f,153
physiological state, 98
precipitate composition, 87,92t
structure with precipitates, 92–94
sulfated excipients
vs. solution stability, 86–87,88–90f
vs. thermal stability, 86
Bioactive peptides and proteins, 242
Biodegradable injectable nanospheres
composed of poly(ethylene glycol)–
poly(lactic-co-glycolic
acid) copolymers, 267,269–273
Biodegradable polymers
erosion, 244–260

Production: Meg Marshall
Indexing: Deborah H. Steiner
Acquisition: Anne Wilson
Cover design: Tana Powell

Printed and bound by Maple Press, York, PA